Milieus of Creativity

Knowledge and Space

Volume 2

Knowledge and Space

The close interrelation of knowledge and power, knowledge and socio-economic development, the conflicts between orthodox and heterodox knowledge systems, and the economisation of knowledge play a decisive role in society and has been studied by various disciplines. The series "Knowledge and Space" is dedicated to topics dealing with the production, application, spatial distribution and diffusion of knowledge. Science Studies, Actor-Network Theory, research on learning organisations, studies on creative milieus, and the Geographies of Knowledge, Education and Science have all highlighted the importance of spatial disparities and of spatial contexts in the creation, legitimisation, diffusion and application of new knowledge. These studies have shown that spatial disparities in knowledge and creativity are not a short-term transitional event, but a fundamental structural element of economy and society.

The volumes in the "Knowledge and Space" series will cover a broad range of topics relevant for all disciplines in the humanities, social sciences and economics focusing on knowledge, intellectual capital or human capital, e.g. clashes of knowledge, milieus of creativity, Geographies of Knowledge and Science, the storing of knowledge and cultural memories, the economization of knowledge, knowledge and power, learning organizations, the ethnic and cultural dimensions of knowledge, knowledge and action, and the spatial mobility of knowledge. These topics are to be analysed and discussed at an interdisciplinary level by scholars from various disciplines, schools of thought and cultures.

Knowledge and Space is the outcome of an agreement concluded by Klaus Tschira Foundation and Springer in 2006.

For other titles published in this series, go to
www.springer.com/series/7568

Peter Meusburger • Joachim Funke
Edgar Wunder
Editors

Milieus of Creativity

An Interdisciplinary Approach to Spatiality of Creativity

KTS
Klaus Tschira Stiftung
gemeinnützige GmbH

Springer

Peter Meusburger
Department of Geography
University of Heidelberg
Germany

Joachim Funke
Department of Psychology
University of Heidelberg
Germany

Edgar Wunder
Department of Geography
University of Heidelberg
Germany

Technical Editor:
David Antal, Berlin
Davidrantal@aol.com

ISBN 978-1-4020-9876-5 e-ISBN 978-1-4020-9877-2
DOI: 10.1007/978-1-4020-9877-2

Library of Congress Control Number: 2008920580

Printed on acid-free paper

springer.com

Contents

Contributors

Günter Abel, Prof. Dr.
Technische Universität Berlin, Institut für Philosophie, Ernst-Reuter-Platz 7, 10587 Berlin, Germany,
abel@tu-berlin.de

Margaret A. Boden, Prof. Dr.
Sussex University, Department of Informatics, Falmer, Brighton BN1 9QN, United Kingdom
m.a.boden@sussex.ac.uk

Ricarda Bouncken, Prof. Dr.
Ernst-Moritz-Arndt Universität, Lehrstuhl für ABWL und Organisation, Friedrich-Loeffler-Str. 70, 17487 Greifswald, Germany
bouncken@uni-greifswald.de

Jens Förster, Prof. Dr.
School of Humanities and Social Sciences, Integrated Social and Cognitive Psychology, Campus Ring 1, 28759 Bremen, Germany
j.foerster@iu-bremen.de

Martina Fromhold-Eisebith, Prof. Dr.
RWTH Aachen, Geographisches Institut
Templergraben 55, D-52056 Aachen
m.fromhold-eisebith@geo.rwth-aachen.de

Joachim Funke, Prof. Dr.
Psychologisches Institut, Universität Heidelberg, Hauptstr. 47, 69117 Heidelberg, Germany
joachim.funke@psychologie.uni-heidelberg.de

Stephan Günzel, Prof. Dr.
Philosophische Fakultät/Medienwissenschaft, Ernst-Abbe-Platz 8, 07743 Jena, Germany
StGuenzel@aol.com

Scott G. Isaksen, Dr.
Creative Problem Solving Group, 6 Grand View Trail, PO Box 648,
Orchard Park NY 14127, USA
sgiaway@cpsb.com

James C. Kaufman, Prof. Dr.
California State University, Department of Psychology, 5500 University Parkway,
San Bernardino, CA 92407, USA
jkaufman@csusb.edu

Rob Kitchin, Prof. Dr.
National Institute for Regional and Spatial Analysis, Hume Building, NUL,
Maynooth, Co. Kildare, Ireland
Rob.Kitchin@nuim.ie

Hans Lenk, Prof. Dr.
Institut für Philosophie, Kaiserstraße 12, 76128 Karlsruhe, Germany
sekretariat@philosophie.uni-karlsruhe.de

Peter Meusburger, Prof. Dr.
Universität Heidelberg, Geographisches Institut, Berliner Str. 48,
69120 Heidelberg, Germany
peter.meusburger@geog.uni-heidelberg.de

Dean Keith Simonton, Prof. Dr.
Department of Psychology, One Shields Avenue, University of California,
Davis, CA 95616-8686, USA
dksimonton@ucdavis.edu

Robert J. Sternberg, Prof. Dr.
Tufts University, Dean of the School of Arts and Sciences, Ballou Hall,
Medford, MA 02155, USA
Robert.sternberg@tufts.edu

Barney Warf, Prof. Dr.
Department of Geography, University of Kansas, Lawrence,
KS 66045-7613, USA
bwarf@coss.fsu.edu

Edgar Wunder, Dr.
Universität Heidelberg, Geographisches Institut, Berliner Str. 48, 69120
Heidelberg, Germany
edgar.wunder@geog.uni-heidelberg.de

Introduction: The Spatiality of Creativity

Peter Meusburger, Joachim Funke, and Edgar Wunder

The concept of creativity used to be seen entirely as an entity depending on the faculties of individuals. Research on creativity in psychology, philosophy, and art criticism focused on the attributes of geniuses, gifted persons, creative artists and scientists, and creative performance and problem-solving. Eventually, researchers acknowledged that the creative scientist or artist does not work in a social, cultural, and economic vacuum. It was accepted that creative individuals are inspired or impeded by societal and organizational structures and that they depend on evaluators, audiences, and research infrastructure. It was recognized that such people may meet with incomprehension, competition, hostility, and social conflict, that interactions play an important role, and that learning processes are situated in environments and spatial structures. With the ascendence of these new perspectives, creativity began capturing attention in other disciplines as well.

A Brief Retrospective

From Persons to Persons in Situations

When research on creativity was still in its infancy (for an overview, see Albert & Runco, 1999; Simonton, 1999), few scholars found it necessary to include the environment in their considerations. At best, they admitted that talented individuals could not develop their creativity in repressive societies. One of the first

P. Meusburger (✉)
Universität Heidelberg, Geographisches Institut, Berliner Str. 48, 69120 Heidelberg, Germany
e-mail: peter.meusburger@geog.uni-heidelberg.de

J. Funke
Universität Heidelberg, Psychologisches Institut, Hauptstr. 47, 69117 Heidelberg, Germany
e-mail: joachim.funke@psychologie.uni-heidelberg.de

E. Wunder
Universität Heidelberg, Geographisches Institut, Berliner Str. 48, 69120 Heidelberg, Germany
e-mail: edgar.wunder@geog.uni-heidelberg.de

P. Meusburger et al. (eds.), *Milieus of Creativity*, Knowledge and Space 2,

scholars to discuss the influence that external conditions (parents, schools, peers, role models, teachers, political institutions, and scientific policies) have on the scientific achievements and careers of eminent scientists was the German chemist and Nobel Prize winner Wilhelm Ostwald. In his 1909 book *Große Männer* (Great Men), which describes the careers of Humphry Davy, Julius R. Mayer, Michael Faraday, Justus Liebig, Charles Gerhardt, and Hermann Helmholtz, he addressed almost all individual, social, organizational, environmental, and political aspects now known to be capable of affecting creativity and scientific careers. However, this early pioneer did not work in any of the core disciplines of the social sciences. As for psychologists, they concentrated more on intelligence than on creativity, at least before Guilford's (1950) famous presidential address to the Association of American Psychologists. Ostwald's research was therefore largely ignored by the epistemic centers of the social and behavioral sciences of that time.

The environmental road to research on creativity was gradually charted in the 1940s and 1950s, beginning with Stallknecht's (1941) discussion of the relations between environment (reality and actual concrete existence) and consciousness. Osborn (1953) continued this line of thought by underlining the importance of environment for the development of creativity. So did Stein (1953) when he pointed out that there is an interaction between the creative individual, the problem on which he or she is working, and the environment in which that person exists.

> To speak solely of the existence of the stresses and strains in the environment without due consideration of the individual, as some investigators do, or to deal primarily with the stresses and strains in the individual and to overlook the nature of the problem or the environment as other investigators do, is an arbitrary approach which is a consequence of the specialization in our profession today. (p. 312)

> The creative product resonates with the needs or experience of a group. Art works resonate with feeling, while technical inventions find resonance because they fulfill practical needs. (p. 318)

> The creative work must strike a chord or resonate in some manner with the group that accepts it. (p. 321)

The way to the interactional and environmental study of creativity was also prepared by environmental psychologists focusing on the relation between actor, situation, and environment, especially by Barker's (1968) concept of action settings. Management studies, too, became interested in the psychological climate of organizations and found that creative persons are very sensitive and responsive to the attitudes and behavior prevailing within an organization or at their place of work (see Raudsepp, 1958).

Not until the latter part of the 1980s did mainstream research on creativity turn to the impact that situations and environments have on creativity. At that point, scholars increasingly began addressing issues that had been raised 80 years earlier by Ostwald (1909). More and more of these late twentieth-century social and behavioral scientists regarded behavior as a function of the interaction between a person and a situation, and situational determinants of creativity became a research focus of cognitive psychologists. It was accepted that creative individuals are embedded in particular environments capable of either fostering or hindering their creativity and that cognitive processes are guided not only by personal capabilities or intrinsic motivation but also by interactions with and influences of the environment. This alteration in the study of creativity was summarized by two leading researchers of that period:

> There has been a concentration on the creative person, to the exclusion of "creative situations"—i.e., circumstances conducive to creativity. There has been a narrow focus on internal determinants of creativity to the exclusion of external determinants. (Amabile, 1983, p. 5)

> We cannot study creativity by isolating individuals and their works from the social and historical milieu in which their actions are carried out. This is because what we call creative is never the result of individual action alone; it is the product of three main shaping forces: a set of social institutions, or *field*, that selects from the variations produced by individuals those that are worth preserving; a stable cultural *domain* that will preserve and transmit the selected new ideas or forms to the following generations; and finally the *individual*, who brings about some change in the domain, a change that the field will consider to be creative.... Creativity is a phenomenon that results from interaction between these three systems. (Csikszentmihalyi, 1988, pp. 325–326)

> Creativity is a phenomenon that is constructed through an *interaction between producer and audience*. (Csikszentmihalyi, 1999, p. 314)

Whether in anticipation of or in response to this turn, some psychologists developed multilevel models of creativity to distinguish between the creativity of individuals, groups, and organizations (e.g., Woodman et al., 1993). Other psychologists applied a systems perspective of creativity, including contextual variables that influence creativity (Csikszentmihalyi, 1999; Simonton, 1975, 1977, 1988, 1990). All this work drew attention to the processes of problem-solving, the interaction between members of teams, the various phases of a creative process, the spatial diffusion of creative ideas and products, and the contextual or environmental determinants promoting or suppressing creativity. When referring to environmental variables, though, most authors mentioned only organizational, cultural, socioeconomic, or political factors. They disregarded the spatiality of creativity and the role of places and spatial contexts.

Some psychologists hypothesize that multiple components must converge for creativity to occur and that creativity evolves through a confluence of various individual abilities, societal structures, economic resources, political conditions, and cultural values (for an overview see Amabile, 1983; Sternberg & Lubart, 1999). This confluence or convergence is inconceivable without a spatial coincidence or co-presence of these components. Processes of learning and gathering experience are inseparable from interactions with a specific environment and from situational challenges.

Creativity and Space

The constituents of creativity and their interrelations materialize in social macrophenomena called creative environment, milieu, or context (see the chapter by

Meusburger in this volume). Such spatially rooted social macrophenomena are not identical with the sum of their components. A creative milieu is not produced solely by a co-presence of particular constituents. Much more decisive are their interrelations and mutual modifications. A creative milieu is a possibility or potentiality, not an actuality. According to Stallknecht (1941), a possibility or probability can be an efficient cause for action. Possibility directs attention to concrete situations, "and this direction is the mainspring of conscious initiative" (p. 622). Possibility can be an efficient cause only when in contact with mind that acts as a "catalytic agent", so to speak (p. 622). Recognizing a possibility earlier than other people do is an important constituent of creativity and competitiveness.

A creative milieu or environment represents a certain potentiality that must be activated through human communication and interaction. What makes a location attractive is its possible or imagined advantages, not the realized ones. It is the potential to communicate with other highly creative persons that attracts artists and scientists from elsewhere. It acts like a magnet for other creative people and thus enhances the attractiveness of a place. One cannot predict whether and how often this potential for integrating diverse viewpoints and knowledge bases is activated and how the relationships between creative agents develop. Those aspects can be described only after the fact. If potential, possibilities, and resources go unexploited, if agents stagnate, if they cling to dominating networks and do not listen to adherents of other paradigms or exchange knowledge beyond their discipline's borders, then locally available intellectual resources may be of little benefit. The mode and intensity of the interrelations between given components vary in time and space; they are not fixed or predictable.

There is also another reason why spatial context is more than the sum of its parts. Its symbolic meaning, reputation, and attractiveness lie not only in its present merits and achievements but also in those gained previously by agents no longer belonging to the context. A place is like a screen on which possibilities, expectations, benefits, and hopes are projected, a surface that reflects reputation back onto the persons and institutions located there.

Interdisciplinarity

The longest tradition in creativity research stems from discussions by philosophers about aesthetic creativity and from investigations by psychologists into intelligence, problem-solving, and individual creativity. But for many decades, these two pioneering disciplines of creativity research did not have much in common when it came to their concepts of creativity. According to Wittgenstein (1966), "aesthetic questions have nothing to do with psychological experiments, but are answered in an entirely different way" (p. 17). Judgment about a work of art is only remotely connected with laboratory-confirmed creativity. Similar gaps exist between other approaches and disciplines.

Human geography, too, has a long tradition in the study of the generation and spatial diffusion of innovations. However, researchers in this discipline did not enter the field of creativity research until the 1990s after first detouring through several other areas of inquiry. Some of these scientists studied spatial disparities of educational achievement, the migration of highly skilled labor, and the importance of co-presence and face-to-face contact for the generation and transfer of scarce and valuable knowledge. Others sought reasons for the spatial concentration of high-level decision-makers, the disparities of knowledge between the center and the periphery, and the role of networks and clusters in the accumulation of knowledge (see Meusburger, 2008). Human geographers began looking into subject-oriented action theory, cognitive processes, relations between structure (environment) and agency (Werlen, 1995, 1997), and theoretical concepts of space. The more they delved into these topics, the more geographical research moved from the macro- and mesoscale (spatial structures and processes) to the microscale (human agency). The deeper they probed, the more their focus shifted from spatial units to individuals and the more they had to incorporate theories and research results from sociology, psychology, and philosophy. As they progressed, they built more and more bridges between geography and the other social and behavioral sciences.

Each discipline that is engaged in creativity research has its strengths in certain aspects and its weaknesses in others. An ever-present danger is the tendency of unidisciplinary researchers "to view a part of creativity as the whole phenomenon" (Sternberg & Lubart, 1999, p. 4). Another hazard is that their narrowed vision of creativity seduces them into downplaying the research questions and methodologies of other disciplines. Human geographers, for their part, are not greatly concerned with analyzing the characteristics of creative persons and with ascertaining the creativity of individuals or work groups with psychometric exactitude. That research agenda falls to psychologists, who have developed various experimental processes for those purposes. Geographers pursuing the topic of creativity focus mainly on the role and impact that milieus, contexts, or environments have on creativity, on the spatial distribution, disparities, and diffusion of creative ideas and products, on the factors constituting creative environments, and on the spillovers of knowledge from science parks and universities.

Geographers examine creative milieus from a variety of angles. In one strand of argumentation, places, locales, and areas are ascribed a constitutive role in the generation of career paths (Pred, 1986; Thrift, 1983). Just as certain age cohorts or time periods offer different opportunities and risks, certain locales and spatial contexts offer different learning opportunities, role models, value systems, challenges, social networks, opportunities for professional careers and vertical social mobility, and face-to-face contact with high-level decision-makers of various fields. From this point of view, a locale is a "meeting place of social structure and human agency, substantive enough to be the generator and conductor of structure, but still intimate enough to ensure that the 'creature-like aspects' of human beings are not lost" (Thrift, 1983, p. 38). A location influences the aspirations, motivations, and interaction of individuals and organizations disposing of the skills, prior knowledge, and resources to exploit these chances.

Economic geographers and regional economists have contributed to the interest in creative milieus by studying the spatial distribution of technical and organizational innovations, innovative products and processes, patents, and research input and output and by analyzing the impact of clusters and networks. Taking a different route, other students of creativity retrospectively explore its spatial disparities by analyzing the careers, professional achievements, and social mobility of elites and the performance of outstanding scientists and artists. This biographical material serves as background information about a creative person, the conditions of his or her early socialization, and the chances and challenges that contributed to that individual's creative career. The emphasis falls on the interrelations of factors and the influence that various spatial contexts and path dependencies have on creativity and scientific careers. Such research on creativity thus complements and amplifies the work done in this area by other social and behavioral sciences.

The attention that creativity has received in an increasing number of disciplines has enriched the work on this subject and has broadened scholarly horizons. The researchers from each field of inquiry bring their own specific ideas, core competencies, and main interests to the task. At the same time, this expansion of research has been problematic. The scales, methodologies, theories, definitions, and indicators of creativity used in research differ from one discipline to the next (and even from author to author within the same discipline). Recognizing that elucidation of a lengthy creative process requires resources other than the description of a creative environment, scholars agree that an individual's creative performance must be measured, analyzed, and explained with resources and techniques that diverge from those used to study the spatial distribution of creative products. In short, the resulting variety complicates interdisciplinary discourse and sometimes dilutes concepts of the core disciplines.

Although innovation, invention, and the generation of scientific knowledge are closely related to creativity, surprisingly few economists and economic geographers have taken notice of the results reported in science studies, psychology, and the geography of knowledge. Until recently, psychologists have similarly disregarded the vast amount of relevant work in science studies. This aglossia results partly from the fact that the concepts, definitions, and methodologies in these disciplines differ from those in economics and economic geography. But it might also be due to parochialism that leads publishers and readers to assume that the most innovative ideas, theories, and results appear in a few journals of one or two disciplines. Until recently, the exchange of ideas and concepts across disciplinary borders left much to be desired.

Goals and Content of This Book

The very appearance of this book in a series entitled "Knowledge and Space" indicates one of the goals behind this enterprise: to raise awareness that spatial disparities of creativity exist and that spatial contexts are important in knowledge generation and creative processes. Are societal factors spatially footloose? What is the point in focusing on places, spatial structures, and spatial relations in creativity research? How should the term *environment* be conceptualized? Are only social factors relevant for the development of creativity or should one also include material artifacts and resources in its definition? How can relationships between environment, cognitive processes, and action be explained without falling victim to geodeterminism? Environmental psychology, human ecology, social geography, semiotics, and actor-network theory offer at least some ways to link between nature (material objects) and society (humans) and thereby find out how sociomaterial things act upon humans and what meaning "materiality [has] in the course of knowledge production" (Jöns, 2006, p. 559).

Yet gaps and contradictory results of the continuing inquiry into creativity remain. Another goal of this book is, hence, to address at least a few of them and to promote an understanding of the approaches taken in other disciplines and at other levels of analysis. In the first six chapters the authors review the most fundamental results of research on creativity from the perspectives of psychology, philosophy, and geography. Psychologist Joachim Funke (Chapter 1) focuses on possible definitions, the methods of analysis, and known determinants of the construct called *creativity*. Robert Sternberg (Chapter 2), drawing on his "investment theory of creativity," argues that creativity is not the same across different domains (e.g., art and science) and that knowledge is one crucial variable explaining why creativity is domain-specific. To be a creative individual in a given domain, one must at least know what the state of the art in that domain is. But knowledge is by no means sufficient for creativity. The third psychologist, Dean K. Simonton (Chapter 3), focuses on scientific creativity, trying to predict creative performance in science by using combinatorial models.

The philosophers Günter Abel (Chapter 4) and Hans Lenk (Chapter 5) deal with possible typologies of creativity, analyzing the typical structures of creative processes. Both authors highlight the importance of symbolizing signs in that approach, the relationship between creativity and rules, and the use of creative metaphors to help overcome limits of human understanding and explanation. The geographer Peter Meusburger (Chapter 6) discusses fundamental concepts of creativity research from the viewpoint of their applicability to human geography. Asking why highly creative individuals are not evenly distributed over time and space, he points out the crucial role of particular milieus in which individuals are raised, trained, and embedded.

Chapters 7–15 delve into rather specific problems and case studies in an investigation of the role that milieus, contexts, and social spaces have in the emergence of creativity. James Kaufman (Chapter 7) is concerned with the relationship between creativity and intelligence, which seems to be amazingly varied across different cultures and ethnicities. To understand the factors that support or hinder the creativity of individuals of differing problem-solving styles, Scott Isaksen (Chapter 8) examines how those people rate their working climates. Similarly, the aim of Ricarda Bouncken's study (Chapter 9) is to explore the effects that national culture has on teamwork and innovation in global teams. The results indicate that cultural values have unequal effects on teamwork and creativity in the innovation process. Martina Fromhold-Eisebith (Chapter 10), an economic geographer, is concerned with the problem of why innovative actors agglomerate and how local contexts sustain economic creativity. On the basis of social cognitive theories, the psychologist Jens Förster (Chapter 11) conducts an experiment with a special priming procedure. He finds that exposing participants to the name of a city they regard as a creative place enhances their performance on a subsequent creativity test. Margaret Boden's research (Chapter 12) centers on conceptual spaces perceived as culturally accepted styles of thinking. She understands creativity to mean the process of moving through such conceptual spaces as one tries to transform one or more dimensions of the space. Rob Kitchin (Chapter 13) exemplifies this theoretical reasoning by highlighting the creative potential of science-fiction literature. According to Barney Warf (Chapter 14), the contingent nature of social reality not only serves as an infinite resource for creativity but also compels a retheoretization of the role that time and space have in the constitution and unfolding of social life. In the final essay of this book (Chapter 15), Stephan Günzel introduces the term "Geophilosophies" to designate fundamental modes of geographical thinking. He also argues that the notion of creative milieus can help researchers reevaluate the origins of geophilosophies in their historical contexts.

As this introduction to the book points out, creative processes on the spatial microscale and the interaction between the environment and the creative individual (or work group) have been studied extensively by psychology and other social sciences. However, less is known about why certain university departments, research units, or scientific cultures have been more successful in producing prominent scientists than others. Even more obscure is the answer to the question of how to explain macroscale spatial disparities of creativity. Why were Florence (fifteenth and sixteenth centuries), Prague (about 1600), Manchester (about 1800), Paris and Vienna (about 1900) such creative places? What cultural, social, economic, and political contexts and what spatial relations enabled Vienna to accomodate between 1890 and 1930 Josef Hoffmann, Hans Klimt, Oskar Kokoschka, Koloman Moser and Egon Schiele in the arts; Alfred Adler and Sigmund Freud in psychoanalysis; Rudolf Carnap, Otto Neurath and Karl Popper in philosophy, the philosophy of science, and mathematics; Ludwig Boltzmann, Philipp Frank and Ernst Mach in physics and philosophy; Julius Wagner-Jauregg, Robert Bárány and Theodor Billroth in medicine; Alban Berg, Johannes Brahms, Anton Bruckner, Josef Matthias Hauer, Gustav Mahler, Arnold Schönberg, Johann Strauss jun., Anton Webern, Hugo Wolf and Alexander Zemlinsky in music; Walter Gropius, Carl Hasenauer, Adolf Loos, Joseph Maria Olbrich and Gottfried Semper and Otto Wagner in architecture; Robert Musil, Arthur Schnitzler and Franz Werfel in literature; Karl Kraus in literary criticism; Friedrich August von Hayek, Carl Menger, Ludwig von Mises, and Joseph Schumpeter in economics; Hans Kelsen in legal doctrine; and many other eminent scholars in other disciplines (for details see Beller, 1993; Brix, 2003; Hanák, 1993; Janik, 1986)? How are the regional systems of knowledge production (Rheinberger, 2003) and the regional conditions of excellence defined? Why did other world cities of comparable size not boast such creative minds?

How can one open the black box and avoid the tautology that someone produces creative ideas or products because he or she is a creative person working in an environment conducive to creativity (Choi, 2004, p. 187). Ambrose (2006),

Gardner (1988), and Thiessen (1998) argue that insights from multiple disciplines are necessary in order to understand the intricate complexities of creativity, prevent intellectual stagnation, and avoid dogmatic insularity in creativity studies. The preexisting knowledge of an expert or a single scientific discipline can become a corset that stifles novel ideas so that thinking leads only to the production of tried-and-trusted, correct answers (Cropley, 2006, p. 402). We editors hope that the co-presence of different and even contradictory approaches and provocative questions in one book will encourage readers either to question some of their beloved paradigms and scientific worldviews or to clarify their assumptions and elaborate their models in increasing detail.

We are very grateful to the Klaus Tschira Foundation for funding our enterprise. We are equally thankful to Christiane Marxhausen and Melanie Kudermann (Department of Geography, Heidelberg University), who are in charge of organizing our symposia, and to David Antal, who does an excellent job as technical editor.

References

Albert, R. S. & Runco, M. A. (1999). A history of research on creativity. In R. J. Sternberg (Ed.), *Handbook of creativity* (pp. 16–31). Cambridge, England: Cambridge University Press.

Amabile, T. M. (1983). *The social psychology of creativity*. New York: Springer.

Ambrose, D. (2006). Large-scale contextual influences on creativity: Evolving academic disciplines and global value systems. *Creativity Research Journal*, *18*, 75–85.

Barker, R. G. (1968). *Ecological psychology: Concepts and methods for studying the environment of human behavior*. Stanford, CA: Stanford University Press.

Beller, S. (1993). Who made Vienna 1900 a capital of modern culture? In E. Brix & A. Janik (Eds.), *Kreatives Milieu, Wien um 1900. Ergebnisse eines Forschungsgespräches der Arbeitsgemeinschaft Wien um 1900* (pp. 175–180). Wien, Austria: Verlag für Geschichte und Politik.

Brix, E. (2003). Wesen und Gestalt kreativer Milieus (Essence and gestalt of creative milieus). In W. Berka, E. Brix, & C. Smekal (Eds.), *Woher kommt das Neue? Kreativität in Wissenschaft und Kunst* (pp. 99–115). Wien, Austria: Böhlau.

Choi, J. N. (2004). Individual and contextual predictors of creative performance: The mediating role of psychological processes. *Creativity Research Journal*, *16*, 187–199.

Cropley, A. (2006). In praise of convergent thinking. *Creativity Research Journal*, *18*, 391–404.

Csikszentmihalyi, M. (1988). Society, culture, and person: A systems view of creativity. In R. J. Sternberg (Ed.), *The nature of creativity* (pp. 325–339). Cambridge, MA: Cambridge University Press.

Csikszentmihalyi, M. (1999). Implications of a systems perspective for the study of creativity. In R. J. Sternberg (Ed.), *Handbook of creativity* (pp. 313–335). Cambridge, England: Cambridge University Press.

Gardner, H. (1988). Creativity: An interdisciplinary perspective. *Creativity Research Journal*, *1*, 8–26.

Guilford, J. P. (1950). Creativity. *American Psychologist*, *5*, 444–454.

Hanák, P. (1993). Social marginality and cultural creativity in Vienna and Budapest (1890–1914). In E. Brix & A. Janik (Eds.), *Kreatives Milieu, Wien um 1900. Ergebnisse eines Forschungsgespräches der Arbeitsgemeinschaft Wien um 1900* (pp. 128–161). Wien, Austria: Verlag für Geschichte und Politik.

Janik, A. (1986). Kreative Milieus: Der Fall Wien (Creative milieus: The case of Vienna). In P. Berner, E. Brix, & W. Mantl (Eds.), *Wien um 1900. Aufbruch in die Moderne* (pp. 45–55). München, Germany: Oldenbourg.

Jöns, H. (2006). Dynamic hybrids and the geographies of technoscience: Discussing conceptual resources beyond the human/non-human binary. *Social & Cultural Geography, 7*, 559–580.

Meusburger, P. (2008). The nexus between knowledge and space. In P. Meusburger (Series Ed.) & P. Meusburger, M. Welker, & E. Wunder (Vol. Eds.), *Knowledge and space: Vol. 1. Clashes of knowledge* (pp. 35–90). Dordrecht, The Netherlands: Springer.

Osborn, A. F. (1953). *Applied imagination*. New York: Scribner's.

Ostwald, W. (1909). *Grosse Männer* [Great men]. Leipzig, Germany: Akademische Verlagsgesellschaft.

Pred, A. (1986). *Place, practice and structure*. Cambridge, England: Polity Press.

Raudsepp, E. (1958). The industrial climate for creativity: An opinion study of 105 experts. *Management Review, 47*, 4–8 and 70–75.

Rheinberger, H.-J. (2003). Historische Beispiele experimenteller Kreativität in den Wissenschaften (Historical examples of experimental creativity in the sciences). In W. Berka, E. Brix, & C. Smekal (Eds.), *Woher kommt das Neue? Kreativität in Wissenschaft und Kunst* (pp. 29–49). Wien, Austria: Böhlau.

Simonton, D. K. (1975). Sociocultural context of individual creativity: A transhistorical time-series analysis. *Journal of Personality and Social Psychology, 32*, 1119–1133.

Simonton, D. K. (1977). Eminence, creativity, and geographic marginality: A recursive structural equation model. *Journal of Personality and Social Psychology, 35*, 805–816.

Simonton, D. K. (1988). *Scientific genius: A psychology of science*. Cambridge, England: Cambridge University Press.

Simonton, D. K. (1990). Political pathology and societal creativity. *Creativity Research Journal, 3*, 85–99.

Simonton, D. K. (1999). Creativity from a historiometric perspective. In R. J. Sternberg (Ed.), *Handbook of creativity* (pp. 116–133). Cambridge, England: Cambridge University Press.

Stallknecht, N. P. (1941). Mind and its environment: Toward a naturalistic idealism. *The Journal of Philosophy, 38*, 617–623.

Stein, M. (1953). Creativity and culture. *Journal of Psychology, 36*, 311–322.

Sternberg, R. J. & Lubart T. I. (1999). The concept of creativity: Prospects and paradigms. In R. J. Sternberg (Ed.), *Handbook of creativity* (pp. 3–15). Cambridge, England: Cambridge University Press.

Thiessen, B. L. (1998). Shedding the stagnant slough syndrome: Interdisciplinary integration. *Creativity Research Journal, 11*, 47–53.

Thrift, N. J. (1983). On the determination of social action in space and time. *Environment and Planning D: Society and Space, 1*, 23–57.

Werlen, B. (1995). *Sozialgeographie alltäglicher Regionalisierungen. Band 1: Zur Ontologie von Gesellschaft und Raum* [Social geography of everyday regionalizations: Vol. 1. On the ontology of society and space] (Erdkundliches Wissen No. 116). Stuttgart, Germany: Steiner.

Werlen, B. (1997). *Sozialgeographie alltäglicher Regionalisierungen. Band 2: Globalisierung, Region und Regionalisierung* [Social geography of everyday regionalizations: Vol. 2. Globalization, region, and regionalization] (Erdkundliches Wissen No. 119). Stuttgart, Germany: Steiner.

Wittgenstein, L. (1966). *Lectures and Conversations on Aesthetics, Psychology and Religious Belief* (C. Barrett, Ed.). Oxford, England: Blackwell.

Woodman, R. W., Sawyer, J. E., & Griffin, R. W. (1993). Toward a theory of organizational creativity. *The Academy of Management Review, 18*, 293–321.

Chapter 1
On the Psychology of Creativity

Joachim Funke

Creative thinking—this combination of words raises the question of whether thinking is possible without creativity, and whether creativity can occur without thinking. But one might also ask: Is this miraculous ability called creativity compatible with the rational act of thinking? Are not irrational elements more important in explaining creativity? Are creative processes accessible with scientific methods at all? Has every human being a creative potential? Instead of providing answers to these questions directly, I structure my paper around the following lead questions:

1. Which methods of analysis are available to researchers working in the field of creativity? What is the source of researchers' knowledge about this issue?
2. What does creative thinking look like, and how does it manifest itself?
3. What are known determinants of creative thinking?
4. Why is there a need for creative thinking?
5. What can be done to improve creative thinking?

Space limitations preclude detailed answers to all these questions, but after reading this article you should feel a bit more informed about the above-mentioned topics.

According to Simonton (2000), creativity is present in all fields of human activity. For example, the building in which you are now was designed by an architect; the clothes you wear were designed by a fashion designer; the chair you are sitting on was designed in a perfect way (hopefully ergonomically); and the book you are reading was designed and produced. Behind each of the things around you, which are normally called *artifacts*, is a person who has created these things with a specific intention in mind.

This omnipresence of creative products in the environment contrasts the comparatively small amount of research that has been conducted on creativity. For many centuries, creative activities were seen as something miraculous, something that comes over a person and needs no further explanation. With the advent of empirical psychology at the end of the nineteenth century, those assumptions

J. Funke (✉)
Universität Heidelberg, Psychologisches Institut, Hauptstr. 47–51, D-69117, Heidelberg, Germany
e-mail: Joachim.Funke@urz.uni-heidelberg.de

P. Meusburger et al. (eds.), *Milieus of Creativity*, Knowledge and Space 2,

about mysterious creative acts slowly changed. An impressive increase in research took place in the 1960s and 1970s, stimulated by an important paper written by Guilford (1950), who argued the need for more and better research on this creativity. But besides Guilford's call for research by the scientific community, there was an event outside academia having at least the same importance or even more: space flight and the endeavor to discover the moon and outer space. Historically, creativity research gained huge impetus from the "Sputnik shock" of the Americans (see Amelang & Bartussek, 1997, p. 260). On October 4, 1957, a small satellite started from the Russian space shuttle platform *Baikonur* and orbited the world—a shock for the Americans who believed their nation to be the leading technological force in the world.

Out of concern that the United States was not producing enough creative scientists, large programs (for example, "Headstart") were immediately launched, an effort that helped identify and support gifted people. With the advent of this research, many new insights about creative processes were gained and came to form psychologists' current point of view definitively. Before going into more detail, I first have a look at the research methods for assessing creativity.

What Types of Creativity Assessment Are Available?

A psychometrically sound assessment of a person's creativity is a difficult enterprise. However, many psychologists have tried to meet this challenge. A comprehensive survey of tests for the measurement of creativity is found in Krampen (1993). In general, there is a distinction between language-based and language-free procedures. Language-based procedures require performances that result in verbal utterances. For example, Guilford's concept of divergent thinking (see below) produced a test labeled "Unusual Uses," which required respondents to name as many uses for a given object as possible. The common brick, for instance, can serve as material for building houses but also as a bedwarmer (after heating it), as a weight in a car's luggage trunk to keep the vehicle from skidding on slippery roads, as a weapon against enemies, or as part of a bed made out of bricks. Flexibility of thinking shows up not only in the simple quantity of different uses named but also in the number of different categories like building material, storage medium, weight, or weapon. Aside from flexibility and fluency, there is also an interest in the originality of responses. Using the brick as a sponge is not obvious to everyone and is therefore a more original idea than its proposed standard usage for building.

Another language-based measurement of creativity, the Remote Associates Test, was proposed by Mednick (1962). The task for the respondent is to find a common link between three stimulus words with a low associative link between them. For example, the common link between *humor*, *pitch*, and *night* is the color black. This procedure measures the flexibility of associations. For sure, one can ask whether this procedure really tests creativity. The procedure described assesses the availability of conceptual structures in semantic memory, which is not unimportant for creative processes, but creative processes are not sufficiently described by this conceptual availability.

Language-free tests for assessment of creativity rely mainly on drawing activities required of the respondent. On the Torrance Test of Creative Thinking (Torrance, 1966), given pictures are to be either completed, newly combined, or produced. Figure 1.1 shows an example for each of the three tasks.

Neither language-based nor language-free assessment procedures have really been able to stand up to criticism. Hussy (1986) went as far to as say that "those measurement instruments for the assessment of creative processes have to be qualified as ineffectual" (p. 78). Even though the psychometric assessment of creativity is not possible by means of reliable and standardized diagnostic procedures, there do exist experimental procedures, which should be mentioned briefly.

Important insights based on experimental studies come from the area of analogical transfer (see Gentner & Stevens, 1983; Holyoak, 1985, 2005). The main question is whether respondents detect the structural equivalence between two different domains spontaneously or rather by means of the experimenter's help. For example, think of the analogy between the solar system and the atomic system (given by the fact that in both systems single elements run on a circular curve around the core and are attracted by that core). Analogical reasoning helps transfer some aspects from the source domain to the target domain. Of course, this facet of creativity is not the whole picture. Results of analogical transfer show the difficulty people have when trying spontaneously to detect the parallels in the deep structure of two domains

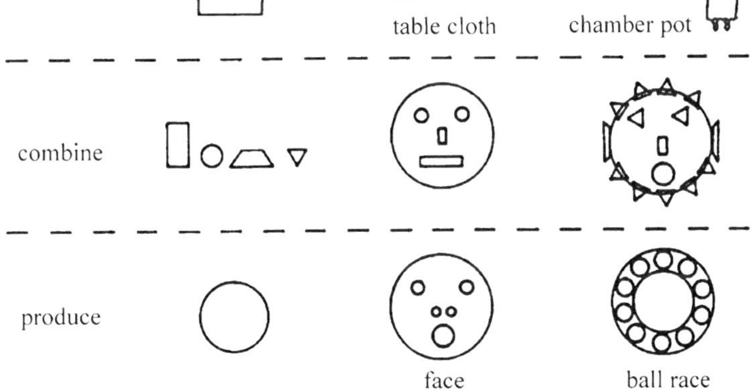

Fig. 1.1 Three examples of creative tasks (completion, combination and production) with non-verbal elements and two types of solution (creative and noncreative)

that are dissimilar on their surface. If hints about the similarity are given, attributes of the source problem can often be used for the target domain. An example is the transfer of solution strategies from one domain to another.

Within the context of scientific discoveries, the principle of induction, which is used in analogical problem-solving, comes into conflict with the principle of falsification. Scientists who want to discover things by means of induction and the use of as many analogues as possible simultaneously have to follow the principle of falsification, which requires strong tests of hypotheses (Bredenkamp, 1980).

The above-mentioned test procedures and the assessment procedures from the area of analogical transfer are not the only instruments and techniques available for research in creativity. Hocevar and Bachelor (1989) mention the following additional techniques: (a) personality inventories, which allow the assessment of certain psychological attributes; (b) biographical inventories, which ask for background information about a person and his or her developmental conditions (the role of biographical analyses as sources for creativity research is stressed by Gardner [1993]); (c) scales for the assessment of attitudes and interests, which ask for specific preferences; (d) person-related assessments by teachers, peers, and supervisors who have seen the rated person for long time periods and know the person well; (e) eminence ratings (e.g., citation frequency, space in biographical texts, and awards); (f) checklist-based self-reports about special performances; and (g) ratings of creative products.

Each of these methods has its pros and cons, so the search for a single type of creativity assessment is misleading. There is no one simple measurement of creativity. Only through a combination of different approaches can a reliable picture emerge. How this picture looks is explained in the next section.

What Is Creative Thinking?

There is the already mentioned popular assumption that creative thinking might be the result of a sudden inspiration, that the solution to a problem suddenly appears in front of the mind's eye (Boden, 1991). Contrary to that popular assumption, psychological research as early as Wallas (1926) indicates that creative solutions are the result of an enduring and long process (Weisberg, 1989). At least the following five phases of creative processes are traditionally mentioned.

Phase 1: Preparation

It is difficult to have a good idea without having worked intensively in the domain under question. Creative inventors know the most important principles of their discipline, and all creative artists have dealt with the products of their predecessors and contemporaries. Creative scientists not only have a long history of ideas behind them but have also reached a high degree of expertise in their domain (Ericsson, 1996). Intensive preparation is a necessary ingredient for important discoveries and creations. Among expertise researchers (e.g., Ericsson, 1996), a saying goes that, roughly speaking, someone who has spent more than 10,000 h on a special topic can be called an expert.

Phase 2: Incubation

Sometimes it is helpful to stop working on a problem for which a creative solution is needed. During the phase of not working on the problem, the brain nevertheless is at work. Incubation becomes strong after the previous phase has laid the groundwork for a kind of "mental infection."

For a long time, it was unclear what happened during this incubation phase. The dynamic of human memory is responsible for the processes of change in associative connections between ideas and representations over time (Finke et al., 1992). The processes during the incubation phase remain below the level of consciousness of the creative person and cannot be influenced actively. But research on the cognitive unconsciousness has provided experimental data showing that intuitive information-processing occurs (Dorfman et al., 1996; Smith et al., 1995; Ward et al., 1997).

Phase 3: Insight

At a certain point in time, a recombined association passes through the threshold of consciousness and produces a flash of insight—the illumination. Gestalt psychologists have called this moment the " 'Aha!' experience." Occurring after appropriate preparation and after some time of incubation, it is the moment of the creative product coming to mind. In medical terms, one has reached the "crisis."

Phase 4: Evaluation

The creative insight has to be evaluated—not all creative insights are really useful. Evaluation is the place for norms and values, which help decide whether a new idea should be discarded or propagated.

Phase 5: Elaboration

From the first idea of an electric light bulb to its first prototype, a long journey had to be taken. Thomas Edison is often quoted for his statement, "Genius is 1 percent inspiration and 99 percent perspiration"—meaning that a lot of force is necessary to make a creative idea work. On the way from the first idea to the final product (a picture, a technical product, a text), a lot of surprises and changes can occur as well.

The above-mentioned five stages of a creative process represent normal stages of information-processing. The opening question, "What is creative thinking?" now has its first answer, which I want to enrich by one further idea. This idea comes from the differentiation between convergent and divergent thinking introduced by Guilford (1967). By *convergent thinking* Guilford means logical procedures, which analytically lead to a certain solution. Divergent thinking, which is predominant in creative processes, is characterized by unusual associations, a shift of perspectives, and the enlargement of the horizon. Normally, a problem-solving process starts with the generation of a sizeable number of ideas (divergent thinking), from which one or more are later selected for elaboration (convergent thinking).

A further conceptual classification differentiates between productive and reproductive thinking. Even if a strong demarcation is not possible, one can describe the endpoints of the scale precisely. With *reproductive thinking* psychologists describe cognitive processes that have only to be reproduced for solving specific problems. Suppose, for example, you want to multiply two numbers, say, 369×264. Even if the exact operation with those two figures has been never done before, the way of solving the problem (the application of the multiplication process) is known and can be reproduced. By contrast, *productive thinking* means that a new way of arriving at a solution has to be found. It is this productive aspect that makes creative processes similar to problem-solving processes. Both constructs have indeed much in common, especially when it comes to complex problem-solving (Funke, 2006).

What Are Known Determinants of Creative Thinking?

Historically, there are three different perspectives on creativity research: (a) the creative person, (b) the creative process, and (c) the creative product. Because some insights about the creative process were mentioned in the previous section, I now go into more detail on the creative persons and their environment. Some statements about the creative product are also made.

The Creative Person

Is it necessary to have extraordinary intelligence for producing creative products? This question was answered by Galton (1869) from the genetic point of view and by Terman (1925) from the perspective of gifted persons (see also Subotnik & Arnold, 1994). Sternberg (1995) concludes, "Bright but not brilliant" (p. 366), which should be read as follows: Above a certain threshold of intelligence, an increase in this ability has no further implications for creative performances. Getzels and Jackson

(1962) have set this threshold at an IQ of 120. The assumption underlying one's conception of intelligence should not be that there is one single general intelligence but that there are multiple intelligences (verbal, logical, spatial, musical, motoric, personal), as formulated in Gardner's (1983) conception or in Sternberg's (1996) ideas on successful intelligence consisting of analytical, creative, and social competencies.

Besides intelligence, there is the more general question concerning the predictive value of personality traits of creative persons. Martindale (1989) and Simonton (1999), for example, point to the importance of variables such as independence, nonconformism, unconventional behavior, broad span of interests, openness for new experiences, risk-taking, and cognitive and behavioral flexibility. Also, the old debate on genius and madness finds some support because creativity is linked to a certain degree to psychopathology (see Eysenck, 1995; Ludwig, 1995). But those pathological behaviors are not necessarily conditions for creativity—on the contrary, sometimes the creative person demonstrates how psychological deficits can be used in an adaptive way (see Csikszentmihalyi, 1997; Ludwig, 1995; Rothenberg, 1990).

With respect to age, it is often said that creativity has a peak when a person is between 20 and 30 years old and decreases thereafter (e.g., Lindauer, 1993). As far as researchers know today, such a pessimistic statement seems unjustified because many factors help maintain creative production at a high qualitative and quantitative level (for gender differences, see Kämmerer, 2000).

The Creative Environment

Life-span oriented research demonstrates that creativity does not always grow where the best and optimal conditions exist. On the contrary, in many cases challenging experiences seem to increase the creative abilities of a person (Simonton, 1994). This finding is interesting because it shows the importance of a creative environment in addition to the creative person. The environment consists of other persons who are creative in a similar way in the same domain. Martindale (1990), for example, shows that writers orient themselves to what other writers (and selected critics) do. These structures were found by Martindale also in art and music. This research shows that it is not enough to concentrate on a single creative person when trying to understand the creative product.

Aside from the influence of environment, there is also a sociocultural influence (Zeitgeist) that belongs to the creative environment. In history, many countries have experienced a flowering of creativity upon the introduction of democracy and liberal systems, as was the case in ancient Greece. According to Simonton (1994), this pattern may be attributed to tendencies to anchor heterogeneity instead of homogeneity. Cultural diversity seems to be an important factor for improving creative environments. Historiometric analyses of creative products seem to support this view (see Simonton, 1984).

The Creative Product

With respect to the creative product, which is a result of creative thinking, two criteria are seen as important: (a) novelty and (b) the usefulness of a particular solution to a problem. Perceived novelty depends on both the evaluating person and social consensus; a creative innovation can have novelty even if it turns out later that this invention has already been made elsewhere. In this vein, Boden (1994) refers to the difference between personal novelty (P-creativity) and historical novelty (H-creativity).

The second criterion, usefulness, ensures that not everything new is simultaneously labeled a creative product. For a product to be called creative, some of the constraints posed by the problem have to have been solved in an optimal way. For example, if one wants to illuminate a dark room in a building, large mirror systems seem less useful than the electric light bulbs used nowadays.

Besides those main criteria, Lubart (1994, p. 291) mentions three subsidiary ones: (c) quality, (d) importance, and (e) history of discovery. With these additional criteria the gradations of product creativity can be conveyed. For example, it makes sense to say that a qualitatively outstanding new product is better than a half-baked product. The importance of a product is also related to its scope. For instance, a new car-alarm system that distinguishes between animal and human contact with a vehicle and thereby avoids false alarms has a lower scope than a new method for cooking with solar energy. Lastly, the history of discovery can change an evaluation if one learns that the invention came about by pure chance instead of hard work. Normally, respect for creative products increases if they are known to have resulted from a very ambitious long-term effort.

The evaluation of a creative product depends not only on historical context but also on the social reference group. This perspective produces a large span of different evaluations of the same creative product. According to Lubart (1994), different background experiences are responsible for that diversity. Art teachers, for example, who have seen many pictures, evaluate a picture by a child more critically than do the child's parents, who are totally enthusiastic about the first products of their son or daughter but who have no real comparison available. Also, different weighting may be responsible for this phenomenon. Depending on the emphasis given to the different criteria, the resulting span of evaluation can be explained.

Why Is Creative Thinking Needed?

The necessity of creative thinking is not open to question if one ponders the continuation of this world. Even though some products of that creativity confront humanity with the greatest ever potential for self-destruction, creative human activity is also precisely what is important for the survival of the human race. Is it necessary for experts to take lessons in creativity? For sure, because experts, especially, can become blind to new ideas (*déformation professionelle*). As early as 1942, Luchins demonstrated with his water-jug problems that human respondents develop certain strategies very quickly and subsequently keep using them even under conditions where easier methods are available.

Gestalt psychologists labeled this effect *functional fixedness* and *Einstellungseffekt*. Frensch and Sternberg (1989) demonstrated its influence in an interesting experiment in which bridge players representing different levels of expertise were pitted against controlled computer opponents. One half of the games were played under normal game conditions; the other half, under either superficially or fundamentally changed rule structures. It turned out that the experts suffered from fundamental rule changes more than the novices did but that even then the experts were better and faster than the novices. Nevertheless, these results show that experts have difficulties adjusting their knowledge to new conditions. Sometimes it might be better to know less (see also Gigerenzer, 2006).

By contrast, Krems (1995) describes a series of experiments in which novices and experts (interns, mechanics, and programmers) had to build hypotheses and draw conclusions from given symptoms. Across all analyzed domains it was consistently found that (a) experts modified their hypotheses much more often than novices did when searching for causes, (b) experts were less prone than novices to verification (i.e., more intense attendance to supporting information than to falsifying information), and (c) the ability to change hypotheses flexibly was based more on case-based knowledge than on rule-based knowledge and was therefore bound to certain domains of knowledge and the experience that one had therein. If one looks into these results, the flexibility of experts might be better than was indicated after the experimental study by Frensch and Sternberg (1989).

The necessity of creative thinking is due not only to the potential blindness of experts when solving complex problems. In a world in which the provision of food and water to an exponentially growing human population is becoming more and more important, in which the military potential for destruction is enough to kill this planet more than once, and in which anthropogenic emissions are increasingly interfering in Earth's very sensible natural cycles (see Wissenschaftlicher Beirat Globale Umweltveränderungen, 1999), the necessity of human creative potential is that it seems to be the only ray of hope. Had it not been for creative processes, the whole history of humankind would not have taken the course that researchers have been able to reconstruct.

Therefore, it is important not only to study the conditions of creative activities but also to look for active improvements in creative thinking. Parents, teachers, schools, and universities are in a certain sense institutions of socialization and can do much to improve creative behavior. The final section deals with this training potential.

What Can Be Done to Improve Creative Thinking?

Many programs have been developed for the improvement of creative thinking. Even though there are researchers who believe that creative potential is given to only a small proportion of humans, a larger group of creativity researchers believes that every person can do something to develop his or her creativity. Amabile (1983,

1996) points to the importance of freedom to decide, unexpected rewards, a positive climate for renovation, and a stimulating milieu as factors that improve creativity. On the other side, she names pressure from colleagues or from evaluation as factors that decrease creativity.

According to Sternberg and Lubart (1991), individual and environmental factors have to be combined. Sternberg (1995, pp. 363–364) formulates several recommendations and attitudes in order to increase creative output:

1. Develop a high motivation for being creative in a certain domain. Do not let yourself be captivated by extrinsic motivation (e.g., money) as reward for creative productions—money corrupts! In general, the motivation for creative acts should come from within a person (intrinsic motivation).
2. Show a certain degree of nonconformism; rules that hinder your creativity may be disregarded. But not all rules and habits are bad. With respect to your own performance, the highest expectations and strong discipline with respect to production are necessary.
3. Be convinced fully of the value and importance of your creative action. Criticism and deprecation from others should not bother you. Self-critique should monitor your own progress and how to improve it.
4. Carefully choose the topics on which you focus your attention—look especially for those not highly appreciated by others.
5. Use analogies and divergent thinking as much as possible. But creative thinking also always has an eye on old traditions, if only to disagree with them.
6. Look for colleagues who help you fight against convention and test new ideas. Search for comrades-in-arms who encourage you to take risks.
7. Assimilate as much knowledge about your domain as possible. This strategy helps prevent you from inventing the wheel for the hundredth time. Try not to be absorbed by these data.
8. Make the strongest commitment to your creative enterprise.

As this list shows, no one factor is made responsible for creative activities; they arise from a broad bundle of conducive conditions. In addition to a creative environment, knowledge, personality, intellectual processes, and intrinsic motivation are necessary ingredients. Sternberg and Lubart (1991, 1995) have labeled their concept "Investment theory of creativity," suggesting that a creative individual "buys low and sells high." Buying low means picking up and creatively developing an idea underestimated by one's contemporaries. Selling high means maximally exploiting the developed idea (financially and otherwise) if you convince other persons of its value.

Yet another part of improving creativity is an important aspect of many training programs, namely, that of putting evaluation on hold in the phase of generating ideas. This objective helps prevent summary rejection of original ideas. If evaluation comes into play too early, it can be a strong barrier against innovation. Ahrens (2000) describes the negative consequences of that premature evaluation has on innovation at British universities. Postponing evaluations is a central part of a method called "brainstorming" developed by Osborn (1953). A small group of persons (6–8) is given the task of generating ideas for 60 min. During this period no critique or discussion is allowed. Afterwards the noted ideas are checked and three questions are asked: Is the idea immediately ready to use? How much do we have to develop the idea? Is the idea useful in principle? The distinction between production and evaluation made by Osborn has been very successful in the context of creative processes (see Taylor, 1964) and has been enriched by many variations (see Seiffge-Krenke, 1974, pp. 264–265). Brainstorming is still a very popular technique used in many companies (see Farr, 1990). Time will soon tell whether "electronic brainstorming" (Roy et al., 1996) is as useful as the older technique.

The history of science demonstrates that creativity depends not only on persons but also on available knowledge within a certain domain. As soon as basic ideas become well-known in a "young" discipline, there is an explosion of creative ideas in that domain. If, after some time, knowledge has increased drastically and the gaps in that knowledge have narrowed, creative inventions also decrease. The domain develops from a positively accelerated development (increasing processes) into a negatively accelerated type of development (breaking face) where the ceiling is reached.

Concluding Remarks

The ideas presented in this chapter explain the necessity of seeing creative thinking as an interaction between a creative personality and a creative environment. The ideas show also that creative performance cannot be prescribed, that it is a treasure to be carefully cultivated, especially in schools and universities. Given the entire accumulation of problems on planet Earth, a major movement is necessary to concentrate humanity's forces on positive goals. Especially with respect to the psychology of creativity, people have to accept that such an endeavor cannot be sustained by individual geniuses.

References

Ahrens, R. (2000). Eine Gefahr für die Universitäten? Forschungsevaluation in Großbritannien [A danger for the universities? Research evaluation in Great Britain]. *Forschung & Lehre, 7*(4), 182–184.

Amabile, T. M. (1983). *The social psychology of creativity*. New York: Springer.

Amabile, T. M. (1996). *Creativity in context: Update to The social psychology of creativity*. Boulder, CO: Westview.

Amelang, M. & Bartussek, D. (1997). *Differentielle Psychologie und Persönlichkeitsforschung. Vierte, überarbeitete und erweiterte Auflage* [Differential psychology and personality research: Fourth, revised, and expanded edition]. Stuttgart, Germany: Kohlhammer.

Boden, M. A. (1991). *The creative mind: Myths & mechanisms*. New York: Basic Books.

Boden, M. A. (1994). What is creativity? In M. A. Boden (Ed.), *Dimensions of creativity* (pp. 75–118). Cambridge, MA: MIT Press.

Bredenkamp, J. (1980). *Theorie und Planung psychologischer Experimente* [Theory and planning of psychological experiments]. Darmstadt, Germany: Steinkopff.
Csikszentmihalyi, M. (1997). *Creativity: Flow and the psychology of discovery and invention*. New York: Harper Collins.
Dorfman, J., Shames, V. A., & Kihlstrom, J. F. (1996). Intuition, incubation, and insight: Implicit cognition in problem solving. In G. Underwood (Ed.), *Implicit cognition* (pp. 257–296). Oxford, England: Oxford University Press.
Ericsson, K. A. (Ed.) (1996). *The road to excellence: The acquisition of expert performance in the arts and sciences, sports, and games*. Mahwah, NJ: Erlbaum.
Eysenck, H. J. (1995). *Genius: The natural history of creativity*. Cambridge, England: Cambridge University Press.
Farr, J. L. (1990). Facilitating individual role innovation. In M. A. West & J. L. Farr (Eds.), *Innovation and creativity at work: Psychological and organizational strategies* (pp. 207–230). New York: Wiley.
Finke, R. A., Ward, T. B., & Smith, S. M. (1992). *Creative cognition: Theory, research, and applications*. Cambridge, MA: MIT Press.
Frensch, P. A. & Sternberg, R. J. (1989). Expertise and intelligent thinking: When is it worse to know better? In R. J. Sternberg (Ed.), *Advances in the psychology of human intelligence* (Vol. 5, pp. 157–188). Hillsdale, NJ: Erlbaum.
Funke, J. (2006). Komplexes Problemlösen [Complex problem-solving]. In N. Birbaumer, D. Frey, J. Kuhl, W. Schneider, R. Schwarzer (Series Eds.), & J. Funke (Vol. Ed.), *Enzyklopädie der Psychologie, Themenbereich C: Theorie und Forschung, Serie II: Kognition, Band 8: Denken und Problemlösen* (pp. 375–445). Göttingen, Germany: Hogrefe.
Galton, F. (1869). *Hereditary genius: An inquiry into its laws and consequences*. London: Macmillan.
Gardner, H. (1983). *Frames of mind: A theory of multiple intelligences*. New York: Basic Books.
Gardner, H. (1993). *Creating minds: An anatomy of creativity seen through the lives of Freud, Einstein, Picasso, Stravinsky, Eliot, Graham, and Gandhi*. New York: Basic Books.
Gentner, D. & Stevens, A. L. (Eds.) (1983). *Mental models*. Hillsdale, NJ: Erlbaum.
Getzels, J. W. & Jackson, P. W. (1962). *Creativity and intelligence: Explorations with gifted students*. New York: Wiley.
Gigerenzer, G. (2006). Bounded and rational. In R. J. Stainton (Ed.), *Contemporary debates in cognitive science* (pp. 115–133). Oxford, England: Blackwell.
Guilford, J. P. (1950). Creativity. *American Psychologist*, *5*, 444–454.
Guilford, J. P. (1967). *The nature of human intelligence*. New York: McGraw-Hill.
Hocevar, D. & Bachelor, P. (1989). A taxonomy and critique of measurements used in the study of creativity. In J. A. Glover, R. R. Ronning, & C. R. Reynolds (Eds.), *Handbook of creativity* (pp. 53–76). New York: Plenum.
Holyoak, K. J. (1985). The pragmatics of analogical transfer. *The Psychology of Learning and Motivation*, *19*, 59–87.
Holyoak, K. J. (2005). Analogy. In K. J. Holyoak & R. G. Morrison (Eds.), *The Cambridge handbook of thinking and reasoning* (pp. 117–142). Cambridge, England: Cambridge University Press.
Hussy, W. (1986). *Denkpsychologie. Ein Lehrbuch. Band 2: Schlußfolgern, Urteilen, Kreativität, Sprache, Entwicklung, Aufmerksamkeit* [Psychology of thinking: A textbook: Vol. 2. Reasoning, judging, creativity, language, development, attention]. Stuttgart, Germany: Kohlhammer.
Kämmerer, A. (2000). Kreativität und Geschlecht [Creativity and gender]. In R. M. Holm-Hadulla (Ed.), *Kreativität* (pp. 301–328). Berlin: Springer.
Krampen, G. (1993). Diagnostik der Kreativität [Diagnostics of creativity]. In G. Trost, K. Ingenkamp, & R. S. Jäger (Eds.), *Tests und trends. 10. Jahrbuch der Pädagogischen Diagnostik* (pp. 11–39). Weinheim, Germany: Beltz.
Krems, J. F. (1995). Cognitive flexibility and complex problem solving. In P. A. Frensch & J. Funke (Eds.), *Complex problem solving: The European perspective* (pp. 201–218). Hillsdale, NJ: Erlbaum.

Lindauer, M. S. (1993). The span of creativity among long-lived historical artists. *Creativity Research Journal, 6*, 231–239.
Lubart, T. I. (1994). Creativity. In R. J. Sternberg (Ed.), *Thinking and problem solving* (pp. 290–323). San Diego, CA: Academic.
Luchins, A. S. (1942). Mechanization in problem solving. *Psychological Monographs, 54*(6, Whole No. 248).
Ludwig, A. M. (1995). *The price of greatness: Resolving the creativity and madness controversy*. New York: Guilford.
Martindale, C. (1989). Personality, situation, and creativity. In J. A. Glover, R. R. Ronning, & C. R. Reynolds (Eds.), *Handbook of creativity* (pp. 211–232). New York: Plenum.
Martindale, C. (1990). *The clockwork muse: The predictability of artistic styles*. New York: Basic Books.
Mednick, S. A. (1962). The associative basis of the creative process. *Psychological Review, 69*, 220–232.
Osborn, A. F. (1953). *Applied imagination*. New York: Scribner's.
Rothenberg, A. (1990). *Creativity and madness: New findings and old stereotypes*. Baltimore, MD: Johns Hopkins University Press.
Roy, M. C., Gauvin, S., & Limayem, M. (1996). Electronic group brainstorming: The role of feedback on productivity. *Small Group Research, 27*, 215–247.
Seiffge-Krenke, I. (1974). *Probleme und Ergebnisse der Kreativitätsforschung*. Bern: Hans Huber.
Simonton, D. K. (1984). *Genius, creativity, and leadership: Historiometric inquiries*. Cambridge, MA: Harvard University Press.
Simonton, D. K. (1994). *Greatness: Who makes history and why*. New York: Guilford.
Simonton, D. K. (1999). *Origins of genius: Darwinian perspective on creativity*. New York: Oxford University Press.
Simonton, D. K. (2000). Creativity: Cognitive, personal, developmental, and social aspects. *American Psychologist, 55*, 151–158.
Smith, S. M., Ward, T. B., & Finke, R. A. (Eds.) (1995). *The creative cognition approach*. Cambridge, MA: MIT Press.
Sternberg, R. J. (1995). *In search of the human mind*. Fort Worth, TX: Harcourt Brace College Publishers.
Sternberg, R. J. (1996). *Successful intelligence: How practical and creative intelligence determine success in life*. New York: Simon & Schuster.
Sternberg, R. J. & Lubart, T. I. (1991). An investment theory of creativity and its development. *Human Development, 34*, 1–31.
Sternberg, R. J. & Lubart, T. I. (1995). *Defying the crowd: Cultivating creativity in a culture of conformity*. New York: Free Press.
Subotnik, R. F. & Arnold, K. D. (Eds.) (1994). *Beyond Terman: Contemporary longitudinal studies of giftedness and talent*. Norwood, NJ: Ablex.
Taylor, C. W. (1964). *Creativity: Progress and potential*. New York: McGraw-Hill.
Terman, R. M. (1925). *Mental and physical traits of a thousand gifted children*. Stanford, CA: Stanford University Press.
Torrance, E. P. (1966). *Torrance test of creative thinking: Directions manual and scoring guide*. Princeton, NJ: Personnel.
Wallas, G. (1926). *The art of thought*. New York: Harcourt Brace and Company.
Ward, T. B., Smith, S. M., & Vaid, J. (Eds.) (1997). *Creative thought: An investigation of conceptual structures and processes*. Washington, DC: American Psychological Association.
Weisberg, R. W. (1989). *Kreativität und Begabung. Was wir mit Mozart, Einstein und Picasso gemeinsam haben* [Creativity and giftedness: What we have in common in Mozart, Einstein, and Picasso]. Heidelberg, Germany: Spektrum der Wissenschaft Verlagsgesellschaft.
Wissenschaftlicher Beirat Globale Umweltveränderungen. (1999). *Welt im Wandel: Strategien zur Bewältigung globaler Umweltrisiken. Jahresgutachten 1998* [World in flux: Strategies for coping with global environmental risks. Annual export report, 1998]. Heidelberg, Germany: Springer.

Chapter 2
Domain-Generality Versus Domain-Specificity of Creativity

Robert J. Sternberg

If Einstein had trained as an artist, would he have been as creative a painter as Van Gogh? How about if Van Gogh had trained as a physicist—would he have been as creative as Einstein? One feels, in answering questions such as these, that the answer is probably "no." But why? Certainly there are people who are multitalented. William Blake achieved great renown as both a writer and a painter. Leonardo da Vinci showed great talent in painting and invention. Richard Feynman is known primarily as a physicist, but his popular books achieved great renown. Is creativity the same or different across different domains, and why? This essay addresses the question of the domain-generality versus domain-specificity of creativity through a theoretical analysis of the construct. As argued in the following pages, there is no general answer to the question of whether creativity is domain-general or domain-specific; creativity varies across individuals as a function of three variables.

The greatest challenge in understanding the domain-generality versus specificity of creativity is in understanding the concept of a domain itself. Is literature a domain, or German literature, or modern German literature, or modern German literature in its original language, or what? Is cognitive psychology a domain, or psychology, or behavioral science, or social science? Because no consensual definition of a domain currently exists, it is impossible at this time to have a clear sense of exactly what domain-specificity means. Domains may themselves be defined at varying levels of generality or specificity.

First, the basic argument is that creativity is largely an attitude toward life. This attitude can, but does not necessarily, extend across a variety of domains. That is, someone might adopt the mindsets that lead to creative thinking across domains, but they do not necessarily do so. Whether they do so or not is one factor in determining the extent to which creativity is domain-general for a given individual. For example, one such attitude is that creative ideas do not necessarily sell themselves, and hence it often is necessary to sell one's creative ideas. One might adopt this attitude, say, in one's work, but not in one's personal life, or vice versa. Even within one's

R.J. Sternberg (✉)
Tufts University, Dean of the School of Arts and Sciences, Ballou Hall, Medford, MA 02155, USA
e-mail: Robert.sternberg@tufts.edu

P. Meusburger et al. (eds.), *Milieus of Creativity*, Knowledge and Space 2,

work environment, one might adopt this mindset, say, in one's interactions with objects but not with people, or vice versa. So domain-generality is a function of the extent to which an individual thinks with a creative mindset across domains. More of these mindsets are described below.

Much of the attitudinal effect is captured through what might be called a *legislative style* of intellectual inquiry (Sternberg, 1997b; Zhang & Sternberg, 2006). A person with a legislative style is someone who enjoys coming up with new ideas. The ability to generate new ideas does not necessarily go along with a desire to generate such ideas. Someone may be more comfortable thinking in traditional ways, even if he or she has the ability to think nontraditionally. In this case, the issue is not how well one can think creatively but rather how much one desires to think in this way. This desire may be mediated, in part, by socialization. In some societies and some religions, creativity is discouraged. The individual may come to believe that one's conformity to existing norms is a good test of one's responsibility as a citizen. Extreme right-wing or extreme left-wing governments, for example—or whatever they may call themselves—may encourage extreme conformity to a societal norm, which may or may not be for the common good. In extreme right-wing societies, such as Nazi Germany, dissenters or people of birth deemed to be unsatisfactory by the government were subject to execution. In the United States, under George Bush the president had the power to detain citizens without due process, a development unprecedented in the history of the country and typically associated with fascist states.

A second variable in determining the extent to which creativity is domain-specific is knowledge. Typically, to think creatively in a domain, one has to know what is known in a domain to go beyond what is known. Someone who is knowledgeable has an advantage at being creative in a domain (Sternberg & Lubart, 1995). But the advantage is tempered by several factors.

First, some domains require very intensive knowledge, whereas other domains require more extensive knowledge. For example, to be creative today in neuroscience, one must be intensively knowledgeable about the workings of the brain and about the research that has been done to date on the brain. The research in this domain is expanding at a rapid rate, so one must constantly be working to keep up to date.

But in many domains, the most creative people are those who are broadly knowledgeable and whose expertise is not limited to one domain (Gardner, 1993). For example, in psychology, many of the most creative scientists have been very broadly trained, often initially studying a discipline other than psychology, as was the case, say, for Helmholtz, Freud, Skinner, Piaget, Beck, Simon, and many others among the most well-known contributors to psychology in the history of the field. Even today, John Gabrieli, one of the best-known contemporary neuroscientists, majored in English as an undergraduate. The advantage that broadly trained people have is that they can bring ideas from one field into another. For example, Simon brought ideas from economics into psychology in his concept of satisficing. George Miller borrowed many of his ideas from linguistics. So being broadly as well as deeply educated enables one to enhance one's creative thinking. At the same time, it may mean that the individual's start on contributions to the field is somewhat delayed.

The situation becomes even more complicated, because expertise in a field can impair as well as facilitate creativity (Frensch & Sternberg, 1989). One can become so used to seeing things in a certain way that one becomes less, not more, creative with the development of expertise. So the acquisition of specialized knowledge does not necessarily facilitate creativity. That knowledge can also diminish it. Whether it enhances or diminishes creativity is largely a matter of attitude, as discussed in the succeeding sections.

A third variable that affects the domain-generality of creativity is the extent to which the environments in which one lives support creativity. An individual does not live in one environment but rather in a multitude of environments. For example, one's family may encourage creative thinking, but one's workplace may not, or vice versa. Certain religions are extremely intolerant of deviations from norms, and may even punish such deviations with death. So one might be allowed to be creative in one's work but not in one's religious beliefs, or vice versa if an organization for which one works insists on blind conformity to a set of norms. Someone may have creative attitudes and an extensive knowledge base but not be allowed to use them in order to be creative.

A society may be led to believe that it is encouraging the intellectual growth of its citizens at the same time that it is stunting that growth. This is what is happening in the United States today, and perhaps other countries as well. The introduction of high-stakes testing into schooling has played into the agenda of an extreme right-wing government by fostering a mentality in schools of cramming for tests that measure knowledge but not critical thinking about this knowledge. Such an agenda is typical of extreme right-wing or extreme left-wing governments, which would risk their own downfall if citizens were to think critically about the propaganda their government feeds them, which often would not withstand even a superficial analysis of the alleged "facts." Creativity and critical thinking are anathema to extremist governments because it might expose the intellectual vacuity of their messages.

Creativity is not a single entity. Rather, there are various kinds of creativity (Sternberg, 1999; Sternberg et al., 2001, 2002). Creativity can work within existing paradigms, transform them, or synthesize them. The more radical the form of creativity, the more it expresses a creative attitude toward life. People may fail to be radically creative not because they lack the knowledge but rather because they lack the desire to experience the kind of rejection that radical creativity often brings with it.

What is creativity and how does it vary, if at all, across disciplines? To analyze this construct, I consider the nature of creativity, drawing upon what is known as the investment theory of creativity (Sternberg & Lubart, 1995).

The Nature of Creativity

The field of creativity as it exists today emerged largely as a result of the pioneering efforts of J. P. Guilford (1950) and E. Paul Torrance (1962, 1974). To this day, the Torrance Tests of Creative Thinking (Torrance, 1974) remain the most widely used assessments of creative talent.

Guilford and Torrance had many more agreements than disagreements about the nature of creativity and how to measure it. Both researchers were basically psychometric theorists who believed in the domain-generality of creativity and who conceived of creativity—and attempted to measure it—from a domain-general psychometric standpoint. But both men were broad thinkers, and their conceptions were much more expansive than the operationalizations of these conceptions through their tests. These two psychologists concentrated on divergent thinking as the basis of creativity, and devised tests that emphasized the assessment of divergent thinking.

My colleagues and I (Sternberg & Lubart, 1995, 1996) have chosen to use a confluence approach as a basis for our work on creativity. I discuss the theory underlying our work and some of the empirical work we have done to test our ideas. These theories are part of a more general theory—WICS—of wisdom, intelligence, and creativity synthesized (Sternberg, 2003b, 2005).

The Investment Theory of Creativity

Our investment theory of creativity (Sternberg & Lubart, 1991, 1995) is a confluence theory according to which creative people are ones who are willing and able to "buy low and sell high" in the realm of ideas (see also Rubenson & Runco, 1992, for use of concepts from economic theory). Buying low means pursuing ideas that are unknown or out of favor but that have growth potential. Often, when these ideas are first presented, they encounter resistance. The creative individual persists in the face of this resistance and eventually sells high, moving on to the next new or unpopular idea.

Aspects of the Investment Theory

According to the investment theory, creativity requires a confluence of six distinct, but interrelated, resources: intellectual abilities, knowledge, styles of thinking, personality, motivation, and environment. Although levels of these resources are sources of individual differences, often the decision to use a resource is a more important source of individual differences. Below I discuss the resources and the role of decision-making in each.

Intellectual skills. Three intellectual skills are particularly important (Sternberg, 1985): (a) the synthetic skill to see problems in new ways and to escape the bounds of conventional thinking; (b) the analytic skill to recognize which of one's ideas are worth pursuing and which are not; and (c) the practical–contextual skill to know how to persuade others of—to sell other people on—the value of one's ideas. These skills can be domain-general, but the knowledge on which they operate is not. The confluence of these three skills is also important. Analytic skill used in the absence of the other two skills results in powerful critical thinking, but not creative thinking. Synthetic skill used in the absence of the other two skills results in new ideas that are not subjected to the scrutiny required to improve them and make them work. And practical–contextual skill in the absence of the other two skills may result in societal acceptance of ideas not because the ideas are good but rather because the ideas have been well and powerfully presented.

In several studies my colleagues and I have tested the role of creative intelligence in creativity. In one of them, we presented 80 individuals with novel kinds of reasoning problems that had a single best answer. For example, they might be told that some objects are green and others blue but that still other objects might be grue, meaning green until the year 2000 and blue thereafter, or bleen, meaning blue until the year 2000 and green thereafter. Or they might be told of four kinds of people on the planet Kyron: blens, who are born young and die young; kwefs, who are born old and die old; balts, who are born young and die old; and prosses, who are born old and die young (Sternberg, 1982; Tetewsky & Sternberg, 1986). The task of the participants in this study was to predict future states from past states, given incomplete information. In another set of studies, 60 people were given more conventional kinds of inductive reasoning problems, such as analogies, series completions, and classifications, but were told to solve them. However, the problems had premises that were either conventional (dancers wear shoes) or novel (dancers eat shoes). The participants had to solve the problems as though the counterfactuals were true (Sternberg & Gastel, 1989a, b).

In these studies, we found that correlations with conventional kinds of tests depended on how novel or nonentrenched the conventional tests were. The more novel the items, the higher the correlations of our tests with scores on successively more novel conventional tests. Thus, the components isolated for relatively novel items tended to correlate more highly with more unusual tests of fluid abilities (e.g., that of Cattell & Cattell, 1973) than with tests of crystallized abilities. We also found that when response times on the relatively novel problems were componentially analyzed, some components measured the creative aspect of intelligence better than others did. For example, in the "grue–bleen" task mentioned above, the information-processing component requiring people to switch from conventional green–blue thinking to grue–bleen thinking and then back to green–blue thinking was a particularly good measure of the ability to cope with novelty.

In another study, we looked at predictions for everyday kinds of situations, such as when milk will spoil (Sternberg & Kalmar, 1997). In this study, we looked at both predictions and postdictions (hypotheses about the past where information about the past is unknown) and found that postdictions took longer to make than did predictions. Novel predictions and postdictions are more challenging and time-consuming than simpler ones.

Creativity and simply thinking in novel ways is facilitated when people are willing to put in up-front time to think in new ways. We found that better thinkers tend to spend relatively more time than do poorer reasoners on global, up-front metacomponential planning when the task is about solving difficult novel reasoning problems. Poorer reasoners, on the other hand, tend to spend relatively more time in local planning (Sternberg, 1981). Presumably, the better thinkers recognize that it is better to invest more time up front so as to be able to process a problem more efficiently later on.

Knowledge. On the one hand, one needs to know enough about a field to move it forward. One cannot move beyond where a field is if one doesn't know where it is. On the other hand, as noted above, knowledge about a field can result in a closed and entrenched perspective, resulting in a person's not moving beyond the way in which he or she has seen problems in the past. Knowledge can thus either help or hinder creativity.

In a study of expert and novice bridge players, for example (Frensch & Sternberg, 1989), we found that experts outperformed novices under regular circumstances. When a superficial change was made in the surface structure of the game, the experts and novices were both hurt slightly in their playing, but quickly recovered. When a profound, deep-structural change was made in the structure of the game, the experts initially were hurt more than the novices, although the experts later recovered. The reason, presumably, is that experts make more and deeper use of the existing structure, and hence have to reformulate their thinking more than do novices when there is a deep-structural change in the rules of the game. Thus, one needs to decide to use one's past knowledge.

Thinking styles. Thinking styles are preferred ways of using one's skills. In essence, they are *decisions* about how to deploy the skills available to one. With regard to thinking styles, as mentioned above, a legislative style is particularly important for creativity (Sternberg, 1988, 1997a; Zhang & Sternberg, 2006), that is, a preference for thinking and a decision to think in new ways. This preference needs to be distinguished from the ability to think creatively: Someone may like to think along new lines, but not think well, or vice versa. To become a major creative thinker, it also helps if one is able to think globally as well as locally, distinguishing the forest from the trees and thereby recognizing which questions are important and which ones are not.

Other research (Sternberg, 1997b; Sternberg & Grigorenko, 1995), has shown that legislative individuals tend to be better students than less legislative students if the schools in which they study value creativity. Students at schools that do not value creativity or that devalue it tend to do worse if they are highly legislative. Students were also found to receive higher grades from teachers whose styles of thinking match their own.

Thinking styles could, in theory, be domain-general, but in practice they usually are not (Sternberg, 1997b). Someone who is legislative at work, for example, may not be in domestic situations. Someone may be legislative in his or her own thinking, but prefer nonlegislative thinking in his or her subordinates. Thus, the styles are only as domain-general as an individual chooses to make them.

Personality. Numerous research investigations (summarized in Lubart, 1994, and Sternberg & Lubart, 1991, 1995) have supported the importance of certain personality attributes for creative functioning. These attributes include, but are not limited to, willingness to overcome obstacles, willingness to take sensible risks, willingness to tolerate ambiguity, and self-efficacy. In particular, buying low and selling high typically means defying the crowd, so that one has to be willing to stand up to conventions if one wants to think and act in creative ways (Sternberg, 2003a; Sternberg & Lubart, 1995). Often, creative people seek opposition in that they decide to think in ways that countervail how others think. Note that none of the attributes of creative thinking is fixed. One can *decide* to overcome obstacles, take sensible risks, and so forth.

One study (Lubart & Sternberg, 1995) showed that greater risk-taking propensity was associated with creativity for artwork but not for essays. When we investigated the reason for this finding, we found that some evaluators tended to give lower grades to essays that took unpopular positions. We learned, therefore, that one of the risks people face when they are creative, even in an experiment on risk-taking, is that the evaluators will not appreciate the risks that go against their own beliefs!

Risk-taking is probably not extremely domain-general. Certainly, there is no necessary relation between willingness to take physical risks and willingness to take intellectual risks. And even within the intellectual domain, one's willingness to take risks may be conditioned by what one perceives as the reward structure for taking the risks.

Motivation. Intrinsic, task-focused motivation is also essential to creativity. The research of Amabile (1983) and of Deci and Ryan (1985) has shown the importance of such motivation for creative work, and has suggested that people rarely do truly creative work in an area unless they really love what they are doing and focus on the work rather than the potential rewards. Motivation is not something inherent in a person: One *decides* to be motivated by one thing or another. Often, people who need to work in a certain area that does not particularly interest them decide that, given the need to work in that area, they had better find a way to make it interest them. They then look for some angle that makes the work they need to do appealing rather than boring to them.

Intrinsic task-focused motivation is highly domain-specific. People are intrinsically motivated to do some things rather than others. So this aspect of creativity may be highly channeled into some activities but not others.

Environment. Finally, one needs an environment that is supportive and rewarding of creative ideas. One could have all of the internal resources needed in order to think creatively, but without some environmental support (such as a forum for proposing those ideas), the creativity that a person has within him or her might never be displayed.

Environments typically are not fully supportive of the use of one's creativity. The obstacles in a given environment may be minor, as when an individual receives negative feedback on his or her creative thinking, or major, as when one's well-being or even life is threatened if one thinks in a manner that defies convention. The individual therefore must *decide* how to respond in the face of the virtually omnipresent environmental challenges that exist. Some people let unfavorable forces in the environment block their creative output; others do not.

Part of the environment is determined by who is doing the evaluating. In one study (Lubart & Sternberg, 1995), we had creative products of individuals of different ages rated for their creativity by raters from different age cohorts. We found informal evidence of cohort matching—that is, raters tended to rate as more creative products of creators of roughly their own age cohort. For example, people often tend to prefer the popular music of the generation in which they grew up as early adolescents more than the popular music of the generation in which their parents or children grew up. Thus, part of what may determine growth patterns of creativity (Simonton, 1994) is in the changing criteria for evaluations of creativity on the part of raters.

Environments may be generally supportive or nonsupportive of creativity, but more often than not, they are mixed. In China today, for example, great creativity is encouraged in the economic domain, but creativity is not encouraged in the political domain and may even be hazardous. In general, repressive governments do not encourage creativity in any domain that they perceive as threatening their own existence. In recent years the United States government politicized science to an extent never before seen in American history, actively supporting researchers perceived as buying into its extreme right-wing agenda and not supporting many of those who did not buy into it. Scientific reports were edited to make them conform to the sociopolitical agenda of the government. Protests by the scientific community were unheeded.

Confluence. Concerning the confluence of these six components, creativity is hypothesized to involve more than a simple sum of a person's level on each component. First, there may be thresholds for some components (e.g., knowledge) below which creativity is not possible regardless of the levels on other components. Second, partial compensation may occur in which a strength on one component (e.g., motivation) counteracts a weakness on another component (e.g., environment). Third, interactions may also occur between components, such as intelligence and motivation, in which high levels on both components could multiplicatively enhance creativity.

Creative ideas are both novel and valuable. But they are often rejected when the creative innovator stands up to vested interests and defies the crowd (see Csikszentmihalyi, 1996). The crowd does not maliciously or willfully reject creative notions. Rather, it does not realize, and often does not want to realize, that the proposed idea represents a valid and advanced way of thinking. Society often perceives opposition to the status quo as annoying, offensive, and reason enough to ignore innovative ideas.

Evidence abounds that creative ideas are often rejected (Sternberg & Lubart, 1995). Initial reviews of major works of literature and art are often negative. Toni Morrison's *Tar Baby* received negative reviews when it was first published in 1981, as did Sylvia Plath's *The Bell Jar* in 1963. The first exhibition in Munich of the work of Norwegian painter Edvard Munch opened and closed the same day because of the strong negative response from the critics. Some of the greatest scientific papers have been rejected not just by one but by several journals before being published. For example, John Garcia, a distinguished biopsychologist, was immediately denounced when he first proposed that a form of learning called classical conditioning could be produced in a single trial of learning (Garcia & Koelling, 1966).

From the investment view, then, the creative person buys low by presenting an idea that initially is not valued and then attempting to convince people of its value. This attitude may be as domain-general as the individual wishes. After convincing others that the idea is valuable, which increases the perceived value of the investment, the creative person sells high by leaving the idea to others and moving on to another idea. People typically want others to love their ideas, but immediate universal applause for an idea often indicates that it is not particularly creative.

The Role of Decision Making

Creativity, according to the investment theory, is in large part a decision. The view of creativity as a decision suggests that creativity can be developed. Simply requesting students to be more creative can render them more creative if they believe that the decision to be creative will be rewarded rather than punished (O'Hara & Sternberg, 2000–2001).

To be creative, one must first *decide* to generate new ideas, analyze these ideas, and sell the ideas to others. In other words, a person may have synthetic, analytical, or practical skills but not apply them to problems that potentially involve creativity. For example, one may decide to follow other people's ideas rather than synthesize one's own. Alternatively, one may decide not to subject one's ideas to a careful evaluation. Or, expecting people to listen to one's ideas, one may decide not to try to persuade others of the value of these ideas. Skills are not enough: One first needs to make the decision to use them.

For example, ability to switch between conventional and unconventional modes of thinking is important to creativity. One aspect of switching between conventional and unconventional thinking is the decision that one is willing and able to think in unconventional ways—that one is willing to accept thinking in terms different from those to which one is accustomed and with which one feels comfortable. People show reliable individual differences in willingness to do so (Dweck, 1999). Some people (whom Dweck calls "entity theorists") prefer to operate primarily or even exclusively in domains that are relatively familiar to them. Other people (whom Dweck calls "incremental theorists") seek out new challenges and new conceptual domains within which to work.

I have proposed a number of different decisions by which one can develop one's own creativity as a decision (Sternberg, 2001): (a) redefine problems, (b) question and analyze assumptions, (c) do not assume that creative ideas sell themselves: sell them, (d) encourage the generation of ideas, (e) recognize that knowledge can both help and hinder creativity, (f) identify and surmount obstacles, (g) take sensible risks, (h) tolerate ambiguity, (i) believe in oneself (self-efficacy), (j) find what one loves to do, (k) delay gratification, (l) role-model creativity, (m) cross-fertilize ideas, (n) reward creativity, (o) allow mistakes, (p) encourage collaboration, (q) see things from others' points of view, (r) take responsibility for successes and failures, (s) maximize person–environment fit, and (t) continue to allow intellectual growth.

These decisions vary in their domain-specificity versus generality. For example, finding what one loves to do is quite domain-specific, but encouraging collaboration may be quite domain-general. If this section of the essay has had any single point to make, it is that creativity is not domain-general or domain-specific. Rather, *aspects* of it are predominantly domain-specific or domain-general, and people individually may differ in the extent to which they show domain-generality in the attitudes that lead to creative thinking.

Evidence Regarding the Investment Theory

Research within the investment framework has yielded support for this model (Lubart & Sternberg, 1995). This research has used tasks such as (a) writing short stories using unusual titles (e.g., "The Octopus's Sneakers"), (b) drawing pictures with unusual themes (e.g., the earth from an insect's point of view), (c) devising creative advertisements for boring products (e.g., cufflinks), and (d) solving unusual scientific problems (e.g., how to tell whether someone had been on the moon within the past month?). Our measures have the same goal as do Torrance's, but we attempt to use tasks that are more oriented toward what individuals do in school and in the real world when they think creatively. This research has shown creative performance to be moderately domain-specific and to be predictable from a combination of certain resources, as described below. The exact blend of resources and the success with which these resources are blended may vary from one culture to another. For example, Niu and Sternberg (2001) found that both American and Chinese evaluators rated two distinct artistic products (collages and science-fiction characters) of American college students to be more creative than products of Chinese college students roughly matched for conventional intelligence (Niu & Sternberg, 2001). This finding held up regardless of whether the raters were American or Chinese.

One concern we have is whether creative skills can be measured in a way that is distinct from the way one measures *g*-based analytical skills (i.e., the kinds of skills measured by conventional tests of general intelligence) and the practical skills that, together with the analytical and creative ones, inform my theory of successful intelligence.

In one study (Sternberg et al., 1999), we used the so-called Sternberg Triarchic Abilities Test (STAT—Sternberg, 1993) to investigate the relations among the three abilities. A total of 326 high school students, primarily from diverse parts of the United States, took the test, which comprised 12 subtests in all. One group of four subtests measured analytical abilities; a second group, creative abilities; and a third group, practical abilities. For each type of ability, there were three multiple-choice tests and one essay test. The multiple-choice tests, in turn, involved verbal, quantitative, and figural content, respectively.

We found that a confirmatory factor analysis on the data supported the triarchic theory of human intelligence, yielding separate and uncorrelated analytical, creative, and practical factors. The lack of correlation was due to the inclusion of essays as well as multiple-choice subtests. Although multiple-choice tests tended to correlate substantially with multiple-choice tests, their correlations with essay tests were much weaker. We found the multiple-choice analytical subtest to load highest on the analytical factor, whereas the essay creative and performance subtests loaded highest on their respective factors. Thus, measurement of creative and practical abilities probably ideally should be accomplished with other kinds of testing instruments that complement multiple-choice instruments. In sum, creative skills could be measured separately from analytical and practical ones.

In a different study (Sternberg & the Rainbow Project Collaborators, 2006), open-ended performance-based measures were used to assess creativity. These performance tasks were expected to tap an important part of creativity that might not be measured by multiple-choice items alone, for open-ended measures require more spontaneous and free-form responses.

For each of the tasks, participants were given a choice of topic or stimuli on which to base their creative stories or cartoon captions. Although these different topics or stimuli varied in their difficulty for inventing creative stories and captions, these differences are accounted for in the derivation of IRT ability estimates.

Each of the creativity performance tasks were rated on criteria that were determined a priori as indicators of creativity.

1. Cartoons. Participants were given five cartoons purchased from the archives of the *New Yorker*, but with the captions removed. The participants' task was to choose three cartoons, and to provide a caption for each cartoon. Two trained judges rated all the cartoons for cleverness, humor, originality, and task appropriateness on five-point scales. A combined creativity score was formed by summing the individual ratings on each dimension except task appropriateness, which theoretically is not a measure of creativity *per se*.
2. Written Stories. Participants were asked to write two stories, spending about 15 min on each, choosing from the following titles: "A Fifth Chance," "2983," "Beyond the Edge," "The Octopus's Sneakers," "It's Moving Backwards," and "Not Enough Time" (Lubart & Sternberg, 1995; Sternberg & Lubart, 1995). A team of six judges was trained to rate the stories. Each of six judges rated the stories for originality, complexity, emotional evocativeness, and descriptiveness on five-point scales.
3. Oral Stories. Participants were presented with five sheets of paper, each containing a set of 11–13 images linked by a common theme: keys, money, travel, animals playing music, and humans playing music. There were no restrictions on the minimum or maximum number of images that needed to be incorporated into the stories. After choosing one of the pages, the participant was given 15 min to formulate a short story and dictate it into a cassette recorder, which was timed by the proctor for the paper assessments and by the internal computer clock for the computer assessments. As with the written stories, each judge rated the stories for originality, complexity, emotional evocativeness, and descriptiveness on five-point scales.

In a sample of 793 first-year college students from around the United States at colleges ranging from not selective at all to very selective, there emerged a separate creativity factor that differentiated the creative performance tests from the other tests. We also found that adding our creative measures to analytical as well as practical measures roughly doubled the predictive value of the SAT for our sample in predicting grades for first-year college students (Sternberg & the Rainbow Collaborators, 2006). The measures also served to *decrease* ethnic differences between groups. In a related study (Stemler et al., 2006), we found that measuring creativity in tests of college-level achievement (in psychology and statistics) also decreased ethnic-group differences. These reductions resulted presumably because groups may differentially emphasize the development of analytical versus creative skills in the socialization of their children.

Conclusion

Creativity is as much a decision about and an attitude toward life as it is a matter of ability. Creativity is often obvious in young children, but it may be harder to find in older children and adults because their creative potential has been suppressed by a society that encourages intellectual conformity. It is neither domain-general nor domain-specific; it has elements of each. We can increase its domain-generality by carrying with us the attitudes that support it across a variety of domains.

In our research, we have found that the domain-specificity of creativity depends in part upon knowledge base, in part upon developed skills for accessing the knowledge base, but particularly upon the decisions that one makes regarding how to use the knowledge base. Knowledge is by no means sufficient for creativity (see Weisberg, 1999). Indeed, our data suggest that knowledge can both impede and facilitate creative thinking (Frensch & Sternberg, 1989). The important element of creativity is the socialization that prepares one to think "outside the box" and thus use one's knowledge in a creative fashion (Simonton, 1988a, b). Children can be taught in ways that emphasize creative thinking, and when they do think creatively, school achievement improves on average (Sternberg et al., 1998a, b). Indeed, measuring creativity also improves prediction of performance in university settings (Sternberg & the Rainbow Project Collaborators, 2006). Parents and schools alike therefore need to rethink how they school children, emphasizing not just the importance of what children are taught but also the importance of how they are taught to think about it.

References

Amabile, T. M. (1983). *The social psychology of creativity*. New York: Springer.

Cattell, R. B. & Cattell, A. K. S. (1973). *Measuring intelligence with the Culture Fair Tests*. Champaign, IL: Institute for Personality and Ability Testing.

Csikszentmihalyi, M. (1996). *Creativity: Flow and the psychology of discovery and invention*. New York: Harper Collins.

Deci, E. L. & Ryan, R. M. (1985). *Intrinsic motivation and self-determination in human behavior*. New York: Plenum.

Dweck, C. S. (1999). *Self-theories: Their role in motivation, personality, and development.* Philadelphia, PA: Psychology Press/Taylor & Francis.

Frensch, P. A. & Sternberg, R. J. (1989). Expertise and intelligent thinking: When is it worse to know better? In R. J. Sternberg (Ed.), *Advances in the psychology of human intelligence* (Vol. 5, pp. 157–188). Hillsdale, NJ: Erlbaum.

Garcia, J. & Koelling, R. A. (1966). The relation of cue to consequence in avoidance learning. *Psychonomic Science, 4*, 123–124.

Gardner, H. (1993). *Creating minds: An anatomy of creativity seen through the lives of Freud, Einstein, Picasso, Stravinsky, Eliot, Graham, and Gandhi.* New York: Basic Books.

Guilford, J. P. (1950). Creativity. *American Psychologist, 5*, 444–454.

Lubart, T. I. (1994). Creativity. In R. J. Sternberg (Ed.), *Thinking and problem solving* (pp. 290–323). San Diego, CA: Academic Press.

Lubart, T. I. & Sternberg, R. J. (1995). An investment approach to creativity: Theory and data. In S. M. Smith, T. B. Ward, & R. A. Finke (Eds.), *The creative cognition approach* (pp. 271–302). Cambridge, MA: MIT Press.

Morrison, T. (1981). *Tar baby.* New York: Alfred A. Knopf.

Niu, W.-H. & Sternberg, R. J. (2001). Cultural influences on artistic creativity and its evaluation. *International Journal of Psychology, 36*, 225–241.

O'Hara, L. A. & Sternberg, R. J. (2000–2001). It doesn't hurt to ask: Effects of instructions to be creative, practical, or analytical on essay–writing performance and their interaction with students' thinking styles. *Creativity Research Journal, 13*, 197–210.

Plath, S. (1963/1971). *The bell jar.* New York: Harper & Row.

Rubenson, D. L. & Runco, M. A. (1992). The psychoeconomic approach to creativity. *New Ideas in Psychology, 10*, 131–147.

Simonton, D. K. (1988a). Creativity, leadership, and chance. In R. J. Sternberg (Ed.), *The nature of creativity* (pp. 386–426). Cambridge, England: Cambridge University Press.

Simonton, D. K. (1988b). *Scientific genius: A psychology of science.* Cambridge, England: Cambridge University Press.

Simonton, D. K. (1994). *Greatness: Who makes history and why.* New York: Guilford Press.

Stemler, S. E., Grigorenko, E. L., Jarvin, L., & Sternberg, R. J. (2006). Using the theory of successful intelligence as a basis for augmenting AP exams in psychology and statistics. *Contemporary Educational Psychology, 31*, 344–376.

Sternberg, R. J. (1981). Intelligence and nonentrenchment. *Journal of Educational Psychology, 73*, 1–16.

Sternberg, R. J. (1982). Natural, unnatural, and supernatural concepts. *Cognitive Psychology, 14*, 451–488.

Sternberg, R. J. (1985). *Beyond IQ: A triarchic theory of human intelligence.* New York: Cambridge University Press.

Sternberg, R. J. (1988). Mental self-government: A theory of intellectual styles and their development. *Human Development, 31*, 197–224.

Sternberg, R. J. (1993). *Sternberg Triarchic Abilities Test.* Unpublished test.

Sternberg, R. J. (1997a). *Successful intelligence.* New York: Plume.

Sternberg, R. J. (1997b). *Thinking styles.* New York: Cambridge University Press.

Sternberg, R. J. (1999). A propulsion model of creative contributions. *Review of General Psychology, 3*, 83–100.

Sternberg, R. J. (2001). Teaching psychology students that creativity is a decision. *The General Psychologist, 36*, 8–11.

Sternberg, R. J. (Ed.) (2003a). *Psychologists defying the crowd: Stories of those who battled the establishment and won.* Washington, DC: American Psychological Association.

Sternberg, R. J. (2003b). *Wisdom, intelligence, and creativity synthesized.* New York: Cambridge University Press.

Sternberg, R. J. (2005). WICS: A model of positive educational leadership comprising wisdom, intelligence, and creativity synthesized. *Educational Psychology Review, 17*(3), 191–262.

Sternberg, R. J. & Gastel, J. (1989a). Coping with novelty in human intelligence: An empirical investigation. *Intelligence, 17*, 187–197.

Sternberg, R. J. & Gastel, J. (1989b). If dancers ate their shoes: Inductive reasoning with factual and counterfactual premises. *Memory and Cognition, 17*, 1–10.
Sternberg, R. J. & Grigorenko, E. L. (1995). Styles of thinking in school. *European Journal for High Ability, 6*, 201–219.
Sternberg, R. J. & Kalmar D. A. (1997). When will the milk spoil? Everyday induction in human intelligence. *Intelligence, 25*, 185–203.
Sternberg, R. J. & Lubart, T. I. (1991). An investment theory of creativity and its development. *Human Development, 34*, 1–31.
Sternberg, R. J. & Lubart, T. I. (1995). *Defying the crowd: Cultivating creativity in a culture of conformity*. New York: Free Press.
Sternberg, R. J. & Lubart, T. I. (1996). Investing in creativity. *American Psychologist, 51*, 677–688.
Sternberg, R. J., Torff, B., & Grigorenko, E. L. (1998a). Teaching for successful intelligence raises school achievement. *Phi Delta Kappan, 79*, 667–669.
Sternberg, R. J., Torff, B., & Grigorenko, E. L. (1998b). Teaching triarchically improves school achievement. *Journal of Educational Psychology, 90*, 374–384.
Sternberg, R. J., Grigorenko, E. L., Ferrari, M., & Clinkenbeard, P. (1999). A triarchic analysis of an aptitude–treatment interaction. *European Journal of Psychological Assessment, 15*(1), 1–11.
Sternberg, R. J., Kaufman, J. C., & Pretz, J. E. (2001). The propulsion model of creative contributions applied to the arts and letters. *Journal of Creative Behavior, 35*, 75–101.
Sternberg, R. J., Kaufman, J. C., & Pretz, J. E. (2002). *The creativity conundrum*. New York: Psychology Press.
Sternberg, R. J. & The Rainbow Project Collaborators (2006). The Rainbow Project: Enhancing the SAT through assessments of analytical, practical, and creative skills. *Intelligence, 34*, 321–350.
Tetewsky, S. J. & Sternberg, R. J. (1986). Conceptual and lexical determinants of nonentrenched thinking. *Journal of Memory and Language, 25*, 202–225.
Torrance, E. P. (1962). *Guiding creative talent*. Englewood Cliffs, NJ: Prentice-Hall.
Torrance, E. P. (1974). *Torrance tests of creative thinking: Norms-technical manual*. Lexington, MA: Personnel.
Weisberg, R. W. (1999). Creativity and knowledge: A challenge to theories. In R. J. Sternberg (Ed.), *Handbook of creativity* (pp. 226–250). New York: Cambridge University Press.
Zhang, L. F. & Sternberg, R. J. (2006). *The nature of intellectual styles*. Mahwah, NJ: Erlbaum.

Chapter 3
Scientific Creativity as a Combinatorial Process: The Chance Baseline

Dean Keith Simonton

The goal of this chapter is to formulate a theory of creativity that uses parsimonious assumptions and logical derivations to obtain comprehensive explanations and precise predictions with respect to the most secure empirical results regarding the phenomenon. In short, the plan is to get the most with the least. The specific formulation is founded on a two-part argument. First, I argue that combinatorial models fulfill these strict requirements. That is, models based on combinatorial processes make the fewest assumptions and by logical inferences explain the widest range of established facts as well as make the most precise predictions with respect to those data. Second, I argue that even if combinatorial models are incomplete from the standpoint of one or more criteria, such models must still provide the baseline for comparing all alternative theories. That is, rival theories must account for whatever cannot be accounted for by chance alone—or what exceeds the chance baseline. This position closely parallels the concept of null hypothesis significance testing in statistics, in which researchers must demonstrate that the discovered effects, whether mean differences or correlations, exceed what could be expected by chance alone (Simonton, 2007). The rationale for this view follows from the standard scientific principle known as "Ockham's razor," or the "law of parsimony." Scientists should prefer a simple explanation over a complex explanation when the former suffices to explain the data.

Because a full treatment of the combinatorial models would require a monograph rather than a chapter, I narrow the focus. In particular, I concentrate on developing combinatorial models of creativity in science (see Simonton, 2004a). This choice is governed by the fact that a great deal is known about scientific creativity, owing to the large body of research done in the psychology and sociology of science (e.g., Feist, 2006). But later I briefly discuss how the model can be extended to all forms of creativity, scientific or otherwise.

D.K. Simonton (✉)
Distinguished Professor, Department of Psychology, One Shields Avenue,
University of California, Davis, CA 95616-8686, USA
e-mail: dksimonton@ucdavis.edu

P. Meusburger et al. (eds.), *Milieus of Creativity*, Knowledge and Space 2,

Assumptions

The combinatorial model begins by defining the three essential components of creativity in science (Csikszentmihalyi, 1990). The first is the individual scientist or inventor, who is the primary agent for the origination of new ideas. The second is the domain in which that scientist operates. The domain consists of the ideas, concepts, theories, techniques, methods, questions, and issues that define a particular scientific discipline. The third component is the field, that is, the set of colleagues who are active in the same domain. Given these definitions, it is now possible to proceed to the six assumptions underlying the combinatorial model.

Assumption 1

During the course of education and training, each individual obtains a random or quasi-random sample of ideas defining his or her chosen domain. Individual domain samples are heterogeneous rather than homogeneous because each person takes a different set of courses at different times; from different instructors; with different textbooks; and with varying backgrounds, interests, and capacities. Given the chaotic nature of the diverse forces that determine the acquisition of expertise in a given domain, one can assume that the samples are random, or at least quasi-random. Nevertheless, it is also clear that the samples will overlap sufficiently and that all of the individuals can be said to belong to the same domain. That is, they represent the same field.

Assumption 2

Within each field individuals will vary in the size of the samples drawn from the domain. A few individuals will have large samples; others, very small samples, with most having samples in the middle range. More specifically, suppose that the sample sizes for any field are distributed according to the normal "bell-shaped," or Gaussian, curve. This supposition reflects the fact that most of the intellectual and dispositional traits that influence expertise acquisition tend to be normally distributed. An obvious example is general intelligence.

Assumption 3

Fields vary widely in their size. Some fields may consist of only a small number of active researchers, whereas other fields may attract dozens of investigators, perhaps even hundreds. At times it is even possible to speak of a field that consists of only one individual, as when there appears a "lone wolf" who opens up a new area of scientific inquiry.

Assumption 4

Each individual subjects the ideas in his or her domain sample to a random or quasi-random combinatorial process. This assumption is consistent with some introspective reports concerning the creative process in scientific geniuses (e.g., Hadamard, 1945; Poincaré, 1913/1921). Even so, it has to be recognized that certain constraints are imposed on this combinatorial procedure. For instance, individuals engaged in what Kuhn (1970) called "normal science" (p. 10) will operate under more constraints than those who can be considered scientific revolutionaries. Similarly, scientists who create in paradigmatic disciplines will have the combinatorial process more constrained than those who create in nonparadigmatic disciplines (Simonton, 2004b).

Assumption 5

The ideas produced by the combinatorial process vary tremendously in quality. That is, they will have differential fitness with respect to various scientific criteria, such as logic and fact. Even those ideas that satisfy minimal standards for "publishability" will vary greatly in quality. Indeed, it is likely that high-impact ideas are extremely rare relative to more ordinary ideas. And unpublishable ideas will outnumber publishable ideas.

Assumption 6

Publishable ideas must be communicated to the larger field of scientists who are active in the same domain. Once published, the idea provides a new element to the domain and can thus become part of the domain sample of subsequent researchers. Even so, the rate at which new ideas enter into the domain pool of ideas varies greatly. In particular, the rate of assimilation depends on several factors, such as communication practices (journals vs. books; least-publishable units), gate-keeping procedures (peer review, editorial policies), publication lags (first- vs. second-tier journals), and diffusion to secondary sources (e.g., introductory texts and popularizations). Hence, the rate at which new ideas enter future individual domain samples will vary across time and discipline.

Implications

To work out the implications of these assumptions, one has to recognize that scientific ideas emerge in two distinct contexts. On the one hand, scientific ideas emerge out of the careers of individual scientists. In this sense, scientific productivity is a phenomenon. On the other hand, ideas also appear within communities of scientists. From this perspective, scientific creativity is a sociological phenomenon.

Scientific Careers: Output

The most conspicuous aspect of any scientist's career is the items that he or she lists in the publication section of his or her curriculum vitae. Although these publications can adopt many forms, the most important and most prominent of them are the articles that appear in professional journals. In any case, two features of these publication lists must be highlighted, namely, individual variation and longitudinal change.

Individual variation. According to Assumptions 1 and 2, each individual acquires a sample of the ideas making up a particular domain, but the size of these samples is normally distributed among the individuals making up the field. According to Assumption 4 each individual applies a combinatorial procedure to these ideas. Because the number of combinations that can be generated from a given set of ideas grows roughly exponentially with the number of ideas in the individual's domain sample, the implication is that the cross-sectional distribution of available combinations will *not* be normally distributed (Simonton, 2003). Instead, the distribution will be highly skewed, with a very long upper tail (Huber, 1999; Shockley, 1957; Simon, 1955). In other words, a very small percentage of the individuals will produce a disproportionate number of the ideational combinations. This skewed distribution is precisely what has been discovered in the empirical literature. Typically, those in the upper 10% in total output publish about 50% of all of the work, whereas those in the lower 50% of the distribution contribute only about 15% of the work (Dennis, 1955). Most scientists tend to cluster at the bottom of the distribution rather than in the middle of it, so the modal output is a single publication. This extremely skewed cross-sectional distribution has been expressed in terms of two laws.

1. According to the Lotka Law, the number of scientists responsible for T publications is inversely proportional to the square of T (Huber, 2001; Lotka, 1926). Stated more formally, $f(T) = k/T^2$, where k is a constant that depends on the particular scientific discipline being examined. By taking logarithms of both sides, one gets $\log f(T) = \log k - 2 \log T$. This version of the law shows that if the distribution is graphed on a log–log plot (i.e., a presentation in which both vertical and horizontal axes are log-transformed), one should get a straight line with a negative slope. This expectation holds very well, at least as a first approximation.

2. The Price Law takes a different approach to specifying the distribution (Price, 1963). Let N represent the number of scientists who are actively publishing in a particular discipline. Then, according to the law, 50% of all of the publications can be credited to an elite consisting of only $N^{1/2}$ scientists. For instance, if 100 researchers are working in an area, then just 10 will account for half of all the publications. One interesting implication of the Price Law is that the output distribution becomes more elitist as a field increases in size. Thus, if there are 100 scientists, then 10% will make half of the contributions, but if there are 1,000 scientists then only about 3% of the scientists will dominate the field (Simonton, 2004a).

Hence, this first implication is amply supported by the data. Nonetheless, according to Assumption 5, ideational combinations vary greatly in quality. Some can be considered breakthroughs, while others will be barely publishable in even second- or third-tier journals. And presumably high-impact combinations are much less frequent than run-of-the-mill combinations. As a consequence, those scientists who generate the most combinations enjoy an elevated likelihood of generating truly important combinations. Put differently, quality should be a positive function of quantity. This prediction of the combinatorial model has been demonstrated in numerous investigations. On the average, those who can list the greatest total number of publications are those who can claim the larger number of high-impact publications (Davis, 1987; Platz, 1965; Platz & Blakelock, 1960; Simonton, 1985; White & White, 1978). This finding can be formally specified as the *equal-odds baseline* $H_i = \rho_1 T_i$, where H_i is the number of "hits" for scientist i, T_i is the total lifetime output for that same scientist, and ρ_1 is the overall "hit rate" characteristic of a given scientific domain. Because both H_i and T_i are ratio-scaled variables with a zero point (i.e., no high-impact works and no publications), the line with a slope of ρ_1 begins at the origin of both the vertical and horizontal axes. It should be obvious that $H_i \leq T_i$.

However, it must be recognized that for any given set of scientists making up a given field, scatter will always exist around the equal-odds baseline (Cole & Cole, 1973; Feist, 1997). Some individuals will have more hits than expected; others, fewer. Yet this outcome is likely given Assumption 1. Ideational samples vary not just in size but also in content. Because scientists are working with different ideas sampled from a particular domain, and because some of these samples may permit more fruitful ideational combinations than do other samples, scientists will not all have identical hit rates. Given that the individual samples have been assumed to be random or quasi-random, this scatter around the baseline can be expressed as a stochastic term. In particular, $H_i = \rho_1 T_i + u_i$, where the last term is a random variable that has a skewed distribution like H_i and T_i, and where $0 \leq u_i \leq T_i(1 - \rho_1)$, so that $H_i \leq T_i$ still holds. If u_i is very large, then scientist i will have a high proportion of hits relative to total output (e.g., "perfectionists"); but if this term is close to zero, the scientist will have a very low proportion of hits relative to total output (e.g., "mass producers").

Longitudinal change. Clearly, total lifetime output is not all concentrated in a narrow period of life but rather is distributed over the course of the career. The question then becomes how productivity is distributed across consecutive years. According to a recent series of investigations, two facts stand out (Huber, 1998a, b, 2000, 2001; Huber & Wagner-Döbler, 2001a, b). First, annual output appears to be randomly distributed across time. That is, there is no evidence for any conspicuous "runs," whether at the beginning, middle, or end of the career. Second, the distribution of output in any given year tends to be best described according to a highly skewed Poisson distribution. More precisely, the probability of having j contributions in a given year is specified by $P(j) = \mu^j e^{-\mu}/j!$, where $e = 2.718\ldots$ and $j! = 1 \times 2 \times 3 \times \ldots \times j$. Yet this is exactly the outcome expected given the combinatorial model (Simonton, 2004a). If each scientist is generating ideational combinations according to a random or quasi-random procedure (Assumption 4), and if the ideas thus produced vary greatly in equality (Assumption 5), then publishable ideas should be randomly distributed across consecutive years of the career. Furthermore, given that the production of a single publishable idea constitutes a relatively rare event for a given year, coming up with two in the same year is even rarer, and contributing three in a particular year is rarer still. Hence, the most frequent amount of output in a given year should be 0; the next most, 1; the third most, 2; and so forth. This kind of distribution is best described as Poisson, the distribution characteristic of rare events—events having such low probabilities of occurrence that they can happen only because there are so many trials.

On the basis of Assumption 5, it should also be apparent that the same two expectations apply not just to quantity of output but also to quality of output. Because high-impact ideas are a subset of publishable ideas, and because the output of the combinatorial process should be variable in quality, it is possible to derive another equal-odds baseline for the output of "hits" for the ith scientist in career year t, namely, $H_{it} = \rho_2 T_{it}$, where ρ_2 is the proportion of hits per total annual attempts and T_{it} is the total output in year t. As before, because the combinatorial process is presumed to be random or quasi-random, the true number of hits per attempts will vary around this chance baseline, requiring the introduction of a stochastic term. The revised prediction then becomes $H_{it} = \rho_2 T_{it} + u_{it}$, where $0 \leq u_{it} \leq T_{it}(1 - \rho_2)$ so that $H_{it} \leq T_{it}$. A considerable body of research supports this theoretical expectation. Those years in which a scientist publishes the most are highly likely to be those years in which that scientist's best work appears (Cole, 1979; Oromaner, 1977; Over, 1988, 1989; Simonton, 1985, 1997). Furthermore, when this prediction is combined with those in the previous paragraph, the expectation is that (a) high-impact publications will be randomly distributed across consecutive years and (b) the annual output of those publications will be Poisson distributed (but with a smaller μ).

Scientific Communities: Multiples

It is now necessary to shift perspectives. Rather than just examine scientific output from the standpoint of the individual, one must also look at discovery and invention from the viewpoint of the scientific community. This necessity is demonstrated in the phenomenon of multiples (Simonton, 2004a), the event in which a particular contribution is made by two or more investigators working independently of one another. Among the famous examples in the history of science and technology are the invention of calculus by Newton in 1671 and Leibniz in 1676; the prediction of the planet Neptune by J. C. Adams in 1845 and Leverrier in 1846; the discovery of the law of the conservation of energy by J. R. von Mayer in 1843, Helmholtz in 1847, and Joule in 1847; the discovery of the periodic law of the elements by DeChancourtis in 1862, Newlands in 1864, L. Meyer in 1869, and Mendeleev in 1869; the proposal of the theory of evolution by natural selection by C. Darwin in 1844 and Wallace in 1858; promulgation of the theory of emotions by W. James in 1884 and Lange in 1887; the development of the principles of marginal utility theory by Jevons, Menger, and Walras in the early 1870s; the invention of the ophthalmoscope by C. Babbage in 1847, Helmholtz in 1851, and Anagnostakis in 1854; and the invention of the telephone by A. G. Bell in 1876 and E. Gray in 1876.

Because such examples are abundant—amounting to hundreds of cases—the multiples phenomenon has often been interpreted as evidence for sociocultural determinism (Lamb & Easton, 1984). At a particular point in time, the "zeitgeist," or spirit of the times, makes a particular discovery or invention absolutely inevitable. The idea is like a ripe fruit ready for the picking. However, the empirical facts underlying the appearance of multiples are not consistent with this interpretation. On the contrary, the multiples phenomenon is actually most consistent with combinatorial models of the creative process in science (Simonton, 2003). Specifically, only such models account for the following four findings: the distribution of multiple grades, the temporal separation of multiple discoveries, the degree of multiple identity, and individual variation in multiple participation.

Distribution of multiple grades. Multiples vary appreciably in the number of independent scientists or inventors claiming the same contribution. This feature has been termed the multiple's *grade* (Merton, 1961). Thus, the invention of calculus can be considered a grade 2 multiple or doublet; the law of conservation of energy, a grade 3 multiple or triplet; the periodic law of the elements, a grade 4 multiple or quadruplet, and so on. If discoveries and inventions are absolutely inevitable, as the sociocultural determinists argue, then high-grade multiples should be very common relative to low-grade multiples, and multiples in general should be more common than "singletons," that is, those ideas that had only a single originator (Merton, 1961). Yet the exact opposite holds: the higher the multiple grade, the fewer the number of cases, and singletons outnumber multiples by a substantial margin. Indeed, the distribution looks just like the distribution of annual output within individual careers. Empirical tests actually show that the frequencies follow the same Poisson distribution (with $\mu \approx 1$). This outcome is precisely what is predicted by the combinatorial model (Simonton, 2004a). Given a field consisting of scientists or inventors subjecting overlapping domain samples to quasi-random combinations (Assumptions 1 and 4), then the odds are nonzero that two or more individuals will chance upon the same or similar ideational combinations. Yet singletons will have a higher likelihood than multiples, and low-grade multiples will have a higher likelihood than high-grade multiples. There is nothing deterministic about this process. The role of the so-called zeitgeist is reduced to providing the ideas that make up the domain from which the contributors draw their respective samples.

Temporal separation of multiple discoveries. Some multiples are virtually simultaneous, whereas others may take decades or even centuries to appear (Merton, 1961). For instance, Bell and Gray showed up on the same day to patent their respective telephones, whereas Mendel's laws of genetic inheritance were not rediscovered until after a 35-year delay. Again, sociocultural determinists draw big inferences about "inevitability" from those cases in which multiples are nearly simultaneous. Yet this argument is also invalid. In line with Assumption 6, it is possible to construct a combinatorial model that includes a "negative contagion process" that accommodates the fact that it takes a finite amount of time for a new idea to enter the domain (Brannigan & Wanner, 1983a; Simonton, 1986). Once an idea becomes part of future domain samples, then it can no longer generate new multiples. At present, no one is working on inventing the wheel. Given this conception of the phenomenon, the distribution of temporal separation will be very similar to that of multiple grades. Most frequently, the independent contributions will occur within a single year, next most frequently within 2 years, and so on. Multiples that take a decade or more to unfold will be very rare and will represent extreme inefficiencies in the communication of scientific or technological innovations. Furthermore, as the diffusion of knowledge has become more efficient over historical time, multiples should more closely approximate simultaneity. Likewise, those disciplines that feature more effective modes of diffusion will produce multiples that more closely approximate simultaneity. Both of these expectations are supported by empirical research (Brannigan & Wanner, 1983b). In addition, when multiples must appear at shorter time intervals, the opportunity for high-grade multiples is also curtailed. That pattern, too, is borne out (Brannigan & Wanner, 1983b). Indeed, with the advent of modern electronic communication, multiples beyond grade 2 have become increasingly rare.

Degree of multiple identity. One significant fact often overlooked by social determinists is that the various contributions said to constitute a multiple are seldom identical (Schmookler, 1966). Indeed, at the time that the ideas first appeared they may not have even been considered the same. Only long after the fact have historians of science applied some generic category to what can actually be considered distinct discoveries or inventions. This phenomenon can be understood in the context of Assumptions 1 and 4. Because scientists are subjecting their respective domain samples to random or quasi-random combinatory processes, and because those samples are themselves a random or quasi-random subset of the ideas making up the domain, the probability is very small that two scientists or inventors will arrive at absolutely identical combinations. Instead, their contributions will converge on some components and diverge on others. Furthermore, the combinatorial model predicts that the magnitude of multiple similarity will display a distinctive distribution. The vast majority of discoveries and inventions will be singletons sharing nothing with other discoveries and inventions. Next in frequency will come those instances of multiples that share just one feature. And rarer still would be those multiples that share two components. And so forth. Some evidence from disputes over patent infractions suggests that this backward-J curve has empirical validity (Schmookler, 1966).

Individual variation in multiple participation. Though multiples constitute comparatively rare events, they are not randomly distributed across scientists and inventors. On the contrary, some individuals tend to be involved in more multiples than do other individuals. The combinatorial model actually specifies who is most likely to participate in multiples (Simonton, 2004a). There are two factors.

First, according to Assumption 2, individuals vary in the size of the samples drawn from the domain, and sample size directly determines the total number of combinations that can be generated over the course of the career. Those who generate more combinations than others are more likely to conceive combinations whose ideas overlap with the ideational combinations of others working in the same domain. Hence, one would predict, on the average, that the most prolific scientists and inventors tend to be involved in more multiples. A single-idea investigator would have the lowest probability of multiple involvement.

Second, according to Assumption 3, domains vary in the size of their corresponding fields. At one extreme, a domain may have only a couple of investigators working with samples drawn from that domain (e.g., new areas of research), while at another extreme there may be dozens even hundreds of investigators creating combinations from overlapping samples from the same domain (e.g., "hot" topics). On average one would anticipate that those who are working in unpopular areas with few colleagues will produce fewer multiples than those who are working in popular areas with numerous colleagues. A lone wolf who opens up new territory would be the least likely to get involved in multiple discovery or invention.

Both of the foregoing predictions have been confirmed in empirical research (Hagstrom, 1974; Simonton, 1979).

Elaboration

I have to impose a qualification on the earlier empirical conclusion that creative output is randomly distributed across the career. That conclusion pertains only to annual productivity at the individual level for the overwhelming majority of scientists. A different picture emerges if two forms of data aggregation are simultaneously imposed (Simonton, 2004a). First, one can aggregate the output counts into larger time intervals, such as consecutive 5- or 10-year periods. Second, one can aggregate across individuals who vary in lifetime output, including those who are the most prolific with the one-idea creators. This twofold aggregation yields a very different result: The output of scientific ideas tends to be a curvilinear function of age (Simonton, 2004a). Productivity rises quickly to a career maximum and then gradually declines thereafter. Both the specific location of the peak and the slope of the post-peak decline are contingent on the nature of the domain (Simonton, 1991, 1997). In some domains, such as mathematics, the peak comes relatively early in the career and the decline is somewhat steep, whereas in other domains, such as geology, the peak comes later and the decline is more gradual.

It is possible to accommodate these aggregate results by a modest elaboration of the combinatorial model (Simonton, 2004a). In effect, one of the model's assumptions was that the ideas generated by the combinatorial process were communicated directly to the scientific community. This simplistic conception is replaced with a two-step procedure. Each scientist begins with a sample of domain ideas that defines his or her *creative potential*. In the first step, these ideas are then subjected to an *ideation* process that generates the raw combinations. In the second step, the resulting combinations are subjected to an *elaboration* process that puts them into publishable form. In short, creativity takes time. Given this two-step process, one can derive a formula that predicts output per time unit as a function of *career age*, that is, the number of years that have elapsed since the scientist began subjecting his or her domain sample to the combinatorial mechanism. That formula is expressed as $p(t) = abm(b - a)^{-1}(e^{-at} - e^{-bt})$, where $p(t)$ is ideational output at career age t (in years), e is the exponential constant (~2.718), a is the typical ideation rate for the domain ($0 < a < 1$), b is the typical elaboration rate for the domain ($0 < b < 1$), and m is the individual's creative potential (i.e., maximum number of ideational combinations in indefinite lifetime). On those occasions in which $a = b$, then $p(t) = a^2mte^{-at}$, an equation that yields a very similar age curve. It is then assumed that the number of contributions T_{it} is proportional to p, the proportionality constant depending on the "least-publishable-unit" characteristic of a discipline.

The curve generated by the foregoing equation correlates in the mid- to upper 0.90s with actually aggregate tabulations in a wide range of data sets (Simonton, 1997). The equation does well even in the prediction of longitudinal output of highly prolific creators. For instance, the correlation between predicted and observed patent output of Thomas Edison is 0.74 (Simonton, 2004a). In addition, by adjusting the ideation and elaboration rates (a and b), the combinatorial model can easily accommodate interdomain contrasts in the career trajectories (Simonton, 1997). For example, the parameters for chemistry are 0.042 and 0.057, whereas those for geology are 0.024 and 0.036. The model can also handle individual differences in trajectories for those scientists or inventors working within the same domain (Simonton, 1997). This explanatory power is based on individual differences in (a) creative potential (m) and (b) the age at career onset (i.e., chronological age at $t = 0$) that enable the model to make a large number of unique predictions that cannot be generated by any rival model (Simonton, 2004a). For instance, the two-step model predicts that the correlation between output levels in consecutive decades of a career will be described by a single-factor model rather than by a simplex or quasi-simplex structure indicative of an autoregressive process (Simonton, 1997). Or to offer another illustration, the model predicts that the correlation between age at first major contribution and age at last major contribution will become zero when age at the best contribution is partialed out (Simonton, 1991). Because these predictions have been empirically confirmed, the model imposes strong constraints on any rival model. To illustrate, the trajectory must be a function of career age rather than of chronological age and cannot be dictated by any form of accumulative-advantage process (Simonton, 2004a).

Conclusion

I have just shown that a combinatorial model of scientific creativity can account for the key features of the phenomenon. Moreover, it does so using fewer assumptions than any alternative theory, and it generates empirically confirmed predictions that cannot be accommodated by rival theories. At present, however, it must appear to be rather abstract and remote from the nitty-gritty details of discovery and invention. Nevertheless, it is possible to ground the combinatorial model in the particulars of scientific practice (see Simonton, 2004a). That is, it is possible to base Assumptions 1 and 4 on (a) the cognitive and dispositional characteristics of creative scientists (e.g., openness to experience, broad interests, and unconventionality), (b) the specific structure of their research programs (e.g., the simultaneous execution of multiple, but interrelated, inquiries), (c) the degree and type of involvement in professional activities (e.g., attendance at professional meetings), (d) the attributes of high-impact research collaborations (e.g., collaborators who are heterogeneous with respect to background, training, and status), and (e) the concrete operation of peer review (e.g., the heterogeneity of referee evaluations). Taken together, these features of the scientific enterprise provide the basis for more or less unrestricted ideational combinations.

Although the emphasis of this chapter is on scientific creativity, it is relatively easy to extend the model to encompass other forms of creative activity, such as that seen in the arts (Simonton, 2004a). This extension requires only the recognition of two facts. First, artistic domains are defined somewhat more broadly than are scientific domains. Artistic domains (e.g., the thematic content of painting and sculpture) may not only overlap more than scientific domains do, such domains may include elements from everyday life (e.g., semiautobiographical novels and self-reflective poetry). Second, the constraints on artistic creativity are less well defined and more weakly enforced than is the case in the sciences. For instance, the definition of a "publishable poem" is far less precise than that of a "patentable invention." One repercussion of these differences is that artistic "multiples" are extremely rare, in contrast to multiple discovery and invention. Another consequence is that artistic creativity can be considered even more strongly combinatorial than scientific creativity. This greater combinatorial freedom is consistent with what is known about the personal characteristics of creative artists (Feist, 1998). Hence, combinatorial models probably provide the optimal approach to a general theory of creativity—a theory that applies to all creative domains.

References

Brannigan, A. & Wanner, R. A. (1983a). Historical distributions of multiple discoveries and theories of scientific change. *Social Studies of Science*, *13*, 417–435.

Brannigan, A. & Wanner, R. A. (1983b). Multiple discoveries in science: A test of the communication theory. *Canadian Journal of Sociology*, *8*, 135–151.

Cole, S. (1979). Age and scientific performance. *American Journal of Sociology*, *84*, 958–977.

Cole, S. & Cole, J. R. (1973). *Social stratification in science*. Chicago, IL: University of Chicago Press.
Csikszentmihalyi, M. (1990). The domain of creativity. In M. A. Runco & R. S. Albert (Eds.), *Theories of creativity* (pp. 190–212). Newbury Park, CA: Sage.
Davis, R. A. (1987). Creativity in neurological publications. *Neurosurgery, 20*, 652–663.
Dennis, W. (1955, April). Variations in productivity among creative workers. *Scientific Monthly, 80*, 277–278.
Feist, G. J. (1997). Quantity, quality, and depth of research as influences on scientific eminence: Is quantity most important? *Creativity Research Journal, 10*, 325–335.
Feist, G. J. (1998). A meta-analysis of personality in scientific and artistic creativity. *Personality and Social Psychology Review, 2*, 290–309.
Feist, G. J. (2006). *The psychology of science and the origins of the scientific mind*. New Haven, CT: Yale University Press.
Hadamard, J. (1945). *The psychology of invention in the mathematical field*. Princeton, NJ: Princeton University Press.
Hagstrom, W. O. (1974). Competition in science. *American Sociological Review, 39*, 1–18.
Huber, J. C. (1998a). Invention and inventivity as a special kind of creativity, with implications for general creativity. *Journal of Creative Behavior, 32*, 58–72.
Huber, J. C. (1998b). Invention and inventivity is a random, Poisson process: A potential guide to analysis of general creativity. *Creativity Research Journal, 11*, 231–241.
Huber, J. C. (1999). Inventive productivity and the statistics of exceedances. *Scientometrics, 45*, 33–53.
Huber, J. C. (2000). A statistical analysis of special cases of creativity. *Journal of Creative Behavior, 34*, 203–225.
Huber, J. C. (2001). A new method for analyzing scientific productivity. *Journal of the American Society for Information Science and Technology, 52*, 1089–1099.
Huber, J. C. & Wagner-Döbler, R. (2001a). Scientific production: A statistical analysis of authors in mathematical logic. *Scientometrics, 50*, 323–337.
Huber, J. C. & Wagner-Döbler, R. (2001b). Scientific production: A statistical analysis of authors in physics, 1800–1900. *Scientometrics, 50*, 437–453.
Kuhn, T. S. (1970). *The structure of scientific revolutions* (2nd ed.). Chicago, IL: University of Chicago Press.
Lamb, D. & Easton, S. M. (1984). *Multiple discovery*. Avebury, England: Avebury.
Lotka, A. J. (1926). The frequency distribution of scientific productivity. *Journal of the Washington Academy of Sciences, 16*, 317–323.
Merton, R. K. (1961). Singletons and multiples in scientific discovery: A chapter in the sociology of science. *Proceedings of the American Philosophical Society, 105*, 470–486.
Oromaner, M. (1977). Professional age and the reception of sociological publications: A test of the Zuckerman-Merton hypothesis. *Social Studies of Science, 7*, 381–388.
Over, R. (1988). Does scholarly impact decline with age? *Scientometrics, 13*, 215–223.
Over, R. (1989). Age and scholarly impact. *Psychology and Aging, 4*, 222–225.
Platz, A. (1965). Psychology of the scientist: XI. Lotka's law and research visibility. *Psychological Reports, 16*, 566–568.
Platz, A. & Blakelock, E. (1960). Productivity of American psychologists: Quantity versus quality. *American Psychologist, 15*, 310–312.
Poincaré, H. (1921). *The foundations of science: Science and hypothesis, the value of science, science and method* (G. B. Halstead, Trans.). New York: Science Press (Original work published 1913).
Price, D. (1963). *Little science, big science*. New York: Columbia University Press.
Schmookler, J. (1966). *Invention and economic growth*. Cambridge, MA: Harvard University Press.
Shockley, W. (1957). On the statistics of individual variations of productivity in research laboratories. *Proceedings of the Institute of Radio Engineers, 45*, 279–290.
Simon, H. A. (1955). On a class of skew distribution functions. *Biometrika, 42*, 425–440.

Simonton, D. K. (1979). Multiple discovery and invention: Zeitgeist, genius, or chance? *Journal of Personality and Social Psychology, 37*, 1603–1616.
Simonton, D. K. (1985). Quality, quantity, and age: The careers of 10 distinguished psychologists. *International Journal of Aging and Human Development, 21*, 241–254.
Simonton, D. K. (1986). Multiple discovery: Some Monte Carlo simulations and Gedanken experiments. *Scientometrics, 9*, 269–280.
Simonton, D. K. (1991). Career landmarks in science: Individual differences and interdisciplinary contrasts. *Developmental Psychology, 27*, 119–130.
Simonton, D. K. (1997). Creative productivity: A predictive and explanatory model of career trajectories and landmarks. *Psychological Review, 104*, 66–89.
Simonton, D. K. (2003). Scientific creativity as constrained stochastic behavior: The integration of product, process, and person perspectives. *Psychological Bulletin, 129*, 475–494.
Simonton, D. K. (2004a). *Creativity in science: Chance, logic, genius, and zeitgeist*. Cambridge, England: Cambridge University Press.
Simonton, D. K. (2004b). Psychology's status as a scientific discipline: Its empirical placement within an implicit hierarchy of the sciences. *Review of General Psychology, 8*, 59–67.
Simonton, D. K. (2007). Chance. In N. J. Salkind (Ed.), *Encyclopedia of measurement and statistics* (Vol. 1, pp. 129–133). Thousand Oaks, CA: Sage.
White, K. G. & White, M. J. (1978). On the relation between productivity and impact. *Australian Psychologist, 13*, 369–374.

Chapter 4
The Riddle of Creativity: Philosophy's View*

Günter Abel

Who Does Not Know What Creativity Is?

The Value of Creativity

Whether in the arts, science, religion, technology, the media, or everyday life, creativity clearly has high value and a positive connotation. This esteem holds across different cultures. Intellectually and aesthetically, creativity is highly prized in persons, processes, and products ranging from an individual's reputation to Nobel prizes. Although—or perhaps because—no one knows what happens as creativity occurs, people take great pleasure in creative persons, processes, and products, indeed, in creativity itself. Creativity is elusive, an intimate stranger, an inscrutable close friend.

Meaning and Sense of the Word

Who does not know what creativity is? Who does not know the difference between creative and noncreative persons, processes, and products? Yet when it comes to spelling out what creativity really is, no one seems to know the answer any longer.

Creativity has to do with bringing something new into being, into the world. It refers not to mere novelty but to something genuinely new, something that did not use to exist. Imagine that I am taking off my watch, putting it on the sheet of paper just in front of me, and writing the prime number 7 on the upper right corner of this sheet. Those acts are probably something new, something never been done before. But it is rather uninteresting and without value (at least for the time being).

G. Abel (✉)
Technische Universität Berlin, Institut für Philosophie, Ernst-Reuter-Platz 7, 10587 Berlin, Germany
e-mail: abel@tu-berlin.de

* Revised version of G. Abel (2006).

P. Meusburger et al. (eds.), *Milieus of Creativity*, Knowledge and Space 2,

One could produce an infinite number of novelties just like it. That which is *creative* is always something genuinely new, illuminating, subjectively precious, and unprecedented.

This observation makes an initial crucial distinction: the differentiation between (a) creativity as mere novelty (the first appearance of something that results from uniquely combining antecedent elements according to given rules) and (b) creativity as genuine, or radical, creativity (the act of bringing about something fundamentally new). Mere novelties come about by combining known elements according to known rules in a way that has not been done before. The improbability of such combinations signals creative momentum. In fact, psychological tests sometimes use the frequency of such improbable combinations as a measure of creativity.

This combination theory of creativity quickly hits its limits, however. First, it can describe only how given elements of a given system are combinable, not what happens when a person modifies, violates, or even abandons the principles and basic rules of the system itself in creating a unique form of organization with its own new principles and rules. In other words, combination theory cannot effectively account for the point at which radical originality is achieved. Second, reference to a new combination does not illuminate how something radically new was made possible and what exactly it consists in. Combination theory presupposes creativity but cannot explain it.

As for radical creativity, one can think of pioneering and style-setting artists, including Beethoven, Cézanne, Picasso, and Michelangelo, or path-breaking research scientists such as Lobachevsky, Copernicus, Einstein, and Heisenberg. It pertains to generative systems (systems that contain, in a mathematical sense, all possible outcomes). Characteristically, radical creativity transgresses and even jettisons the rules and basic patterns of an underlying generative system in order to usher in new rules and principles and thus organize the material in a fundamentally new way. Well-known examples are the transition from Euclidian to non-Euclidian geometry, from the tonal to the atonal system in music, and from the linear to the ring-shaped concept of the benzene molecule within the carbon ring of Kekule.

Is There a Science of Creativity?

If radical creativity is attributed so basic a function, how does one describe and explain the character of the processes of creativity itself? After all, the cognitively and aesthetically enlightening, inspiring, and style-setting features of creative processes arguably:

1. Cannot be reduced to a set of antecedent elements and their principles and rules (which are violated and superseded in radical creativity)
2. Cannot be deduced (or, hence, forecast) logically, causally, or psychologically from such a given set of elements

3. Are linked more to luck than to reason
4. Are known to be quite spontaneous
5. Are unpredictable, and
6. Are characterized by discontinuities

These six aspects strongly suggest that no scientific explanation of creativity is possible in the narrow terminological sense of "science."

Can the semantics of science change to provide for a *scientific* analysis of creative processes, as when positing that creativity itself becomes inherent in scientific objects such as matter and the universe and thereby becomes a new paradigm in scientific research? The question is challenging. At the moment, an alteration of that sort would eventually modify science with regard to law-likeness, predictability, and the projectability of scientific predicates.

This prospect in no way implies that one must understand creative persons, processes, and products as romanticizing mystifications; the entire phenomenon of creativity must be placed beyond this dichotomy. Nor does it in any way imply that creative ideas occur only by chance, that one just has to wait for them. The opposite is the case. Apparently, intimate knowledge of the given field is necessary for a creative thought to leap out. Moreover, creative minds are usually workaholics par excellence. To cross boundaries, one has to know them. Picasso knew and mastered the dominant painting techniques of his time well before he came up with his own style. Schönberg mastered the mechanisms of tonal music before he started producing atonal compositions. Lobachevsky knew Euclidian geometry before he rejected the fifth axiom and thereby made way for non-Euclidean geometry. The cliché of the "lazy genius" is totally misleading.

Weak, Strong and Moderate (Intuitive) Creativity

Against this background, I distinguish three types of creativity: (a) weak creativity (the act of combining existing elements into new arrangements); (b) strong creativity (the transformation, violation, and replacement of old principles, rules, and patterns with new ones; see Hausman, 1998, p. 454); and (c) moderate, or intuitive, creativity (the phenomenon associated with the constitutive role that human imagination plays in human cognition, perception, language use, and representation). Let me elucidate the aspect of moderate, or intuitive, creativity in greater detail.

Perception and imagination. Whenever people successfully individuate the contents of what they perceive, say, and think, there is more to the process than meets the eye (on this point and the following two paragraphs, see Abel, 1999, pp. 145–168). Every actual situation (i.e., every situation existing at the present moment), also has constitutive nonactual components. Perceiving the person on the other side of the street as Uncle Paul means that one must immediately draw on constitutive former perceptions of Uncle Paul. "Imagination" is taken to mean precisely this individual capacity to bring the nonactual components (without which perceiving, speaking, and thinking would be impossible) into a present process of perceiving, speaking, and thinking. Without this genuine imagination, a person would only see someone on the other side of the street but would not recognize the figure as Uncle Paul.

Imagination also enters into perception when one concentrates on the conceptual components of perception. Imagination is operant in the processes of (a) perceiving different objects of the same kind (e.g., recognizing a variety of objects as "chairs") and (b) having different perceptions of the same object (e.g., recognizing the book in front of me to be the book I bought in Berlin yesterday). Hence, both the identity of kind and the identity of concepts always presupposes *intuitive imagination*, which is the key element of intuitive creativity and a basically relevant facet of the other two types of creativity as well. Intriguingly, this constitutive role of imagination in perceptual representation is not represented in perception itself. Nothing about the perceived objects (percepts) reveals their dependence on their imaginative, creative constituents.

Meaning, reference, and imagination. Understanding the meaning and reference of a word or expression, whether natural or artificial, includes grasping the propositional attitudes of the speaker (or hearer) and the intended meaning(s) of the word or expression. But attitude and content are not conveyed by the syntactical occurrence of the word or expression alone. Imagination—and, hence, creativity—is necessary in order to assign and understand meaning successfully. Creativity is not entirely restricted to the extraordinary. It begins at home.

So does "reference." Nothing about a table suggests that it is denoted by the word *table* and that that word refers to the table. Establishing and fixing the relation between the intensional and extensional realm of a denoting sign depends on imagination. But this decisive relation is always presupposed when *table* is used to speak about tables and to refer to them. No successful cognition, meaning, reference, or representation is without imagination, without intuitive creativity. It is hidden but intensely relevant for human routines.

In mental representation, this statement holds independently of whether one is a propositionalist (such as philosophers Jerry A. Fodor and Zenon W. Pylyshyn, conceiving representation as a language-, sentence-, and proposition-like phenomenon) or a pictorialist (such as cognitive psychologist Stephen M. Kosslyn or philosopher Ned Block, conceiving mental representations as quasi-pictorial processes). In both views, imagination and, hence, intuitive creativity are necessary ingredients of cognition. Other clear cases of imagination in the developed sense are (a) memory (e.g., speaking and thinking about past events), (b) reference to possible future situations (e.g., mentioning one's birthday next year), (c) fictional discourse (e.g., understanding sentences like "Don Quixote plays tennis with Pegasus and Mr. Pickwick in Wimbledon"), and (d) counterfactual conditionals (e.g., understanding sentences such as "If Peter hadn't bought the car, he wouldn't have had the accident").

This intermediate, moderate, or, more specifically, intuitive creativity goes beyond weak creativity insofar as it does not merely combine given elements in a new way but rather brings presently nonactual aspects into the actual processes of perceiving, speaking, and thinking. But it is not yet *radical* creativity. Intuitive creativity also involves a form of negativity, which plays an important part in creating something radically new.[1] The questions and problems pertaining to these three types of creativity—weak, strong, and moderate (intuitive)—likewise differ by type. They, too, are about novelties, the nonactual components of actual events, and radically new processes and things.[2]

Creativity in Psychology, Creativity in Philosophy

Psychological Features of Creative Persons

There is a difference between the treatment of creativity in psychology and its treatment in philosophy. To put it in a terribly simplified way, the focal interest in psychology lies in the personality features of creative individuals and in their psychic motivational profiles that are conducive to creative actions. Examples are Wallas's (1926) identification of preparation, incubation, illumination, and elaboration as phases of the creative processes, distinctions that in principle go back to Helmholtz (1884) and Poincaré (1908).

[1]Creativity does not thereby become a negational act. Creativity is a positive, positioning act. It is *positio*, to borrow a term from medieval philosophy. Creativity is, however, internally tied to a form of negativity (whose broad sense includes more than logical and grammatical negation) as part of its new positioning activity. Thus, one cannot state positivist grounds for creativity in a strict sense. Consider also the aspects characteristic of radical creativity (see this chapter's subsection entitled "From Possibilities to Potentialities"): the violation of established rules, the establishment of differences, the withdrawal from familiar horizons, and the renunciation of hitherto existing worldviews and established norms and standards. The presence of nonactual components within actual processes of creativity is constitutive for all these processes. In addition, the modal connection between creativity and potentialities and the distinction between "nothing" (*Nichts*), "not-yet-being" (*Noch-Nicht*), and "actual reality" (*Aktual Wirkliches*) might be helpful in elucidating characteristics of creativity's processes. For these reasons, a comprehensive philosophy of creativity must include the relation between creativity and negativity in a systematic way. *Positio*, *negatio* and their interaction are characteristic for creative processes, states, and abilities.

If creative momentum is required in each individuation of things or processes as such-and-such things or processes (in assigning or denying qualities that do or do not belong to the essence of a thing or process), then an element of creativity appears in Spinoza's famous phrase "determinatio negatio est" (determination is negation), which he communicated in a letter of 1674 (see Spinoza, 1977, p. 210). According to Spinoza, the figure (*figura*) is negation in that mere matter is indefinite in itself and therefore without figure. That is, matter in itself does not possess figure (*gestalt*). Hegel (1832/1975) elevated this momentum to the thesis that negativity is inherent in the form. He saw the inner "ground" of the becoming (*Werden*), of the "unrest of self-motion" (*Unruhe der Selbstbewegung*), and of the "pulsation of liveliness" (*Pulsation des Lebendigen*) as being based on this form of negativity (vol. 1, p. 157; vol. 2, p. 61).

[2]At this point, still finer distinctions are necessary: (a) epistemic creativity (concerning the generation of new epistemic objects); (b) semantic creativity (concerning the "semantic innocence" and the new organization of the semantic features of signs, that is, their meaning, reference, and conditions for truth or satisfaction); (c) agent creativity (concerning the bringing into being by human actions); and (d) cognitional creativity (concerning the changes in cognitive horizons, principles, and rules of individuals or other generative systems).

Creative individuals have a number of psychological characteristics, including a well-developed sense of imagination; advanced problem-solving skills; the ability to construe new structures and find regularities quickly in seemingly chaotic situations; a willingness to challenge traditional assumptions, standards, and norms; and the production of surprising visualizations. To arrive at scientific answers to the types of questions pertaining to this make-up, psychologists have required subtle tests, models, survey procedures, and sound interpretations thereof to provide a basis for an explicit psychometrics of creative attitude—an intense field of research in recent decades (see Lenk, 2000; Sternberg, 1999).

As part of that foundation, one aspect of the psychology of creative persons deserves special mention in this chapter. It is that highly creative individuals, more than other persons, risk becoming overwhelmed by stimulation because they usually expose themselves to a rather unfiltered flow of it. However, they are often able to use this surfeit chaos productively, channeling it into their output. (The fact that this later transfer does not always succeed, as the lives of creative persons show at times, is presumably the ultimate meaning of the observation that being creative can be highly uncomfortable, even mortally dangerous.) Such overflow and the ability to transfer it into acts and products of creativity should be distinguished conceptually and empirically from the kind of overflow evident in cases of clinical psychoses, such as paranoid schizophrenia. That condition is depicted, for instance, in the film *A Beautiful Mind* (2001), the story of John F. Nash, who received the 1994 Nobel Prize in economics for his contributions to establishing the mathematical principles underlying game theory, an examination of the rivalries among competitors with mixed interests.

Psychological and psychiatric models of creativity are sometimes jeopardized by methodological inadequacies stemming from the orientation to "normality" in their tests, models, and survey procedures and by the fact that their authors often unwittingly presuppose the phenomena of creativity rather than explain them. Despite Plato's assertion that genius is a state of divine insanity, Hans J. Eysenck (1993), a researcher in the fields of intelligence and creativity, has clearly shown that a genuinely creative state cannot be equated with a psychotic state or be seen as one of its manifestations. Eysenck has pointed out that schizophrenia kills all creativity. Similarly, Salvador Dalí (1942, p. 349, note) appealingly remarked that the sole difference between him and a madman is that he is not mad.

Philosophical Assumptions About Creativity

The philosophical question of what creativity is does not center on personality traits of creative individuals but rather on phenomenology and the structures of creative processes themselves. This section offers a brief phenomenology of creativity, the elements of which I call "assumptions about creativity"—the set of requirements that one presupposes as given and satisfied in cases where creativity is ascribed to persons, processes, and products. Possible assumptions about creativity might include the following acts:

- Generating multidimensional associations
- Coupling predicates and subjects to unusual judgments
- Building analogies between remote realms or referents only remotely associated with each other
- Producing metaphors and transferring them from one area to another in order to organize material in a new and informative way
- Simultaneously activating two or more ideas, images, or thoughts and having them interact
- Prompting thought experiments
- Breaking common and established patterns of observation
- Modifying and violating traditional ideas
- Risking discontinuities
- Changing cognitive perspectives
- Raising new points of view
- Juxtaposing methods of one discipline with those of others
- Risking category mistakes
- Switching between different systems of description
- Constructing new epistemic objects as the subjects and entities of cognitive and scientific research
- Modifying hidden collusions
- Modifying, transforming, and violating established rules, principles, patterns, and worldviews

These and other features touch on the phenomenologically and structurally intrinsic features of processes of creativity itself. They also mark the difference between a bounded and a free, creative mind. Of course, merely having some of these features does not guarantee that creativity gets off the ground. They are assumptions about creativity, not criteria of its occurrence. Individually, none of the aforegoing assumptions is either sufficient or necessary for creativity and its ascription. But together they can be sufficient, and each of them can separately be necessary.

Computational Psychology of Creativity

Within computational psychology and cognitive sciences, creativity is usually explained and defined as "the mapping, exploration, and transformation of structured conceptual spaces" (Boden, 1994, p. 84, 2004). From this perspective, "conceptual spaces" can be modeled best in computational terms. Computational psychology pertaining to creativity rests on concepts used in research on artificial intelligence, that is, on the question of what kind of human skills one wants to teach computers (e.g., perceiving, speaking, thinking, and—in robotics—rudimentary movements).

The issue of the relation between a computer's skills and creativity is interesting. On the one hand, every activity of a computer is based on human creativity in the sense that the computer has been programmed by humans. On the other hand, it is possible that computers can be creative in the weak sense introduced above. Boden (1994, p. 85) poses four questions to which common sense would immediately answer "No!" but to which the computational psychologist would answer "Yes!":

1. Can computational concepts help people understand how *human* creativity is possible?
2. Can computers (now or in the future) ever do things that at least appear to be creative?
3. Can computers ever appear to recognize creativity—say, in poems written by human beings or in its own novel ideas about science and mathematics?
4. Can computers themselves ever really be creative (as opposed to offering performance whose originality is merely apparent and due wholly to the human programmer)?

Boden's responses to the first three questions are "*Yes, definitely*; *Yes, up to a point*; and *Yes, necessarily (for any program that appears to be creative)*. In short, computational ideas can help us to understand how human creativity is possible" (p. 85). This is the case because a computer, understood as a generative system, can do only what its program allows it to do. That is why computers are so ideal and interesting with respect to the question of creativity.

I criticize this basic assumption of computational psychology on two counts: (a) the fundamental failure of computer functionalism as a comprehensive theory of mind, especially of the creative mind, and (b) the basic understanding of a generative system. The prevalent idea that the human mind is like a computer[3] posits that mental—hence, also creative—states and processes can be compared to functional and logical states of computers and that they can be described and explained as one would the functioning of computers (Putnam, 1975, pp. 325–451, articles 16–22). In this conceptualization the creative mind is part of the software. Accordingly, it is possible to formulate conditionals. For example, when the generative system and the mind are in a definite state, then a definite input leads to a definite output and therewith to a new definite state. This model is highly attractive because it does not tie mental states and processes to neurobiological realizations and reductions.

But a key objection to it comes from Putnam (1988) himself, the very founder of computational functionalism. He points out that functionalist models of the mind and those based on calculus hit their limits as soon as interest turns to having a comprehensive theory of human cognition, especially of the creative mind. (Though Putnam does not explicitly refer to the *creative* mind, it is clearly implicit in his considerations.) People cannot individuate the contents of their beliefs and ideas

[3]For a more detailed description and critique of this model, see Abel (2005, especially pp. 12–18).

within conceptual space without considering aspects located outside the human brain conceived of as a computer. This impossibility especially applies to the individuation of the contents of creative processes, states, and phenomena. To clarify the semantic and representational features of the creative signs involved in these processes, one must consider many aspects of the relations between signs, time, situations, attitudes, and contexts that are not based on calculus. The creative mind cannot be analyzed or individuated by means of calculus. Above all, computational psychology's model holds that the process of breaking rules and principles, which is characteristic for radical creative thinking, must be regarded as calculable in principle. But obviously, the creative violation of established rules does not follow any meta-rule. That fact is the point of radical creativity. And insufficient intelligence is certainly not the reason for the absence of proof that human creativity is based on calculus.

The second reason for criticizing the basic assumption underlying computational psychology is the concept of a generative system, specifically, the boundedness of creativity to signs and systems. The next section deals with this aspect of my refutation.

Creativity and Signs I: Creative Processes as Signo-Interpretational Processes

Presumably, the most basic characteristic of the human mind, particularly the creative one, is to use and understand symbolizing signs (see Abel, 2005, especially pp. 20–23). Mental creative processes are performed as signo-interpretational processes. The idea is not that mental and cognitive operations are simply operative manipulations of given inner signs (which is the view characteristic of cognitive science and computational psychology). Nor is it that they can be characterized by the new combination of preexisting elements, as claimed in the combination theory of creativity.

It does not mean merely the human mind's dependence on external signs, either—at not least in the sense that a creative mind needs intermediary signs in order to articulate, present, and communicate its forms and contents to other persons. One must take one step more. In the words of Peirce (1960): "We have no power of thinking without signs" (no. 5.265). He even goes so far as to say: "When we think, then, we ourselves, as we are at that moment, appear as a sign" (no. 5.283). Pushing the point, I add that there are no creative minds without signs, with the expression *sign* being understood broadly as including a holistic sense of cognition and intuition. Given the internal relation between sign and interpretation, creative thinking can inherently be described as a signo-interpretational process. People think and are creative by *virtue* of signs, not by means of them. A creative mind is an individual mind capable of channeling these elemental processes in a new, rule-setting, and style-forming way. A creative mind uses given signs in a new way. It invents new signs and interpretations and implements new rules and patterns for their functioning. These three capacities are the signo-interpretational trio of human creativity.

Generative System and Tacit Knowledge

When it comes to system relatedness in the computational and empirical sciences, a generative system always accompanies tacit knowledge. To me, that understanding of tacit knowledge is insufficient. In this context the term really means something entirely different from the sense it has generally had ever since Ryle (1949) and Polanyi (1958, 1966) distinguished between "knowing-*that*" and "knowing-*how*." According to both authors, tacit knowledge refers to those forms of nonpropositional knowledge that cannot be articulated by a *that*-clause and that refer instead to knowing-*how* (as in knowing how to open a bottle of wine).

In connection with generative systems, the expression *tacit knowledge* appears in the sense used by the strand of language philosophy in which language is said to be a generative system—a system that, because of its orthographic symbols, contains, in a mathematical sense, all sentences ever possible (see, for example, Davies, 1986; Dummett, 1973; Evans, 1982; Miller, 1997). Accordingly, a competent speaker possesses the tacit knowledge with which to understand and compose every possible sentence. Ultimately, a quasi-axiomatic structure of language is posited, a structure that defines all possible sentences and inferential relations of that language. Tacit knowledge, therefore, refers to this presupposed structure of a language. It also offers an answer to the question of how it is possible for speakers or listeners of a natural language to produce or understand the infinite number of sentences they have never built or heard before. To put it another way, how is it actually possible to learn a natural language?

The link between a pretended complex of tacit knowledge and the question of creativity is readily apparent. Within formal and ordinary languages the functioning of tacit knowledge and creativity is thought to be about variations of techniques. The character of those languages is thought to be modeled on the patterns of formal semantic theories. The compositionality of meaning, the principle according to which larger linguistic units are composed of smaller ones, is obviously important in this context. It seems that this notion addresses the complex of problems concerning creativity. However, it does so only in the framework of combination theory, the limitations of which I have identified above. If meaning is made up of the elements of a sentence, then it seems as though new combinations of those elements will lead to new meanings and could therefore be called creative.

In my opinion, neither the quasi-axiomatic understanding of language as a system nor the notion of meaning's compositionality is able to explain what radical creativity really comes to. The phenomenon of radical creativity reveals the limits of formal semantics in attempts to explain ordinary creative language. Obviously, creative speaking, thinking, and acting cannot (Wittgenstein, 1980) be described as the result of a calculus with determined rules. The point is not just to arrange elements in a hitherto unknown way but rather to modify or violate established rules and principles, even to supersede them by setting up new ones.

How Does Creativity Proceed?

Creativity and Signs II: The Space of Arbitrariness and Polysemiosis as the Space of Creativity

I have just advanced three theses: (a) Creative processes can be conceived as signo-interpretational processes, (b) There is no creative mind without signs, and (c) Creativity is not to be understood as a mere operative (i.e., merely instrumental) manipulation of given signs with their own exact and delimited meanings. The additional aspects identified in this section bring out what I call the *potentiality space of creativity*.

Every actual use of signs can be labeled "new" in the sense that using and understanding a sign is not simply a replication or recursion of an established convention. The time interval alone precludes the use of any sign in exactly the same way twice. This fact, too, opens up the space of creativity. Language, like any other network of symbols, is not a conventional system with predefined structures that humans first learn and then apply to given situations and contexts. The codes of natural languages are not guaranteed in advance and once and for all. Their lives are "à la merci du lendemain," at tomorrow's mercy (de Saussure, 1957, p. 72; see also Frank, 1991, pp. 55–57).

Creativity in natural languages and other symbol systems makes use of this openness and of the arbitrariness and polysemous nature of signs and languages. Radical creativity is manifest in transforming, violating, and, if necessary, replacing the previous and established structures determining a use of language and other signs. Wit, irony, and metaphors are examples in this context. The human ability to build new sentences and new chains of signs suggests that the meaning of words and symbols are due to the compositionality principle. As for the semantics of sentences and signs, compositionality can account only for the combining of separate elements, that is, for weak creativity. It cannot appropriately explain strong or radical creativity. Radical creativity is not just a matter of content consisting in a combination of single units. It pertains to content consisting of fundamentally new sign formation (including the assignment of its semantic features) that goes beyond the meanings of its parts. It ranges up to inventing new signs and features and to construing new epistemic objects and states of affairs. Hence, the formation embodies more than the sum of the elements composing it. However, radical creativity typically modifies and violates a sign's hitherto existing form and content, overtakes it, realigns it, or even replaces it with newly invented signs and their semantic characteristics (meaning, reference, and conditions of truth or satisfaction).

Signs and sign systems have two aspects that are all-important to creativity space. One is their arbitrariness, "the displacement of the relation between the signifier and what is signified" (de Saussure, 1957, p. 6). It opens up a space of noncalculable and indeterministic characteristics and, hence, possibilities of meaning and reference. The other aspect is their *polysemiosis*. The word is derived from *polysemy*, meaning the multiple meaning and reference of a sign or a sign system. I regard it as a term that enlarges and changes Peirce's (1960) and Morris's (1971) notion of *semiosis* as a process of effectiveness in using and understanding signs. Polysemiosis, too, opens up space for determining meaning and reference. Creativity in signs and languages may in some sense be tied to syntactic and grammatical features. But the central thing is that creative processes always have the power to push the established features, rules, and principles beyond their conventionally prescribed ends. Strong creativity uses and fills this indeterminate space of arbitrariness and polysemiosis, which one might call the signo-interpretational space of creativity.

Given the key role that the arbitrariness and polysemiosis of signs play in creativity space and given the fact that a creative mind having no signs has no power either, the relation between a sign and its successor sign becomes pivotal. Signs are usually followed by other signs. And these successor signs either perpetuate the use and understanding of their preceding signs or they interrupt this continuity (as when one has a problem in using and understanding a sign because its interpretation has become disputable). This point is paramount because it dismisses a particular model of creative language and nonverbal signs, one according to which the successful creative use of language and signs involves processes that render implicit structures explicit and places inferential semantics at the bottom.[4] The model can answer neither the question of creativity nor that of many other phenomena (such as the diverging, metaphorical, ironical, or fictitious use of signs).

The relation between a sign and its successor sign is neither an inferential nor a deterministic relation (logically or causally). It is a *free* relation that, in cases of successful communication, is nonetheless characterized by the fact that the "right" or "fitting" successor sign is directly understood. Take, for example, a successor word in a conversation or a successor line or successor image in a poem. Or think of a flash of inspiration in mathematics that solves a heretofore unsolvable problem. Obviously, the creativity of inventing new and directly understood successor signs is essential in such cases. It is relevant above all when the rules of using signs have been revised, violated, or even replaced by new ones. This matter touches on the question of the relation between creativity and rules.

---

[4]This understanding of the functioning of languages, especially that of creative languages, differs fundamentally from Brandom's (1994) view, which is being widely discussed at present.

Creativity and Rules: Rule-Following, Rule Violation, Rule Invention

Logical or causal determination is absent not only in violating a rule but also in customary rule-following in normal language.[5] Hence, intuitive creativity is at work even in normal rule-following creativity. The operating rules of actual speaking and thinking do not determine the future use of a word, sign, or thought, just as past usage does not determine the current one. These two aspects are part of the space of linguistic and nonlinguistic creativity (e.g., pictorial, graphic, musical, gestural, or mimic creativity) as well.

This concept of rule-following reinforces the manner in which the determination and reorganization of the semantic characteristics of signs depend on interpretation. As already underlined, creativity with regard to language and signs consists essentially in using this space or scope of indeterminacy. And radical creativity consists in reorganizing this space by implementing new rules or principles, changing, and, if necessary, violating the established ones. Instances of these processes are wit, poetic language, and the creation of metaphors. Another example is the diachronic semantic characteristics of the words and signs used, say, for epistemic objects—such as the terms *atom*, *galaxy*, or *gene*.

It may be that only the *conscious* violation of established rules bring something genuinely new into the world. But creative rule violations are not intended to achieve a state of chaos at all. Kant stressed this point when he emphasized that the ability to create and set new rules is to be called *ingenious* and is crucial for art. The creative mind is rule-setting, not a mind of self-satisfying rule destruction. Nietzsche had a name for those who destroy rules but who lack the power to create new ones: "decadents."

To characterize an ingenious mind as a rule-setting mind (or, in the arts, a style-setting mind) means much more than to say that it grasps concepts. Rules cannot be modeled as concepts. In the arts, for instance, even a newly established rule or style cannot be used as a principle for the production of future works of art, although imitators and mannerists would have people believe otherwise. The point about *creative violation of rules* is twofold: (a) rule violation does not follow a meta-rule, and (b) one opens up new and deep insights by establishing new and revealing rules with the greatest of ease. Something of the deep correlation between truth and creativity flashes in this realization (see Abel 2009, in press). Truth (in the broad sense of the word) is located in this open space of indeterminacy, and creativity is instrumental in ferreting out truth.

[5] This idea of linguistic rule-following must be extended to the entire field of *nonlinguistic* signs. It contradicts the view that language is a quasi-axiomatic system having rules that a speaker first acquires and then applies to given cases. That concept misses the central point of what it means to speak a language and be in relation to other persons and to the world by virtue of a language. The sense of "rule-following" that I address here is the one Wittgenstein elaborated (see especially Wittgenstein, 1980, nos. 198–242).

This view has a bearing on the relation between creative thinking and the concept of "rule." Given the difference between strong and weak creativity as introduced at the beginning of this chapter, one must also distinguish between different kinds of rules. Hintikka (1997) has distinguished two types: (a) definitory rules, which in chess, for example, establish what counts as an allowed or as an unallowed move with the pieces; and (b) strategic rules, which, as developed in game-theory, "specify what a given player should do in every possible situation that can arise in the course of a play of the game" (p. 68). In Hintikka's view, the definitory rules are not linked with creativity. To him, creativity "is a matter of strategic rules" (p. 68).

Unsurprisingly, I argue for broadening the types of rules: (a) definitory rules, (b) strategic rules, and in cases of radical creativity (c) regularities that build up in the indeterminate openness, arbitrariness, and polysemiosis of signs and their successor signs and interpretations. Works of art, pioneering scientific theories, and individual forms of life emerge from and move toward this space.

From Possibilities to Potentialities

The fact that new things come into being shows that they are possible. Philosophically, the relation between creativity and possibility is fundamentally relevant. It transcends the combining of given logical possibilities (weak creativity), including the idea of possible world semantics; it encompasses creative dispositional potentialities (strong creativity). The realm of potentialities is much more extensive than the realm of logical possibilities, which consists in the principle of avoiding self-contradictions in biconditional yes–no alternatives. Radical creativity has to do with dispositional potentialities. Obviously, there are possible and real things and events (not to mention forms of life and mental states) that have to be classified as impossible by the criteria of what is logically possible in the narrow sense of possibility. The form and content of paintings by René Magritte or drawings by M. C. Escher, though logically impossible, are directly grasped, pretty real, and not nonsensical. The modal element, understood in its deep sense as potentiality, comes into play within creative processes, persons, and products. More precisely, it is an essential working part of the picture.

The point of a language is not that it consists of the letters of an alphabet and that various combinations of these letters are possible in order to form words, sentences, and whole discourses. The point is rather that language is a potential: "La langue est quelque chose de potentiel, la parole est du réalisé" (de Saussure, 1957, p. 20). The distinction that Aristotle drew between potentiality and actuality can help make decisive differentiations between kinds of creativity, namely, Nothing, Not-Yet-Being, and Actual Reality. Inasmuch as people are not up to considering creativity only as a *creatio ex nihilo* (creation from nothing), this demarcation can be helpful. Technological artifacts—technical machines and systems—are invented and tangibly brought into the world. Technical creativity, artifacts, and technology itself would all be miracles if they were looked upon as *creationes ex nihilo*. Artifacts do not come from nothing.

But I wish to emphasize a critical point about Aristotle's concept of *actus* (actuality) and *potentia* (potential). To Aristotle, the concept describes ontological movement, a transformation of potentiality into actuality. With a rock, for instance, this transformation could be the move from potentiality into a statue by an artist, that is, into the reality of being a statue. But an ontology in which something's unfolding is predisposed threatens to divest strong creativity of its radical character. For strong, or radical, creativity becomes challenging the moment it breaks into ontology, into the assumptions of metaphysics themselves. Neither radical creativity nor radical temporality, which are internally linked, figured in classical metaphysics.[6]

Creativity as Emergence

How does the creative mind fit into a naturalistic and scientific worldview? I propose that creativity can be conceived of as a phenomenon of emergence, a "surprising coming about," with emergence being understood as a given in both the philosophy of mind and the systemic sciences of complexity, such as synergetics, self-organization theory, and chaos theory (on emergence, see Beckermann et al., 1992; Krohn & Küppers, 1992; on the history of emergentism, see Stephan, 1999). Emergent phenomena and creative processes share key characteristics, particularly three to which I have called attention in this chapter to describe the features of creativity: (a) radical newness, (b) unpredictability, and (c) nonreducibility to antecedent elements.

A second response to the question of the relation between the creative mind and a naturalistic, scientific worldview is today's mainstream meta-theoretical perspective on emergence in synergetics, self-organization theory, and research on chaos theory. The central point is that the genuine process character of creative events cannot be described in terms of either the starting or the ending situation. Phenomena of this kind are usually called emergent phenomena. This component of creativity raises the question of how creative processes can be described, characterized, and

[6]Unlike Plato, who saw divine inspiration as the root of creative minds, creative processes, and products, Aristotle conceived of creative processes or acts only as new instantiations of preexisting forms. According to him, creative production can result only in whatever was already predisposed in the antecedent form. In this sense Aristotle's philosophy provides instruments for weak creativity only; it does not suffice for a satisfying treatment of radical creativity. Whereas Plato ultimately jumped too quickly to the idea of divine and irrational inspiration (of course, this assessment is a simplification of Plato's thought), Aristotle came close to reducing the nature of creative processes to little or nothing more than the act of updating antecedent forms. The challenging problem of creativity becomes paramount with the insight that creative processes and products are not reducible to antecedent elements and forms.

interpreted at all when the vocabulary of neither the initial nor final situation is able to articulate the specifically new and creative character of these processes. The problem of describing, presenting, and representing creativity thereby becomes one of the language used to communicate about creativity.

To grasp the difference between a "philosophy of creativity" and "research on the systemic theory of creativity," one must first understand that both fields deal with (a) the relation that the phenomenon has to a generative system and its systemic properties and, more important, (b) the phenomenon's newness cannot be ascribed to systemic properties in a reductionist way. Thinkers engaged in system-based scientific research emphasize the first of these two aspects. Those engaged in philosophical research accentuate the second aspect and try to elucidate creativity in terms of phenomena and concepts, including creative processes, persons, and products. This interface between philosophy and science is intriguing, especially when radical creativity modifies, violates, and even replaces the principles and patterns of the underlying generative system. Theories of systemic complexity will therefore not have the last word in the elucidation of radical creativity. Shifting from complexity theories to creativity theories could make creativity a new paradigm of scientific explanation, too. Such an explicit step has yet to be taken as far as I can see. But something of the sort is already going through people's minds.

The question and phenomenon of creativity is currently felt to be a challenge for science itself. I do not mean only the possibility that "science" may have to surrender in the battle to explain the phenomenon of creativity, be it in neurobiology, modern brain research, or computational psychology. The challenge goes deeper than such renunciation as soon as it becomes necessary to grasp the processes of nature itself as being internally creative. Discussion about creative universes in astrophysics is not the only discourse that goes conspicuously in that direction.

But philosophy is no better off than science when it comes to explaining the phenomenon of creativity. The enormous impact and challenge that the question of creativity has on what philosophers themselves understand philosophy to be is obvious. Ultimately, the phenomenon of creativity leads directly to the question of whether the character of "what there is" is one of metaphysical determinacy or indeterminacy, of permanent pre-established being or radical processual becoming, of an a priori order or of unpredictable and incalculable processes. The entire issue exists within both the smallest and the biggest worlds, ranging from the events among elementary particles to astrophysical processes, from the creativity of human individuals to their interactions with other persons and the world.

The Smallest Contains the Biggest

Ontologically, what one states to be the components or building blocks of nature or the world is crucial. Thus, it makes a key difference whether components or building blocks of nature are posited as things in the sense of material objects occupying space–time–places or rather as processes and events.

I share the view held by those who understand the world as a world of processes, of process things (see Abel, 2004, pp. 222–235). This view is consistent with thinking in modern physics. Indeed it is consistent with the dominant idea within the sciences in general, which is to understand physical objects as sequences of processes or events, which for their part are no longer seen as having the categorical status of "things" (with their paradigm of material objects). Hardly anything is proper for comparing microphysical structures ("particles") with macroscopical objects, which had been the major paradigm of *thing*-ontology. In modern physics, a thing is conceived of as a series of processes or events that are linked in time and identical in type. The physical identity identicalness of individual objects across a time interval is based on the type-identicalness of the processes or events involved.

The assumption that there are process things is required also by the logical form of many kinds of linguistic sentences. In the field of analytic philosophy and after the preliminary work of Reichenbach (see 1947), Davidson (e.g., 1967) showed that the logical form of many of the sentences of our natural language cannot be construed without the assumption regarding processes or events as genuine individuals. Examples of such sentences are those expressing the relations of order in terms of temporal succession, causality, explanation, and action. Take the following sentence, for example: "The creativity conference opened in the main lecture hall of the university and then spread out over different rooms and lasted several days." A person understanding that statement casts processes or events as genuine individuals (in this case, "conferences" spreading out over different spaces and times) and not merely as things to which something happens (e.g., a wooden desk occupying a space–time–place and getting darker in color) (see Abel, 1985, 2004, pp. 222–231).

The shift from this kind of process philosophy to creative processes (and, for instance, to the idea of a creative universe) is readily accomplished. The processes mentioned can be characterized as processes of dynamic variations, of creative development, of dynamic reorganization and new organization—in short, as processes of perpetually creating something new. They can therefore be described as *creative* processes in a broad sense. Every natural process of transformation into a newly organized flow and figure may be understood as a formation of something new (as thought by Whitehead, among others). The spectrum of these processes, then, extends from elementary particles to the formation of new stars in the universe, including the emergence of creative ideas in the minds of individual persons.

How the processes of the continuum between, say, the Big Bang and the creativity of the human mind can be understood is a thrilling question. The relation between creative elementary particles or universes and creative individuals or human minds is *not* that of the relation between the universal and the particular. It is not a discursive relation of having the universal subsume particular cases. It might rather be about the relations in which the universal and the individual were once considered to be of the same nature in principle, as in the tradition of philosophy especially in Cusanus (1440/1977) and Leibniz (1720/1998). From that kind of viewpoint, the universal is conceived of as being determined through infinite degrees down to the individual; and, reciprocally, the individual is conceived of as being determined up to the universal. To elucidate this type of internal relation between the universal and the individual, I offer two phrases from Cusanus (1440/1977) along with my own translation:

- "In qualibet enim creatura universum est ipsa creatura" (for in each creature the universe is this creature itself) (Book 2, Chap. 5, p. 36)
- "Individua vero sunt actu, in quibus sunt contracte universa" (only those individuals are real within whom the universe is present in a contracted way) (Book 2, Chap. 6, p. 46)

In a contracted way, each individual, each finite being, contains the whole universe. The smallest contains the biggest.

If asked today, here and now, astrophysicists would be apt to say that humans would not even be here and that the human mind would not work the way it does if it had not been for the Big Bang. Conversely, one hears that there is a continuum between the lone individual and the universe, a continuum from which the individual issued and to which that individual's perceptions, actions, thoughts, and theories refer. How people see themselves as human beings, as individuals, within that picture and how they do fit into it are two of the most challenging issues at the interface of current philosophy and modern sciences or the scientific worldview. Thus, whatever is brought into existence by an individual in a creative way can be seen as a creative modification of the universe at the same time. And reciprocally, the universe manifests itself within the creative individual and, in turn, is conceived of by that individual as he or she sets up a creative theory of the universe. The creative human mind specifies the whole by casting itself as something different from the universe, that is, in a relation of negation to the whole. It adopts the status of negative, selective attention, which is a precondition for every concept formation (see Kant, 1781/1968, B 156, footnote) and which one may call a creative attitude. This self-recursive or Möbius-strip-like structure is fundamentally relevant, given the position of human beings in nature. The creations of individuals contain the universe, and creative individuals can create new things in the universe. And they do so with remarkable success!

The idea that the universe is a "creative advance into novelty" was developed especially by Whitehead (1929, p. 222). Today this notion is far from being tied solely to his name. One encounters it in the present concepts of astrophysics (see also Kanitscheider, 1993). The difference between Whitehead's approach to the question of creativity and the one put forth in this chapter is that I have not directly assumed an ontology of "actual entities." Instead, I have focused on the presupposition that people's sentences, actions, and thoughts make logical sense. In the context outlined above, one also comes to conceive of nature as process nature and subsequently to conceive of these processes as creative processes in the broad meaning of the term. Creativity is not only an option, it is also a condition of vital truth.

References

Abel, G. (1985). Einzelding- und Ereignis-Ontologie [Ontology of individuals and events]. *Zeitschrift für philosophische Forschung, 39*, 157–187.

Abel, G. (1999). *Sprache, Zeichen, Interpretation* [Language, signs, interpretation]. Frankfurt am Main, Germany: Suhrkamp.

Abel, G. (2004). *Zeichen der Wirklichkeit* [Signs of reality]. Frankfurt am Main, Germany: Suhrkamp.

Abel, G. (2005). Geist – Gehirn – Computer. Zeichen- und Interpretationsphilosophie des Geistes [Mind–brain–computer: Signo-interpretational philosophy of mind]. In R. Dürr, G. Gebauer, M. Maring, & H.-P. Schütt (Eds.), *Pragmatisches Philosophieren. Festschrift für Hans Lenk* (pp. 3–36). Münster, Germany: LIT.

Abel, G. (2006), *Die Kunst des Neuen. Kreativität als Problem der Philosophie* [The art of the new: Creativity as a problem in philosophy]. In G. Abel (Ed.), *Kreativität* (pp. 1–24). Twentieth Deutscher Kongress für Philosophie, Kolloquiumsbeiträge. Hamburg, Germany: Felix Meiner.

Abel, G. (2009, in press). Wahrheit und Kreativität [Truth and creativity]. In G. Abel & M. Ruggenini (Eds.), *Wahrheit, Freiheit, Existenz*. Berlin: Parerga.

Beckermann, A., Flohr, H., & Kim, J. (Eds.) (1992). *Emergence or reduction? Essays on the prospects on nonreductive physicalism*. Berlin/New York: deGruyter.

Boden, M. A. (1994). What is creativity? In M. A. Boden (Ed.), *Dimensions of creativity* (pp. 75–118). Cambridge, MA: MIT Press.

Boden, M. A. (2004). *The creative mind: Myths and mechanisms* (revised and expanded 2nd ed.). London: Routledge.

Brandom, R. B. (1994). *Making it explicit: Reasoning, representing, and discursive commitment*. Cambridge, MA: Harvard University Press.

Cusanus [Nikolai de Cusa] (1977). *De docta ignorantia* (Latin-German 2nd ed.) P. Wilpert, H. G. Senger, & E. Hoffmann (Eds.). Hamburg, Germany: Meiner (Original work written 1440).

Dalí, S. (1942). *The secret life of Salvador Dali*. London: Vision.

Davidson, D. (1967). The logical form of action sentences. In N. Rescher (Ed.), *The logic of decision and action* (pp. 81–95). Pittsburgh, PA: University of Pittsburgh Press.

Davies, M. (1986). Tacit knowledge, and the structure of thought and language. In C. Travis (Ed.), *Meaning and interpretation* (pp. 127–158). Oxford, England: Blackwell.

de Saussure, F. (1957). Cours de linguistique générale, Introduction [General course of linguistics, Introduction]. In R. Godel (Ed.), *Cahiers Ferdinand de Saussure* (Vol. 15, pp. 6–103). Geneva, Switzerland: Librairie Droz.

Dummett, M. (1973). *Frege: philosophy of language*. London: Duckworth.

Evans, G. (1982). *The varieties of reference*. Oxford, England: Clarendon Press.

Eysenck, H. J. (1993). Creativity and personality: Suggestions for a theory. *Psychological Inquiry, 4*, 147–178.

Frank, M. (1991). Individualität und Innovation [Individuality and innovation]. In M. Frank (Ed.), *Selbstbewußtsein und Selbsterkenntnis* (pp. 50–78). Stuttgart, Germany: Reclam.

Hausman, C. R. (1998). Creativity. In M. Kelly (Ed.), *Encyclopedia of aesthetics* (pp. 453–456). Oxford, England: Oxford University Press.

Hegel, G. W. F. (1975). *Wissenschaft der Logik* [Science of Logic] G. Lasson (Ed.). Hamburg, Germany: Meiner (Original work published 1832).

Helmholtz, H. L. F. von (1884). *Vorträge und Reden* [Lectures and speeches]. Braunschweig (Brunswick), Germany: Vieweg.

Hintikka, J. (1997). On creativity in reasoning. In A. E. Andersson & N.-E. Sahlin (Eds.), *The complexity of creativity* (pp. 67–78). Synthese Library, Vol. 258. Dordrecht, The Netherlands: Kluwer Academic.

Kanitscheider, B. (1993). *Von der mechanistischen Welt zum kreativen Universum. Zu einem neuen philosophischen Verständnis der Natur* [From the mechanistic world to the creative universe:

To a new philosophical understanding of nature]. Darmstadt, Germany: Wissenschaftliche Buchgesellschaft.
Kant, I. (1968). *Kritik der reinen Vernunft* [Critique of pure reason] (reprinted 2nd ed. of 1787). In Königliche Preußische Akademie der Wissenschaften (Ed.), *Kants gesammelte Schriften* (Vol. 3). Berlin: Walter de Gruyter (Original work published 1781).
Krohn, W. & Küppers, G. (Eds.) (1992). *Emergenz: Die Entstehung von Ordnung, Organisation und Bedeutung* [Emergence: The formation of order, organization, and meaning]. Frankfurt am Main, Germany: Suhrkamp.
Leibniz, G. W. (1998). *Monadologie* [Monadology] (French-German ed.) H. Hecht (Ed.). Stuttgart, Germany: Reclam (Original work published 1720).
Lenk, H. (2000). *Kreative Aufstiege. Zur Philosophie und Psychologie der Kreativität* [Creative ascents: On the philosophy and psychology of creativity]. Frankfurt am Main, Germany: Suhrkamp.
Miller, A. (1997). Tacit knowledge. In B. Hale & C. Wright (Eds.), *A companion to the philosophy of language* (pp. 146–174). Oxford, England: Blackwell.
Morris (1971). *Writings on the general theory of signs*. The Hague, The Netherlands: Mouton.
Peirce, C. S. (1960). *Collected papers*, Vol. 5 (2nd ed.) C. Hartshorne & P. Weiss (Eds.). Cambridge, MA: Belknap.
Poincaré, H. (1908). *Science et méthode*. Paris: Flammarion.
Polanyi, M. (1958). *Personal knowledge: Towards a post-critical philosophy*. Chicago, IL: Routledge & Kegan Paul.
Polanyi, M. (1966). *The tacit dimension*. Garden City, NY: Doubleday.
Putnam, H. (1975). *Philosophical papers: Vol. 2. Mind, language and reality*. Cambridge, England: Cambridge University Press.
Putnam, H. (1988). *Representation and reality*. Cambridge, MA: MIT Press.
Reichenbach, H. (1947). *Elements of symbolic logic*. New York: Macmillan.
Ryle, G. (1949). *The concept of mind*. New York: Barnes & Noble.
Spinoza, B. De (1977). *Briefwechsel* [Correspondence] C. Gebhardt (Ed.). Hamburg, Germany: Meiner.
Stephan, A. (1999). *Emergenz. Von der Unvorhersagbarkeit zur Selbstorganisation* [Emergence: From unpredictability to self-organization], Dresden, Germany: Dresden University Press.
Sternberg, R. J. (Ed.) (1999). *Handbook of creativity*. Cambridge, England: Cambridge University Press.
Wallas, G. (1926). *The art of thought*. New York: Harcourt, Brace.
Whitehead, A. N. (1929). *Process and reality: An essay in cosmology* (corrected ed.). New York: Macmillan.
Wittgenstein, L. (1980). *Philosophische Untersuchungen 1* [Philosophical investigations]. Frankfurt am Main, Germany: Suhrkamp.

Chapter 5
Creativity: Multidimensional Associative or Chaotic Process? Methodological Comments on Creative Processes and Metaphors in Aesthetics and Innovation

Hans Lenk

Creativity as Mental Propulsion

Combining psychological and methodological aspects in a developmental theory of creativity, Sternberg (2003; see also Sternberg, 1988, 1999) develops a theory of types of creative processes that he calls "the propulsion theory of creative contributions" (2003, pp. 124–143). This new theory of types of creative contributions is intended to replace the investment theory of creative contributions (Sternberg & Lubart, 1995), in which he and his coauthor refer essentially to the decision of a creative person to become and stay creative, to act creatively by engendering ideas beyond or against the usual expectation ("Defy the crowd!"). The main accent of the theory is on being ready to forward, advertise, fight for, and socially innovate new exceptional ideas by which Sternberg describes the intellectual capacities, knowledge, styles of thinking, personality variables, risk acceptance, and the motivation and attitude to overcome obstacles, tolerate ambiguities, and dig into intrinsic motivation and the description of creativity-fostering facts within the social (and natural) environment. The intention is to weave all these components into an integrated set of dynamics and account for certain thresholds and interactions between them. The message sounds rather traditional, like the common theories drawing on well-known and fashionable accounts of capacities and capabilities and on intelligence and personality traits and social factors in an attempt to provide an integrative construct supported by empirical studies and accompanied by educational recommendations like "Sell your creative ideas!" Psychologists like factors that account for deferred-gratification patterns; appropriate willingness to assume risks, engage in conflicts, and save time for creative thinking; practical paragon personalities; and mutual fructification through cross-thinking (Sternberg, 2003, pp. 106–123).

Sternberg's propulsion theory facilitates a much more differentiated consideration of the structural conditions, occasions, causes of events, and stages of progress in typical creative contributions. He distinguishes between eight types of creative

H. Lenk (✉)
Institut für Philosophie, Kaiserstraße 12, 76128 Karlsruhe, Germany
e-mail: sekretariat@philosophie.uni-karlsruhe.de

P. Meusburger et al. (eds.), *Milieus of Creativity*, Knowledge and Space 2,

contribution. The main thesis is that "[c]reativity is by its nature *propulsion*. It moves a field from some point to another. It also always represents a decision to exercise leadership. The creator tries to bring others to a particular point in the multidimensional creative space" (Sternberg, 2003, 125–126). "Leadership, like creativity, is propulsion," too (p. 141). The eight types of contribution constitute a qualitative, nominal classification rather than an ordinal one. Yet "certain types of creative contributions probably tend, on average, to be greater in amounts of novelty than are others" (p. 126). Creativity is notably instigated and characterized by the rather fundamental novelty of what results from creative processes and work involving high-quality assessments and judgments.

Sternberg (2003) tries to group the types of creative contributions into a few major categories, such as those of accepting, perpetuating, or rejecting current paradigms and those of trying to integrate multiple current paradigms that feature a kind of combinatorial *meta*type. The whole enterprise revolves around the dynamic development that the field of creativity undergoes by virtue of the contributions the creative individual makes to bring the field closer to a kind of goal state or new direction. The first, not really genuine, type is mere *replication* that does not change the field at all. The second one, too, *redefinition* of a field or problem, is at most a new perspective but does not change the field. The third type, *forward incrementation*, is typical for what one can call "small creativity" (see also Koestler, next section). It gives rise to meaningful solutions (of a largely combinatorial kind) that fall under the current that is moving in the field anyhow.

By contrast, the fourth type of creative contribution, *advance forward incrementation* or *accelerated forward motion*, involves consciously and explicitly changing the field by transgressing the status quo: "The creator accelerates beyond where others in his or her field are ready to go—often 'skipping' a step that others will need to take" (p. 134). The fifth type, *redirection*, even changes the direction of the extant development from the given state at present. It thus includes a deviation from the past and from the actual general line and strategies. The sixth type is a combination of *reconstruction and redirection* in which the creator moves the field back to a previous point and then moves it in a different direction. "The work is judged as creative to the extent that the individual is judged as correctly recognizing that the field has gone off track and to the extent that the new direction is viewed as a useful one for the field to pursue" (p. 136).

The seventh type of creative contribution is *reinitiation*, a new direction and reorientation. It represents "a major paradigm shift" in which "the contributor suggests moving in a different direction from a different point in the multidimensional space of contributions [by a new] start-over" (p. 138) after having criticized traditional hypotheses, suppositions, premises, and so on. The last, the eighth (actually the genuine seventh), type is *integration*, in which the contributor or "creator puts together two types of ideas previously seen as unrelated or even as opposed. Formerly viewed as distinct ideas, they now are viewed as related and capable of being unified" (p. 139).

Sternberg (2003) presents case studies on all these types of creative contribution, mostly from the psychology of intelligence and capability and from social psychology, but also from music, arts, literature, and science. For instances of new paradigmatic reinitiations, he cites Leon Festinger's theory of cognitive dissonance in psychology (see Festinger & Carlsmith, 1959, for example) as well as Marcel Duchamps's new paradigms in arts and John Cage's in music. The integration type of creative contribution may be illustrated by a new theoretical combination of quantum theory and the general theory of relativity. The types are meant to improve the explanation and representation of changes, differentiations, cultural takeovers, and social innovations without prejudging the quality or heights of creativity and without claiming that just one type is relevant and fitting for a given creative contribution or case. These analytical types overlap, a relationship emphasizing that more than one process, procedure, or encompassing type is relevant. Creativity cannot be characterized by just one trait, process type, or paradigm structure. It is a multifactor phenomenon, and "types of creative contributions do not immediately translate into levels of creative contributions" or other paradigms (Sternberg, 2003, p. 140). Thus, "the propulsion model may help explain several creativity-related phenomena, although it does not provide a unique explanation" (p. 141). Moreover, it may classify software developments and new programs in computers as a sort of replication and may lead to differentiated answers to the long-standing question "of the extent to which creativity is domain-specific or domain-general" (p. 141). Whereas "successful forward incrementations may be largely domain-general,... the ability to perform a reinitiation may be quite a bit more domain-specific, requiring a sense or even feeling for a field that goes well beyond the kinds of more generalized analytical abilities measured by conventional tests" (p. 142).

Sternberg (2003) thinks his classification of types might not be exhaustive yet, admitting it to be "unlikely that there is any one 'right' model of types of creative contributions" (p. 143). In addition, creators and even children

> will need to decide for themselves ... how they wish to unlock and express their creative potential.... [T]hey will decide, because creativity is a decision. How can one encourage people to decide for creativity? According to the view of creativity as a decision, fomenting creativity is largely a matter of fomenting a certain attitude toward problem solving and even toward life. (p. 143)

His model is in fact also a bit integrationist and largely one-dimensional. One can certainly deal with more than two different creative inputs of theories, as would be the case in integration (the eighth type of creative contribution). A creatively dividing development, differentiating refinement, and parsing of factorial components as *another* model is also conceivable. A further possibility—which I think very important—is the methodological ascent to higher theoretical metalevels or even to comprehensive interdisciplinary metatheories, metalanguages, and metaperspectives leading to new higher level insights that I have called "creative ascents" (*Kreative Aufstiege*; see Lenk, 2000a, pp. 59–64, 164–165).

Sternberg's (2003) paradigms, typological classifications, and differentiations seem meaningful, however. They allow one to characterize different forms of creative progressive developments, the movement of the fields of creative procedures or paradigms shifts, and all incremental progress in a manner even more systematic than the interrelationships Sternberg describes. (It must be considered whether, methodologically speaking, these model types are of an ideal-type character in

Max Weber's sense: conceptualizations that are to be more or less clearly separated in a reality but that are frequently overlapping and open to interpretation or to combination with one another.) They might be interesting for describing jumps and shifts across the traditional fields in arts, music, and literature and for projecting or transferring new creative or provocative paradigms from one field of human culture to another. Creative hybrids in multimedia arts; in the sciences and medical technologies; between the inorganic and organic research fields; between artificial intelligence, virtual reality, transgenic manipulation of organisms, and neuroimplantation; between social processes and mass suggestion, artificial worlds, world representations, and "Ways of World-making" (Goodman, 1978): they will also be topics of future creativity research on aesthetics and life in general (because aesthetic processes and products, too, are changing people's lives all the time). Transgressing borders, frontiers, and the restrictive fields of cultural and social life seems to be the indicative mark of progress-oriented neotenic society seeking to move beyond all the traditional boundaries between old realities and new virtual, artificial worlds and realities. This feature is bound to pose great challenges to a future philosophy of creative processes, designs, and developments. Whitehead *redivivus*? Whitehead reshifted, virtually virtualized, represented on a metalevel—Whitehead artificially refined and alienated? These are truly thrilling ontological and methodological questions. Existing psychology and philosophy of creativity seem to be a bit behind the times in dealing with such pressing and acute topics as artificial world; artificial life; artificial intelligence; and computer design in arts, science, and technology. Think only of molecular design; artificial cloning; computer-aided, fMRT-guided surgery; any multimedia technology; or any mixture of different art fields and modalities.

As yet, there is no deeply rooted philosophical anthropology of creativity and creative developments of such hybrid modes and fields of creative phenomena and boundary-crossing developments and strategies. Plessner's (1928) law of artificial naturalness of the human being is too gross and too vague, even paradoxical, to really offer explanations. It just covers deeply interacting and co-evolving factors and effects in a superficial formula, however right that formula may basically be. Postmodern aggregations and collages, quotations of old-fashioned paradigmatic or basic styles, and quasi-ironic self-disassociation from one of them by using all of them (especially by simultaneously contrasting and incorporating outdated styles) seem due to similar oversimplification, though new creative insights and types of postmodernist development do exist in arts and aesthetics.

In this chapter creativity is dealt with in general as a multidimensional process of association that carries novelty and originality for persons and creative processes profiting from ideal delineations such as Sternberg's (2003) types. The persons and creative processes meant are those that are still walking a tightrope between adopting traditional methods, developing new original approaches to fields of creative activity, and spanning different modalities and realms of cultural life and even technologies. Unquestionably, this balancing act is also characteristic of aesthetic creative processes, persons, and new perspectives. Sternberg (1988) talks about creative persons, processes, products, and places. In Lenk (2000a, pp. 91–93)

I added many other topoi (features) such as creative potentials, problem-provoking challenges, production activities, partnerships, populations, cultural preferences, and priorities of values.

Is Creativity a Pluridimensional Associative Process?

Traditionally speaking, it seems characteristic of creativity and creative persons that they tend to oscillate between originality and traditional methods, experiencing a state of suspense and/or an "optimal mix" between "iconoclasm and traditionalism" (Simonton, 1988, p. 413). This condition sounds quite paradoxical, but sustained productive tension still seems indispensable for originality that leads to creative outcomes.

Another attribute of creative innovations seems to be the cross-fertilization of different areas, disciplines, and, sometimes, diverse capacities and opportunities. This interaction, however, often eventually affords creative persons a semiexternal or marginal vantage point within their own discipline. They sometimes depend on that perspective for their creativity. Such individuals might not even be discovered as truly creative instigators, inventors, or discoverers until very late, if at all, as with Gregor Mendel (1822–1884) regarding hereditary statistics and with Julius Robert von Mayer (1814–1878) regarding the relation between heat, energy, and entropy in thermodynamics. The implication is that creative collision, the "fusion" of creativity and innovation, is often characterized by the tension between traditionalism (the established methods and common opinions within a discipline), iconoclastic radical orientation and innovation, and the possibility of transferring what is fundamentally novel from one area to quite another. Thus, it seems that confrontation and struggle between different approaches and areas are conducive to creativity, even necessary for it.

Creativity indisputably also stems from certain cultural and social conditions and from particular psychological dispositions and motivations (see Lenk, 2000a, pp. 76–173). This constellation, however, constitutes necessary, though generally insufficient, conditions for explaining outstanding accomplishments by intuitive or analytic geniuses. Simonton (1988) sees chance intervening at different points and junctures. It figures as essential to the permutation of mental elements in the inception of new innovative ideas, in the comparison of relations between configurations, and in the probabilistic interplay between quantity and quality of the output. Chance especially plays a role in the acceptance of a new idea and in historical development, as in cases of simultaneous discoveries and developments.

Simonton's (1984, 1988) theory of creativity is, however, mainly about combinatorial (normal) creativity. Granted, one must refrain from mere stereotyping by freely permutating the combinations, using them exhaustively, and linking them in new arrangements and configurations of known achievements that seem to be characteristic of what one can call the "reproductive-creative" type. Nonetheless, this theory does not account for the overwhelming creativity of geniuses. Some elements of the personalities, products, stimulations, and inspirations of normal magnitude and of the places and the processes are describable more in historical and methodological terms than in psychological ones. The "four" theories of creativity (personalities, products, places, and processes) seem to be too down to earth to cover the eminent examples of creativity by a genius like Mozart. For lack of repetition, statistical reliability, validity, and generalizability, psychological models and tests dealing with such exceptional personalities have their limits (on Mozart, see Gardner, 1993; Hildesheimer, 1977; Küster, 1991).[1]

Methodologically speaking, the approach taken by Koestler (1966) seems more interesting than that of combinatorial psychological theories of creativity. He compares creative discoveries and developments in science, art, and other creative areas with phenomena of humor and jokes. He does so by focusing on the fusion—or, more exactly, what one can call an associative fusion—entertained in theories of the comical. He emphasizes the association ("bisociation," pp. 25, 36) of different planes, perspectives, and approaches from quite diverse areas. They might be connected in a flash of illumination or inspiration, like an "Aha!" experience—a sudden insight, impulse, or burst of ideas that potentially leads to a specific combination or conjunction of the various factors from different angles and that then culminates in a real fusion of them. This phenomenon is frequent with jokes. Interconnections that had not ordinarily been expected or suspected emerge in this kind of fusing culmination. The explosive comical effect in jokes certainly relies on a confrontation, confounding, or even "con-fusing" of the rules of the game of different realms and planes that are usually alien to one another. Bisociation unexpectedly conjoins them, leading to a collision ending in laughter, new mental or spiritual synthesis, or the differentiated confrontation of parts within an aesthetic experience. Koestler thinks that bisociations may account for all comical, tragical, or spiritually stimulating or inspiring effects (p. 36). Whether or not they are a comical, tragical, or purely intellectual experience of fusion, they illustrate the magical pattern of bisociation.

As with jokes and humor, mutual association typically also characterizes new knowledge, intellectually novel insights, and innovations (Koestler, 1966, pp. 73–74, 105). In most regards these kinds of discovery, too, originate in a bisociation of different planes, dimensions, or areas from relevant perspectives that remain unconnected otherwise. The "spiritually" stimulating effects take center stage in this context. Koestler, however, does not define additional intrinsic features of the differentiation between the comical, the tragical, or the fusing new discovery. He states only that the discoverer has looked around in one or two areas for a long

[1] Weisberg (1986, 1993) denies the very existence of geniuses, the corresponding exceptional personalities, and the extraordinary visions and experiences of *heurēka*. He instead sets store in normal successive acceptance and the continuous development of "elements." He apparently acknowledges only combinatorial creativity and what be called a sort of combinatorial gymnastics (Simonton, 1988). However, he generalizes this standpoint from an insufficient number of single cases (e.g., from Charles Darwin and from the discovery of DNA structure by James Watson and Francis Crick). A mathematician like Srinivasa Ramanujan would go far beyond the scope of combinatorial gymnastics. The same is true particularly in Mozart's case.

time (i.e., the exploratory appetance behavior expounded by ethologists) before the respective bisociation will really fuse. The researcher or thinker searches for ways to state a problem clearly and precisely, to find a clear leading question, and to solve it on a specific plane E_1 (but in vain). At a critical moment a particular interpolation (unlike the merely exploratory extrapolation within E_1) coming from a given plane E_2 orthogonal to E_1 as it were (thus representing an independent dimension) triggers a fusing bisociation, suddenly opening the connection between initially quite different planes or "systems of experience." The revelation that occurs seems to be the wit of a joke, the surprise consisting in the unexpected "lightning bolt" from another plane when routine responses are expected. The comparison to lightning is common sense in humor, in the comical, in sudden novel insights called creative.

Koestler's (1966) bisociation, the fusing creative occurrence of an idea, combines hitherto two unconnected systems of experience, links their respective planes or symbols and approaches. At the intersection of those planes, it leads to what is called a novel idea or the experience of laughter and the comical epitome. (According to Koestler, tragic effect might also be entailed by such bisociation.) The subjective experience is projected onto a connection that has a corresponding objective frame of reference deviating from routine patterns of thought. If successful, that frame of reference acquires a creative combination consisting of two different kinds of dimension.

In a sense, the concept of this bisociation is quantitatively and terminologically too restricted. The model refers either to only two factors or planes of bisociation, thereby ignoring the possibility of multiassociative associations of creative derivation, or to just the "exchange of concepts," thereby merely projecting or simulating "one-track 'digital' associating" (Polet, 1993, p. 298), albeit from two different planes. As pointed out by James (1880), real processes are much more multifactorial and complex. They rely on parallel wiring and multiplex switching. He spoke of the "cauldron of bubbling ideas" in creative processes and of chaotic systems, emphasizing that a multivoiced or multilane configuration typically tends to be involved in the conjoining and associating that takes place in creative processes. These activities are bound to lead to unilateral narrowing of consciousness, but such restriction is only the tip of the iceberg. Beneath it, in the unconsious part of the mind, there is an abundance of rich structures and a chaotic profusion of close interconnections and parallel wirings. This assertion is assuredly right, but it seems to be only implied in Koestler's (1966) model. The idea of approaching bisociation from just two planes or areas is too restricted to cover processes of multiassociation, and such combinatorial approaches easily mislead one to just another "digital" or combinatorial psychological, or now rather methodological, way of dealing with the main aspects of creativity. As with Simonton's theory of creativity, Koestler's concept of bisociation, too, thus seems inadequate for thoroughly dealing with the creativity of extraordinary geniuses.

In any case, Koestler's (1966) approach should not be reduced to merely an extrapolation on one or two planes or an interpolation or a transposition between just two planes or areas, as suggested by the word "bisociation." To me, that kind of sketch oversimplifies the general phenomenon of extraordinary creativity. Instead, it is frequently necessary to deal with multiple collisions, collusions (playing together), confounding phenomena, interconnections, and interstimulations of many kinds and planes. Extraordinary creativity is actually a rather multifarious and mostly unconscious interplay of many factors unconfined by the proverbial narrowness of the conscious mind. It might be almost infinitely many planes cross-cutting each other, flexibly intermingling in confrontation and collision zones and leading to a solution or fusion in the form of an unexpected insight.

Moreover, Koestler (1966) pays little attention to the creative building up of metalevels, which seem to be an immanent facet of theoretical and intellectually abstract insights derived from metalevel models, analyses, and schemas (see Lenk, 1993, 1995b, 2000a). Horizontal bisociations of different disciplines and perspectives are not the only highly typical features of intellectual discoveries (particularly fundamental ones) and of generalizations and overarching insights. So are, to my mind, the creation and shifting up, or raising, of metalevels. The transcending interpretation arrived at through higher levels of perspectives, analyses, interpretations, and consciousness is a decisive feature of intellectual creativity beyond Koestler's concepts of extrapolation, interpolation, transposition, and transformation (which are apparently oriented only to single-level explanations). The creative ascent means going to abstract modeling or to the abstraction of more general concepts. It also means overarching and summarizing translevel concepts on different planes and metalevels.

It seems that surveying and overarching specific levels and planes is particularly important for novel insights of an intellectual and profound kind. In this context one can speak of "transcending" instead of just "transposing" or "transforming." It is about "metatransposing," or even vaulting, to higher levels—of metainterpretations from higher level perspectives (like the approach to a higher order consciousness in the philosophy of mind). Creativity, particularly with respect to intellectual endeavors, insights, and activities, is not in fact restricted to different perspectives on the same plane or level. It is frequently the metainterpretations, the creation of new planes and levels, that are especially creative and characteristic of going beyond mere combinatorial creativity.[2] Perspectives usually do have levels, if not multileveled (i.e., level-overarching) patterns. Not only does it seem necessary to put on a new "thinking cap," (*neue Denkmütze*) as the science historian Herbert Butterfield has labeled it (as cited in Koestler, 1966, p. 255). The mental transpositions within the planes of scientists would not originate simply in new observations and additional data but also, and mainly, from rearrangement of the available data bundle into a totally new system of mutual relations upon receiving a new framework. This process would be the donning of a new thinking cap: NEW THINK!

Therefore, Koestler's (1966) key idea that two different, hitherto unconnected systems of experience are conjoined by a flash of inspiration that metaphorically

[2] According to Kant's (1790/1968, pp. 307–308) theory of creativity and originality of genius, it is characteristic that a genius not only has new insights and findings within a field but that he or she sets or changes the *rules* of new areas in the historical development of the arts (§46). The same is, *mutatis mutandis*, true also of intellectual approaches, in particular the transcending of limits and frontiers between different areas, as in science and philosophy.

combines two or three orthogonal planes in a specific line or point has to be extended or generalized. Although the basic idea of associating different experiential systems (not just through bisociation but rather through multiple sociation) is certainly valid and intriguing as a guide or model for capturing processes and ramifications of creative processes and developments, it does not go far enough. In the case of real creativity, such conjoining or crosscutting cannot be conceived of as simply an aggregating of values and magnitudes. It is about genuinely integrating and structurally establishing the internal mutual effects, interference, and fructification of perspectives, a process that cannot be understood by a model for adding up factors (p. 252).

This criticism applies to Koestler's own approach as well. Many creative bisociations cannot be restricted to the accrual, criss-crossing, or crosscutting of different planes or to a particular way of combinatorially establishing relations. The circumstances surrounding processes of fundamental creativity are usually much more complex and more interesting than the act or fact of just conjoining two planes or factors in a kind of fusing process.

To be sure, Koestler highlights the deeper transformations of perspective and fusion of interpretations by using metaphors, analogies, analogical concepts, comparisons, transformations, cross-comparisons, cross-thinking, cross-interpretations, and certain conflicts between partial perspectives and approaches. He also emphasizes conflicts within the creative personalities themselves (as captured by psychological research and theories, see Gardner, 1993; Simonton, 1984, 1988; Weisberg, 1986, 1993). All these factors are due to exacerbated tensions that sometimes eventuate in blockage but occasionally enhance the probability that such a multiple association or collision of insights will result in a highly creative discovery or mental "strike." One could even speak of a collision of conflict-bound preliminary or initial constellations of factors, of collusion, interplay, or mutual connectedness that comprises the interaction of the different experiential systems and sometimes leads to an associative fusion. One often encounters the exchange of different codes, and at times it even becomes consciousness. Fixed strategies are rendered flexible, a result for which one typically must shift to another framework. Switching and modifying frameworks is very important in fundamental creativity processes.

However, the solution or solubility of a complex multiassociation problem is not predictable. It cannot be causally explained or deduced or combinatorially and mechanistically produced or imposed. Koestler's approach does not offer an explanatory theory but rather amounts to a kind of phenomenological attempt to describe each strike, burst, or explosive fusion. However, such mental lightning bolts or other striking events are not reducible or restrictable to combinatorial gymnastics (a statement also true of Simonton's and Koestler's theories). Bisociation or even multiple association tends to be oriented to combinatorial manipulation of approaches and the access it gives to different experiential systems in a rather systematic combination. Yet frequently, even typically, a random coincidence is triggered by external circumstances. Psychologists and sociologists of science (e.g., Merton, 1957, pp. 12, 103) talk of "serendipity" when such a stimulating experience from the environment or sociocultural vicinity has a fusing effect.

One may try to model these kinds of stimulating experiences from the environment by analyzing the factors that increase the probability of such collusion. It might also be possible to model them by conceiving of a mental strategy for scanning or sampling features in a subjective internal mental map. Koestler (1966) describes wandering around within a "virtual inner landscape" (pp. 167–168) for goal-oriented thinking, as when a person directs the focal beam of consciousness on different parts of the internal map, trying to explore it. But none of these literary devices suffices for thorough theoretical comprehension of the factors and phenomena that creativity entails, let alone for an exacting explanation of them. The metaphors are merely an attempt to circumscribe something that is actually undepictable "from outside." Koestler does, however, refer to the unconscious, to crosswise thinking and interpreting, even "thinking away" or pushing aside (p. 149), precisely what is included in indirect strategies that emerge from autobiographical accounts reported by the mathematician Poincaré (1913). These strategies are intended to induce the necessary associations for solutions to problems by extending the time and circumstances of incubation so as to increase the probability of an essential stroke of insight. As Louis Pasteur once said: "Luck would only hit the prepared mind" (as quoted in Koestler, 1966, pp. 112–113).

In brief, then, Koestler's (1966) model is too simple. It pertains to only two intersecting levels or perspectives and confines the preparation of the creativity situation to combinatorial procedures alone. In particular, it does not actually encompass higher and more abstract levels. Koestler did not see that horizontal bisociation and association are complemented by the existence of a third kind—namely, vertical association, even metalevel multiassociation from metatheoretical and metalinguistic vantage points that afford different perspectives on lower level phenomena. People may also creatively associate vertically. One could even speak of meta-associations and of methods to create them. Thus, it is necessary to generalize and elevate Koestler's methodological model to a multi- and metalevel theory of creative processes. It is about looking for variations, interpolations, extrapolations, transpositions, transformations, selections, and so on to collide within the same level or within two planes and to find overarching wider and higher "superperspectives." It is also about identifying attempts and strategies with which to ascend to higher levels of modeling, abstract structuring, and the flexible use of metaphors and metametaphors, even "creataphors" (see "Toward a Strategy of Creataphors," below).

What about the artist's creativity? Is it only similar to the creativity of the scientists, as proposed by Koestler (1966, pp. 366, 371)? He believes that the development of the creative process and the creative personality is very similar in science and art and that observations about the inception of new ideas are just as valid for the scientist as for the artist. According to Koestler, fundamental novelties will emerge if unforeseen transpositions of awareness occur and if a hitherto ignored part of the spectrum of human existence is emphasized (p. 371). In both science and art such novelties spring from an unexpected connection or even "conwiring" of as yet separated systems by means of bisociation (p. 443). He states that all great discoveries of both science and art stem from such bisociations and associations. It is the fate and privilege of scientists and artists to have to walk the tightrope of these intersecting lines. One can and should extend, generalize, and modify this associative model in the direction of multiassociation and vertical leveling. But there may still be another grain of truth in this similarity between the two areas of creative activities, at least with respect to the really creative processes, developments, and personalities. I suggest that the creative phenomenon in all these areas is of the same structure, that the causes of creative processes and acts seem to be mainly of like structure, and that the motivation of the creative processes appears to be very similar in both fields.[3]

New truths and new beautiful phenomena are gained only through creative acts, and they themselves have a "creativating" effect (a psychic outcome that instigates and enhances creativity). When it comes to pioneering activities and trail-blazing or epochal new effects, perspectives, and approaches, however, it is important to describe truly fundamental creative processes. Mirroring and re-experiencing truth or beauty that is already known are not called creative acts but rather a surreptitious re-experiencing of former creative processes (though such "reliving" of creativity is motivating and important for all normal persons, even highly creative ones, outside their own fields). Originality, the element of novelty, must be added if the process is to amount to genuine creativity.

However, even such perspectives that are legitimate in principle do not suffice. Being genuinely creative also requires inclusion of at least the following four characteristics and (e)valuative perspectives:

1. The principal orientation is to configuration, wholeness, and totality, particularly with respect to especially great creativity (see Polet, 1993, p. 93, 114).
2. Novelty is an essential constituent in principle. It is certainly included in the requirement of originality. But the concept of novelty is still too general; it must contain the notion that the development of new perspectives, new modes of representations and perspectives, new rules, and new fields are indicative of genuinely high-level creativity. Originality does not consist only in elementary extending approaches on the same plane or in the establishment new combinations of already known factors and solutions. Real creativity of high standard necessitates the establishment and inception of new foundations, new fundamental perspectives, and new levels and metalevels of interpretation. In sum, it is new perspectivity and a new perspectivism that count.
3. In keeping with Kant's (1790/1968) concept of the genius (§46), corresponding insights hold for the insights of the creative individual and for the inception of new rules for the creation of interpretations and metainterpretations. These new rules constitute not only a new special or "individual rule of the game" (Koestler,

[3]To explain the scientist's motivation to seek truth and ultimate causes and the artist's experimentation with the ultimate realities of what can be experienced by producing works of art, Koestler (1966) draws on Freud's (1930) idea and critical reinterpretation of Romain Rolland's "oceanic feeling." It is the climax of satisfaction and the most sublime expression of the integrative striving of the human being. Kepler, according to Koestler (1966), also had the intoxicating feeling leading to the experience of wonderful clarity, beauty, and truth simultaneously upon discovering his second law. Similar reports are attributable to H. Poincaré (1913, p. 393).

1966, p. 424). They are also a totally new direction of art (e.g., 12-tone music or the transition from painting on canvass to reliefs and collages that extend art to three-dimensional space and integrate that space with traditional pictures). All these reorientations amount to the establishment and application of new rules or new rules of evaluation, which lead, of course, to radically new styles and subsequently to new developments and offshoots. According to Kant, the genius establishes new rules for himself or herself and thereby henceforth may create new standards of valuation and evaluation in general. This kind of neoregularism or neostandardism can be analyzed and related to the metalevels of analysis and interpretations that are implied in the approach of methodological schema-interpretationism (see Lenk, 1995b, 2000a, for instance).
4. The encompassing phenomenon of creativity and the creative thus reaches across individual areas, producing something rather philosophical. This insight is expressed by the fact that all abstract models and higher levels of interpretation and their respective developments are layered one over the other. The corresponding metaperspectivism might lead to level-transcending creations, to *metacreativity*. This possibility might even result in an interdisciplinary, overarching view informing a philosophy of potential creative activities and concentrating on the quality of and similarity between phases, kinds, structures, and basic motivational factors of creativity and the creative in very different areas.

Is There What Can Be Called a Chaotic Creativity?

Cramer (1994, p. 259; see also Cramer & Kaempfer, 1992) thinks that the beautiful is to be interpreted as a kind of tightrope walk between order (or the ordered) and the chaotic (chaotic phenomena). The intriguingly ordered structures of fractal geometry are especially relevant in this context. They expose relations and correlations between the physics of complex dynamical systems with fractal (chaotic) attractors ("strange attractors") and evolutionary biology. Because all developments in living systems generally depend on the current state at the time of their respective evolving systems, there are formal identities, or at least analogies. Cramer (1989) tries to apply the theory of deterministic chaos to the transitions between order and chaos in the arts, to the reception of the beautiful, and to relevant, notably aesthetic, experiences. "Aesthetic" beauty originates wherever chaos borders on order and order on chaos. Beauty is equal to the open, irrational order of the transition and, in keeping with its own principle, is transitory, fragile, endangered, and unique—as is life itself. Beauty can exist only as "living beauty" (Cramer, 1994, p. 259). This notion is certainly remiscent of Goethe's statement in his lyric cycle "Urworte. Orphisch: Daimon" that beauty can be realized (in a double sense!) only as gestalt that lives, develops, always modifying and renewing itself (*Geprägte Form, die lebend sich entwickelt*; 1885, p. 319).

According to Cramer (1994), fractal geometry's nonlinearity and the mathematics of chaotic systems and phenomena ("procreating the beautiful form"; p. 261) allows one to describe nature more effectively than the Newtonian approach in theoretical physics. "Reality of the cosmos" (p. 261) is nonlinear, whereas linear equations and superpositions of magnitudes and linear combinations of them are just a very simplified model. A similar phenomenon is encountered with works of art:

> Novelty originates in going through chaotic zones. Art creation is an act in the highest possible neighbourhood to 'just not yet chaos', ... The work produced in an artistic tight-rope walk at the edge of chaos would in the truest sense contain the moment of the artist [a climax even conjured up by Lessing, for example], and it is exactly this fact that would render it a work of art, that this moment is fixed so that it can never deny its subtly endangered creative process any more. (p. 280)

The process also shows the orientation at the symmetric and ordinary rule-governed structures as well as the minor deviations having sometimes surprisingly new original variations. Total symmetry is, as a rule, boring (as known from psychological experiments comparing responses to images of natural human faces and responses to images in which one facial hemisphere is the mirror image of the other; see Cramer, 1994, p. 277). In other words, all that would enliven the work of art is the deviation from and modification of the symmetrical and rule-governed structure, including fractal self-similarity.

If the desire is to develop the ideas of self-similarity further toward an aesthetics of an approach based on chaos theory and fractal geometry, one must first ask what such an aesthetics would consist in. Would it depend on the fact that people's re-experiencing of structures is biologically preprogrammed? After all, human neuronal assemblies and their stabilized, though flexible, interconnection in the brain tend to follow such ramifications. A person's brain manifests oscillations and stabilizing oscillatory processes similar to such dynamical systems. Holistic interconnections and feedback processes seem to play a decisive role in both areas. Neurologists contend that brain patterns are stabilized and swung by such oscillations and the corresponding coherence of firing and spiking rates according to "a hire-and-wire" principle of a dynamical oscillatory kind. Researchers such as Freeman and Skarda (1985) try to discover and identify strange attractors that are chaotic and fractally structured attractors within the brain itself. This kind of result could at least in principle be the starting point for, say, an aesthetics based on a fractal basic model structure and on background chaos of brain processes. It could make sense of why such quasi-natural, fractal, very ramified, dynamically complex structures would be evaluated as "beautiful." Cramer (1994, also in 1989, Chapter 6) thinks that chaos research would contribute to a new understanding of the aesthetics of the beautiful and to the interpretations of the arts of different periods, cultures, and schools (see also Briggs & Peat, 1989/1993, p. 28; Cramer & Kaempfer, 1992).

As for genuine creativity, there are two rather more interesting questions: What is the difference between fractal computer-produced shapes and structures on the one hand and highly creative art on the other? What is the difference between computer-generated graphics, or a series of "pictures" drawn from the edge of the Mandelbrot set, and the spiral-shaped seahorse-like structures of some pictures by Picasso or Van Gogh? Briggs (1992/1993, p. 171) claims that a genuine work of art seems very "catching" because it corresponds to the brain's receptivity, but that the greatness consists in resisting this customary tendency of the brain by deviating from the standard form of self-similarity and the expected level in terms of fractal structure. It is about deviating in a more surprising than systematic way. It seems that "a great work of art would provoke in every (novel) encounter in the human brain a new, very strange attractor" (p. 174). A person would therefore experience such a varying and varied creation or pattern again and again in new ways.

Reflectaphors

The exceptionality, the greatness, of a great work of art resides in this ambivalence, which borders on artificial self-similarity (understood in the sense of fractal geometry). That ambivalence is an expression or instantiation of it and its ever-reproducing or repeating patterns and structures, from which the work of art, in turn, deviates notably. In that manner the work of art typically arouses and repeatedly produces a kind of new "reflectaphoric"[4] tension, revealing and reconstituting itself at ever deeper levels with each further development or new encounter. Great works of art do use self-similar forms and colors, but they vary them, deviating from ever-relevant rhythmical regularity. They avoid strict repetition; they do not just mirror the self-same partial structure, though they might self-reflexively go back or feed back on these patterns by creatively modifying and varying the structures. They always create tensions of a new kind, providing stimulating instances of ambivalence, provoking them, alluding to them. Such a new variation of nuances is the factor also informing the new tension and deviations in the use of creative metaphors that Briggs and Peat (1989/1993) call "reflectaphors" (p. 302). These reflectaphors are metaphors or metaphor-like structures deploying a special tension in the interplay of similarity and difference in kind and structure, of harmony and dissonance: This "reflectatoric" (p. 302) or reflectaphoric tension is dynamic. It provokes and produces an ever new kind of vivacity, even in experiencing, perceiving, and sensing. One experiences astonishment or perplexity when entertaining unexpected perspectives and points of view. Therefore, according to Briggs (1992/1993):

> [I]n producing works of art, artists have to find the right distance between the forms of expressions of their own reflectaphors by striving for the right balance between harmony and dissonance in order to create the tension and multifarious ambiguities that an artwork can reveal. This right balance would outstrip the processes of thinking and prevent the process of habituation. For it would improve our understanding to perceive words or forms or sequences of tones as though for the first time, that is, each time in a new way no matter how often we have perceived them before. (p. 174, author's translation)

[4] Briggs (1992/1993) uses the term "reflectaphor" (p. 174) for an artificial juxtaposition with many self-similar forms, instances of ambivalence, and dynamical tendencies—even on several levels of sensing and interpretation. Not only are forms self-similar to one another and mirrored in those as in a metaphor, there is tension between "similar *and* different forms of expressions" (p. 174, author's translation). This "reflectaphoric tension" shakes and moves human understanding with a mixture of amazement, respect, bewilderment, perplexity, and the sentiment of unexpected truth or beauty.

People do not deal only with balance at a single level; forms of tension have contrasting levels and metalevels. Harmony and dissonance at different levels and on different planes also play an overarching role, as previously mentioned with regard to the levels of creativity in intellectual, aesthetic, and humorous productions and activities. By stabilizing and interpreting metabalancing processes (as mentioned with respect to metainterpretations in transitions between levels of interpretation), one might conceive of a creative ascent overarching the single-level balance and extending to a metabalance. In the present context only this creative ascent is to be applied to the reflectaphoric tension and play between different functions of the reception—and creation—of a great work of art. According to Briggs (1992/1993), artists and poets "find the reflectaphoric harmony by trying out the distance between self-similar conditions" and the respective deviations and conscious differentiations "in their own understanding" (p. 174, author's translation). As he asks, does a metaphor lead to a surprising effect even if frequently re-read? If it does, if the metaphor is different within the overall self-similarity of the reflectaphoric tissue, and if its ambiguities do interact with other forms and gestures of the work that are slightly modifying the self-similarity at large, then a work of art is "living and dynamic" (p. 174).

Toward a Theory of Creative Metaphors

Writing about cognitive theory of metaphor, MacCormac (1985) extended metaphoric processes and operations from the linguistic and literary perspective to prelinguistic processes of imaging and thinking that seem to be of special importance for the understanding of creative activities and processes. In his approach, the creation and usage of metaphors must be conceived of as processes taking place at three related levels, not just that of language. They are the speech and "language process" as a "semantic and syntactic process" leading to a linguistic explanation and especially as "a cognitive process set in the context of a larger knowledge evolutionary process" (p. 42). Establishing metaphors is not only understood as a semantic process but also explained "as an underlying cognitive process without which new knowledge might not be possible" (p. 42). Examples he cites are metaphors such as the famous one by Charles Sherrington: "The brain is an enchanted loom where millions of flashing shuttles weave a dissolving pattern" (p. 28). The function of metaphors consists in creating tension between the two *relata* (referents) of the metaphor. That is, they display a "diaphoric quality" that may lead to a new representation, a surprising opposition, in any case to a tension in the adapted or habituated scheme, provoking at times emotional restlessness. The tension comes from "an apparent semantic anomaly rather than from emotional discomfort" (MacCormac, 1988, p. 85). "The psychological tension arises from a semantic tension" (p. 85).

Whenever a metaphor spreads within a language community, the speakers and hearers become accustomed to it. By and large, it thereby loses its semantic and psychological tension and may eventually acquire a new meaning in the dictionary. According to MacCormac (1988) many metaphors start

> their literary lives mostly as diaphorical, that is, as productive or prolific metaphors (though they always have also epiphoric quality). Later on, they gradually become largely epiphoric ones, expressing analogies rather than suggesting potential meanings, and finally wind up as "dead metaphors" within the corpus of normal language. Metaphors die if at least one of their referents adds a new lexical meaning to a dictionary entry. (p. 86, author's translation)

MacCormac's (1985) claim amounts to the idea that metaphors as the basis for conceptual semantic anomalies are engendered by a surprising, more or less conscious opposing activity of the referents or relata whereby the identification of dissimilarity is especially conducive to their transformation. This relationship had not previously occurred to anyone. Through it, "the creation of a new meaning" is established and ensured (p. 50). Creativity lies in the selection of suitable referents that have or produce "enough similarity for recognition" and re-identification and that yield sufficient dissimilarity of "the right kind" in order to create new "hypothetical possibilities" (p. 148), say, for interpretations and research or artistic variations. This thesis applies to the establishment of new metaphors and perspectives in all creative areas of association and imaging as well as to the inception of new basic ideas in scientific research.

The crux of the matter is that the creative production of new hypotheses and comparisons—scientific or not—would be impossible without metaphors and that semantic modifications in language would therefore be drastically restricted. Without metaphors, without the intentional conceptual construction of semantic anomalies, one would hardly able to speculate about or venture into the unknown. Thus:

[M]etaphors perform the cognitive function of creating new meanings through the juxtaposition of referents in language: Without them, humanity would find it difficult to extend its knowledge into the unknown, and language would be largely static. The diaphor offers the possibility of taking a familiar referent and transforming it by juxtaposing it with a referent or referents not normally associated with the familiar referent. The combination of referents that produces semantic anomaly forces the hearer or reader of a metaphor to locate the similarities among the attributes of the referents as well as the dissimilarities. Not only does the recognition of similarities not seen before produce new insights or new meanings, but especially the identification of dissimilarities allows for the possibility of transformation of these dissimilarities into previously unthought of similarities, thereby ensuring the creation of a new meaning. (MacCormac, 1985, p. 50)

Highly creative persons characteristically seem to fashion and frequently use metaphors in language, especially in metaphoric imagination, referring back to deeper processes. Constructing or establishing metaphors is also a process of making new cognitive associations. MacCormac (1988) stresses that the creative formulation of new metaphors expands the imagination best when the most unusual combinations are used (p. 92). How these unusual and vivid combinations of concepts are to be expressed in words remains a secret. "Were he a painter, a poet would not be compelled to dress nonverbal intuitions in words, but since language is his artistic medium he has to express all his concepts in language" (p. 93). According to MacCormac, the poet thus strives to find "metaphors" in order to render "greater suggestive force" (p. 93) to language, probing and proving one of the

> miracles of language, namely, its plasticity and creativity, its capability to grow with, in, and through the mind of a skilled language user. The distance between the imagination of the poet replete with fantasy and the banality of normal speech would determine the battle about artistic moods and ways of expression. A poet would constantly push the limits of normal language beyond the usual framework. Whereas the gap between fantasy and usage becomes narrower whenever the poet creates new metaphors for expression, ironically victory eventually becomes a kind of defeat because the poetic language is no longer fresh and unused. (p. 93)

MacCormac (1988) stresses that poets always have to create new vivid and sparkling visions or the creations will wear out and become commonplace or even vulgar because of their success. It seems to be true dynamics of wearing off and using up the creative potential and semantic visionary content of metaphors. These dynamics have crucial influence on aspiration, fantasy, visionary force, potential, originality, and novelty. In short, new fruitful metaphors setting off creative dynamics that open up new realms and combinations of ideas eventually wear thin. That fate awaits the invention of new styles, the setting of new rules, and the wide circulation of creative productions. The dynamics reach far beyond poetry and the fine arts, affecting creative production in other realms such as the formation of new ideas and new visions in all creative fields, even in philosophy. They extend to the motivation and aspiration of the creative person, to the language, and to the poet designing and grasping new syntheses. They also act on the activities connecting representations and concepts, on the continued development of styles, and on perspectives and modes of experiencing and sensing in the interpretation of the world. These observations are notably true for philosophy, scientific discoveries, technical developments, mental imaging, and, above all, the fine arts.

All creative realms and processes of the above-mentioned associations and multiassociations, the development of new perspectives on higher levels, and the phenomena of creative ascent (not only transpositions on the same plane) correspond to this pattern. This relationship might even be referred to as the interplay of different sense impressions (synaesthesia) and as imagistic or pictorial representations as studied by Kosslyn (1980).[5] Generally speaking, the idea seems very plausible that metaphoric processes are the basis of creative processes and that the conception of the metaphoric is not just restricted to external language and purely syntactic and grammatical forms. It might also be true even if one avoided identifying all metaphors with these creative processes of multiassociative and deep psychological provenance.

A new expression should be coined for this basis, however. I propose *createphor*: a concept of creative cognitive activities that link usually unassociated concepts, representations, or imaginings through contrasts (dissimilarities) and comparisons (similarities) of characteristic features, properties, and modes of experience and that lead to a dynamic development of new perspectives in creative

[5] Kosslyn even utilizes the "mental eye" metaphorically as though it were a kind of television tube. Theories about metaphor are often themselves metaphoric and use metaphors, but this does not necessarily mean that all language use is metaphoric (MacCormac, 1985, 57–72).

activity and knowledge. Instead of the metaphorical consciousness hypostatized by Cohen (1958/1979), what one may call "creataphorical consciousness" would be more specific when one speaks of creative persons and attitudes. The term refers to the fact that consciousness and a distinct dynamical tendency are always necessary in order to use and establish new tension-generating metaphors (reflectaphors) as vehicles of the creative. The creative metaphors that truly lead to novelty are "creative reflectaphors" and, hence, creataphors—innovative creative metaphors of a dynamic provenance. It would certainly be interesting to explore and explain mental and psychical functions of the creataphors and reflectaphors within and corresponding to the creative activity of the artist, poet, scientist, or creative philosophical thinker. Very few pioneering studies in that realm exist.

To summarize in general terms, the development and use of creative metaphors sheds a sort of explanatory, at least plausibility-enhancing, illustrative light on the origin, course, and flow of creative processes and on the conceptions and interpretations entertained by creative persons. Therefore, MacCormac (1985, pp. 50–52) seems to be right when he extends the originally only language-oriented theory of metaphors into a more general theory of creativity pertaining to metaphoric imagining and thinking. However, the theory should also be extended to creative actions and activities. One could and should terminologically distinguish it from its strictly linguistic connotations, perhaps by speaking of "metaphor in the narrow sense" when meaning the linguistic realm. Referring to a general theory of creative processes involving cognitive as well as acting and creative (or, more widely, poietic) metaphors and reflectaphors, one could instead speak of creataphors (dynamic, progressive, far-reaching, far-guiding, creative reflectaphors of representations and imaginations, or even judgments in a Kantian sense).

Creative games and play, such as the playfulness expressed by the classical Latin word *creare* (creating something new in knowledge and cognition or some other area), do not appear in Caillois's (1958/2001) famous list of the kinds of play and games. In fact, Caillois makes no mention at all of what is genuinely creative (nor of the creative play inherent in the capacity of judgment, or *Einbildungskraft*, à la Kant). Creative games (*Kreativspiele*)[6] have to be characterized by another feature: *creativitas* (creativity). That term, however, is not classical Latin but rather neoclassical Latin: *creans* (the creating). *Creativitas* is distinguished from that which is or was created, the *creatum* (after Whitehead, 1978). The simile and metaphor of play and games is obviously a very encompassing phenomenon in human

[6] Are, for instance, Wittgenstein's "language games" (*Sprachspiele*) or "schema games" (*Schemaspiele*, see Lenk, 1995b), as I called them regarding the play of schematized representation and imaginations, really creative games? Or do they represent yet another extended form? They need not necessarily be creative; as a rule they can turn out to be rather conventional. Wittgenstein understood the terms "game" and "play" (The German word, *Spiel*, covers both English terms in a more general but less differentiated concept having many connotations) in a way that makes it a rather vague expression with open borders and dimming or blurred edges (PI §71). Many phenomena may be called games or play: There is no unique thoroughgoing trait of combining or covering all connotations at the same time. Nor is there one for combining or covering all the edges of chaotic phenomena of deterministic chaos theory mentioned above.

life. Even some natural scientists generalize this notion to account for some of the most encompassing phenomena of all. For example, Eigen and Winkler (1975) developed the idea that *Spiel* (play), understood in a rather extended way, is the fundamental principle informing the creation of life and dynamical shapes, almost in Goethe's previously cited sense of a poietic, living, developing form. Playful creations may be products of a quasi-Darwinistic selection principle or a dynamic of self-organization at a rather generalized level of interpretation.

One should, however, proceed by making rather differentiated distinctions. Game and play among conscious humans and rather highly developed animals (like dogs and primates) are certainly different from the "play" of physical or chemical elements in a dissipative, dynamic system of deterministic provenance. In the same vein, popular and scholarly opinion is divided about creation as selection with respect to the concept of creativity. According to Darwinism, selection is but reproduction, descent with modification by natural selection (i.e., selection in a specifically biological, hereditary sense). Valuation and modification tend to enter at random. It is not a controlled interaction and reaction but rather much more a random selection and random modification. By contrast, an intentionally productive and strategic creation much more neatly corresponds to the usual concept of creativity. In intentionally productive and strategic creation there is no selection with just random modification but rather an election with strategic modification, that is, a rather intended and purposeful, telic modification under strategic, at times conscious variations. This production of variants is indeed highly characteristic of creativity in the arts. From these strategic intentionality-guided perspectives, random creativity in a Darwinistic and neo-Darwinistic sense should therefore at least ideal-typically be distinguished from a *designer* or *design creativity*.

Creataphors and the Creataphoric Being

To a large extent, philosophical reflection is a function of ever-changing and, at times, particularly new perspectives. In this sense, it is creative. Genuine philosophizing is not just mirroring (somehow passively reflecting) the given. It always amounts to interpreting, engaging in active conceptual work, or even changing

---

(Unfortunately, a corresponding theory of probabilistic chaotic states and systems has not yet been developed.) Chaos games might be an interesting idea regarding fractal computer-graphics and the question of whether they have aesthetic value, whether they represent art, and whether high art can be grasped from a fractal geometric and chaos-theoretical point of view. Play- and game-like phenomena regarding chaotic phenomena and processes of self-organization (*Selbstorganisationsspiele*) may and should at times be analyzed with a chaos-theoretical approach. Playful appearances on the brink of the chaotic outside and inside the respective strange attractors surely count as games of order but certainly not as games of competition, chance, mimicry, or intoxication in Caillois's (1958/2001) sense. In Caillois's theoretical vein, chaos games or games of self-organization systems could also be counted as kinds of play or games. But Caillois never touched on such alternative interpretations nor, interestingly enough, on genuine games or play of creativity.

perspectives, gaining new vantage points and delimitative experiences, making a transition between levels, or transcending them. Genuine philosophizing is creative. It creatively transcends levels and limits via and in interpretations and conceptual designs. Philosophizing as the activity of transcending interpretation should at least be creative in that sense. Philosophy at its best is a creative, transcending, interpreting activity; it is transinterpreting and metainterpreting. Like people in other creative realms, philosophers, too, are required to take risks, to develop designs, creative activities, and creative acts. They should internalize Weiss's (1992, p. 634) suggestion that every creative activity embodies a characteristic, "unique" creative impulse far beyond the usual areas of creative production like the arts. The creative basic impulse can be grasped only as a kind of theoretical construction or interpretive construct (see Lenk, 1993, 2000a, b, 2003) and need not be described as an ontological real causal entity per se. It is necessary to develop a creative philosophy of creativity itself, incorporating modern methodological insights such as those about the constructive-interpretative constitution of all knowledge, *Erkenntnisse*, and action structures, that is, all phenomena of "grasping" (in a double sense, as in "grasping reality"; see Lenk, 2003).

A useful prompt might be to use the Darwinistic metaphor of evolution and combine it with the activities of shifting and upgrading levels, of transitioning to different levels, of overcoming as well as transcending their limits, and of devising symbolic metainterpretations. Processes of self-organization in the universe obviously show the existence of a structuring tendency to build certain systems with emergent properties (see Lenk & Stephan, 2002) that are the basis of all structures, shapes, and forms stemming from processes of interaction, developments, chance encounters, and interstitions. To that extent, one may uphold Whitehead's (1978) basic pattern (a Darwinistic perspective of sorts, so to speak) without considering creativity in the narrower sense. Whitehead's (1978) principle of originality or Weiss's (1992) factors of excellence and creative ventures are certainly implied. Creativity would then be given only if (a) chance activities are not the only ones in a particular goal-oriented or teleogenic activity, (b) this goal-oriented or teleogenic activity is pursued by a creator, and (c) fundamentally new structures and phenomena are involved.

This approach goes beyond just living up to or living out a creative impulse or drive in works. Conceptual developments, such as theories, new perspectives, approaches, and—last but not least—philosophical conceptualizations and theories of design may be creative, too. Creativity is possible and especially important in transcending limits, levels, and strata of perspectives. The essential, highly creative element in philosophy consists in the activity of transcending metainterpretation, as mentioned above. The transition across levels is possible only through symbolization and the shaping and modification of metaphors. The createphors as tension-maintaining, ever-further-stimulating dynamic metaphors are centers of creative processes and acts.

Creativity in this sense is not characterized only by novelty; possibly (but not always) by goal-orientation and conscious orientation to end states, objectives, or outputs; and by prospective excellence and originality. It is also characterized by a continuous exploratory activity of dynamic curiosity. This added dimension of creativity applies at least to creative philosophers who continuously think ahead, who see and search for new problems, deeper questions, and more overarching perspectives in order to arrive at ever higher levels and strata of interpretations and generalizations, if not universalization. Humans as metainterpreting (see Lenk, 1995a), ever symbolically transcending beings are the creative beings *par excellence*. Human creativity is always *creans*. Expressions like "creative ventures" (Weiss, 1992) and "creative ascents" (Lenk, 2000a) intriguingly reflect this view. Accordingly, one should certainly not fail to foster high creativity in specific ways. Encouraging it is necessary in order to open new perspectives, developmental fields, scope, and alleys for potential creative capacities and people, including opportunities for creativeness. It seems crucial to open and maintain those opportunities through stimulating vantage points, affordances (in Gibson's sense), instigation, and motivation. *Homo semper interpretans, ludens, creans*: The human creature is always the interpreting, metainterpreting, playing, and creative being.

Especially creative reflectaphors consist in seeing and establishing similarities and differentiations from a variety of perspectives on diverse levels and overlapping strata. If stimulation toward new developments is based on transpositions to other perspectives and toward higher levels and strata, then one has a particularly creative (creativity-stimulating) reflectaphor. I proposed a new word for it: *createphor*. Creataphors are also metaphors, but they are special ones that overarch perspectives, that bridge and transform as well as maintain tension within a stimulating play between similarities ("homeotaphors," "syntaphors," or both; see MacCormac, 1985, pp. 38–42) and dissimilarities ("diaphors" as well as dissonances). Creataphors constitute creative play and games and vice versa. A *creataphoric process*, or a creataphoric instead of just a metaphoric *and* reflectaphoric activity, is a rule. It seems to be a rather interesting idea relating back to the human as the creative being that has the capacity to generate not only metaphors and combinatorial creativity but also creative reflectaphors and creataphors. Humans are creative and creataphoric, particularly striking and characteristic attributes of the metainterpreting being. In other words, the constituent creative metainterpretational element of the creataphors characterizes these special capacities of a human being with respect to dynamic, creative representation and creative production (*Gestaltung*)—as opposed to mere usage of symbols or just interpretation restricted to a unique perspective. It is the capacity to transcend special perspectives; to arrive at higher perspectives, levels, and more abstract interpretive strata; and to change approaches and perspectives at the same level.

Moreover, creativity is symbolic authentic activity, *Eigen-activity*. Such a philosophy of being creative simultaneously amounts to a philosophy of an extended personal and authentic activity by human, subjective, social, or artificial interpreting systems. The capacity to design, establish, maintain, and change metaphors, reflectaphors, and creataphors is a kind of characteristic anthropological feature. Only the human being can discover analogies and think in metaphors and all their modifications such as reflectaphors and creataphors in order to develop new creative metaphors allowing human knowledge to be extended into the realm of the hitherto unknown. This creataphoric ability also pertains to higher order representations, metasymbolizations, and abstract metalevels, which are particularly important in philosophy, epistemology, and the methodology of actions and design. Only humans are capable of transcending any position, level, stratum, or perspective to arrive at ever new viewpoints. The drive to be creative, to transcend limits and levels, even if only in a symbolic manner, is characteristic of all creative and aesthetic activity. Innovative human life turns out to be possible only if it is embedded in continually practiced creativity of at least an intermediate range.

It is indeed a kind of creative play with metaphors, namely, reflectaphoric metaphors, especially the *creative* reflectaphoric metaphors (creataphors). Humans (at least creative humans) can even be ascribed a creataphoric consciousness as a specification of metaphoric consciousness by which the human being is understood to be the potential author and agent capable of creating creataphors, of being the specifically creataphoric or "creataphorizing" being. Creativity is a permanent and continuing creative process, a kind of ongoing transformation of creataphors. It is the capacity and motivation to reach beyond old or dying metaphors and reflectaphors by engaging in genuine creativity. The metainterpreting being is the creative and creataphorizing or creataphoric being at the same time. ***Homo meta-interpretans sive homo creataphoricus***.

References

Briggs, J. (1993). *Chaos: Neue Expeditionen in fraktale Welten* [Chaos: New expeditions into fractal worlds] (F. Griese, K. Dürr, U. Mihr, & B. Schweighofer, Trans.). Munich, Germany: Hanser (Original work published 1992).

Briggs, J. & Peat, F. D. (1993). *Die Entdeckung des Chaos* [Turbulent mirror] (C. Carius, Trans.). Munich, Germany: DTV (Original work published 1989).

Caillois, R. (2001). [Man, play, and games] (M. Barash, Trans.). Chicago, IL: University of Illinois Press (Original work in French published 1958 as *Les jeux et les homes: le masque et le vertige*).

Cohen, J. L. (1979). The semantics of methaphor. In A. Ortony (Ed.), *Metaphor and thought* (pp. 64–77). Cambridge, England: Cambridge University Press (Original work published 1958).

Cramer, F. (1989). *Chaos und Ordnung* [Chaos and order]. Stuttgart, Germany: DVA.

Cramer, F. (1994). Das Schöne, das Schreckliche und das Erhabene: Eine chaotische Betrachtung des lebendigen Formprinzips [The beautiful, the dreadful, and the sublime: A chaotic consideration of the living principle of form]. In G. Bien, T. Gil, & J. Wilke (Eds.), *'Natur' im Umbruch* (pp. 259–282). Stuttgart, Germany: Frommann-Holzboog.

Cramer, F. & Kaempfer, W. (1992). *Die Natur der Schönheit* [The nature of beauty]. Frankfurt am Main, Germany: Insel.

Eigen, M. & Winkler, R. (1975). *Das Spiel: Naturgesetze steuern den Zufall* [Games: Laws of nature governing chance]. Munich, Germany: Piper.

Festinger, L. & Carlsmith, M. (1959). Cognitive consequences of forced compliance. *Journal of Abnormal and Social Psychology*, 58, 203–210.

Freeman, W. & Skarda, C. (1985). Spatial EEG patterns, non-linear dynamics and perception: The neo-Sherringtonian view. *Brain Research*, *357*(3), 147–175.

Freud, S. (1930). Das Unbehagen in der Kultur [Civilization and its discontents]. In A. Freud (Ed.), *Gesammelte Werke* (Vol. 14, pp. 419–506). Frankfurt am Main, Germany: Fischer.

Gardner, H. (1993). *Creating minds: An anatomy of creativity as seen through the lives of Freud, Einstein, Picasso, Stravinsky, Eliot, Graham, and Gandhi*. New York: Basic Books.

Goethe, J. W. (1885). *Goethes sämtliche Werke* [Goethe's collected works] (Vol. 1). Stuttgart, Germany: Gotta.

Goodman, N. (1978). *Ways of worldmaking*. Indianapolis, IN: Hackett.

Hildesheimer, W. (1977). *Mozart*. Frankfurt am Main, Germany: Suhrkamp.

James, W. (1880). Great men, great thoughts, and the environment. *Atlantic Monthly, 46*, 451–449.

Kant, I. (1968). *Kants Werke. Akademie Textausgabe: Vol. 5. Kritik der Urteilskraft* [Critique of judgment]. Berlin: De Gruyter (Original work published 1790).

Koestler, A. (1966). *Der göttliche Funke. Der schöpferische Akt in Kunst und Wissenschaft* [The divine spark: The creative act in art and science]. Bern: Scherz.

Kosslyn, S. M. (1980). *Image and mind*. Cambridge, MA: Harvard University Press.

Küster, K. (1991). Der neue Wolfgang Amadé. *Bild der Wissenschaft, 6*, 26–30.

Lenk, H. (1993). *Interpretationskonstrukte* [Constructs of interpretation]. Frankfurt am Main, Germany: Suhrkamp.

Lenk, H. (1995a). Das metainterpretierende Wesen [The meta-interpreting being]. *Allgemeine Zeitschrift für Philosophie, 20*, 39–47.

Lenk, H. (1995b). *Schemaspiele. Über Schemainterpretationen und Interpretationskonstrukte* [Schema-games: Schema-interpretations and constructs of interpretation]. Frankfurt am Main, Germany: Suhrkamp.

Lenk, H. (2000a). *Kreative Aufstiege. Zur Philosophie und Psychologie der Kreativität* [Creative assents: On the philosophy and psychology of creativity]. Frankfurt am Main, Germany: Suhrkamp.

Lenk, H. (2000b). Outline of systematic schema interpretation. In D. O. Dahlstrom (Ed.), *The proceedings of the Twentieth World Congress of Philosophy: Vol. 8 Contemporary philosophy* (pp. 121–132). Bowling Green, OH: Philosophy Documentation Center.

Lenk, H. (2003). *Grasping Reality: An interpretation-realistic epistemology*. Singapore: World Scientific.

Lenk, H. & Stephan, A. (2002). On levels and types of complexity and emergence. In E. Agazzi & L. Montecucco (Eds.), *Complexity and Emergence* (pp. 13–28). Singapore: World Scientific.

MacCormac, E. R. (1985). *A cognitive theory of metaphor*. Cambridge, MA: MIT.

MacCormac, E. R. (1988). Die semantische und syntaktische Bedeutung von religiösen Metaphern [The semantic and syntactic meaning of religious metaphors]. In J.-P. van Noppen (Ed.), *Erinnern, um Neues zu sagen* (pp. 84–175). Frankfurt am Main, Germany: Athenäum.

Merton, R. K. (1957). *Social theory and social structure*. Glencoe, IL: Free Press.

Plessner, H. (1928). *Die Stufen des Organischen und der Mensch. Einleitung in die philosophische Anthropologie* [Stages of the organic and man]. Berlin: Gruyter.

Poincaré, H. (1913). *The foundations of science* (G. B. Halstead, Trans.). New York: Science Press.

Polet, S. (1993). *Der kreative Faktor: Kleine Kritik der kreativen (Un-)Vernunft* [The creative factor: A brief critique of creative reason and unreason]. Bensheim, Germany: Bollmann.

Simonton, D. K. (1984). *Genius, creativity, and leadership: Historiometric inquiries*. Cambridge, MA: Harvard University Press.

Simonton, D. K. (1988). Creativity, leadership, and chance. In R. J. Sternberg (Ed.), *The nature of creativity* (pp. 386–426). Cambridge, England: University Press.

Sternberg, R. J. (Ed.) (1988). *The nature of creativity*. Cambridge, England: University Press.

Sternberg, R. J. (Ed.) (1999). *Handbook of creativity*. Cambridge, England: University Press.

Sternberg, R. J. (2003). *Wisdom, intelligence, and creativity synthesized*. New York: Cambridge University Press.

Sternberg, R. J. & Lubart T. I. (1995). *Defying the crowd: Cultivating creativity in a culture of conformity*. New York: Free Press.

Weisberg, R. W. (1986). *Creativity: Genius and other myths*. New York: Freeman.

Weisberg, R. W. (1993). *Creativity—Beyond the myth of genius*. New York: Freeman.

Weiss, P. (1992). *Creative ventures*. Carbondale, IL: Southern Illinois University Press.

Whitehead, A. N. (1978). *Process and reality*. London: Free Press–Collier Macmillan (Original work published 1929).

Chapter 6
Milieus of Creativity: The Role of Places, Environments, and Spatial Contexts

Peter Meusburger

Although some of the earliest case studies on famous scientists addressed the importance of parents, peer groups, teachers, and fortuitous events for creative persons (e.g., Candolle, 1873; Ellis, 1926; Ostwald, 1909), interactions with the environment did not figure in the first theoretical conceptions of creativity. Researchers claimed that creative persons are gifted with special innate talents and capacities that others lack and that creativity is a gift or innate talent that cannot be acquired or taught (see Boden, 2004, pp. 14–15). This concept eventually raised a number of questions. For example, why are highly creative individuals not evenly distributed over time and space? Why are certain cities and historical periods characterized by great creativity in the visual arts, music, and science, whereas others are not? Why are certain research departments or universities so successful at copiously producing outstanding creative scientists, whereas others are not? Why does the large majority of Nobel Prize winners stem from such a small share of universities? Reflecting growing interest in the social environment as a variable, these questions indicated a change in creativity research.

However, interest in *spatial* disparities of creativity and in the impact of spatial contexts, spatial settings and spatial relations on creativity did not evolve until the late twentieth century. One reason for such belatedness is that new, original, and valuable ideas and topics often encounter resistance because they usually threaten continuities and tradition and may destroy existing paradigms, power relations, and self-efficacy. Both ignorance and the highly valued preexisting knowledge of experts can block novel ideas and can lead merely to the production of tried and trusted correct answers (Cropley, 2006, p. 402).

Eventually, an ever greater number of scholars accepted that creativity is not an innate attribute of a single individual, no matter how intelligent and talented that person might be. It was recognized that creative ideas emerge and develop in complex, dynamic interaction between the creator and his or her environments (see, for example, Amabile, 1983a, 1988, 1996; Amabile et al., 1990, 1996; Csikszentmihalyi, 1988, 1999; Ericsson, 1996; Feldman et al., 1994; Gardner,

P. Meusburger (✉)
Universität Heidelberg, Geographisches Institut, Berliner Str. 48, 69120 Heidelberg, Germany
e-mail: peter.meusburger@geog.uni-heidelberg.de

P. Meusburger et al. (eds.), *Milieus of Creativity*, Knowledge and Space 2,

1988, 1993a, b, 1995; Mayer, 1999; Mumford, 1995; Sternberg & Lubart, 1999). It was realized that creative ideas arise from a large set of well-developed skills and a rich body of domain-relevant knowledge that must be acquired through laborious apprenticeship (Simonton, 2000, p. 152). Creativity, in other words, therefore requires time and certain environmental conditions.

Second, viewing creativity from an interactional perspective accentuates the relevance of situational, contextual, and cultural determinants and various layers of existential dimensions. Place matters, because a stimulating environment and a talented individual must come together and interact before a creative process can occur (see Sternberg & Lubart, 1991). That process requires preparation through learning, gaining experience, and identifying and solving problems. It takes cognitive skills and results from complex and dynamic interaction between the actor and his or her surroundings. This individual potential for creativity has to be actualized and cultivated by the family, the school environment, role models, organizational structures, challenges, disciplinary cultures, and chance opportunities in professional careers. As this viewpoint suggests, interaction of this sort is not mechanistic. Creative, talented people are not just raised, trained, and embedded in particular milieus. In their careers they tend to be attracted to certain institutions and places where they can develop their abilities and ideas, have the occasions to interact with other knowledgeable agents, procure the necessary support, be inspired, tackle challenges, and command the necessary resources. "Complex problem solving implies the efficient interaction between a solver and the situational requirements of the task and involves a solver's cognitive, emotional, personal, and social abilities and knowledge" (Frensch & Funke, 1995, p. 18). In short, the interactional perspective posits that the social and material environment, with its ability to promote or hinder such development, is an important constituent of creativity.

A third reason for the turn to the spatiality of creativity is that early problem-finding and problem-solving depend on perceptual discernment and environmental sensitivity, that is, on "the ability to be aware of and to correctly identify events within one's environment" (Carlozzi et al., 1995, p. 366). A number of studies support the hypothesis that creative persons have a heightened perceptual awareness (Stamm, 1967, p. 93) and that they are likely to be more sensitive to environmental stimuli than are less creative individuals (Barron, 1969; Carlozzi et al., 1995, p. 371). Because of the great sensitivity, keen attention, and prior knowledge that creative individuals have, they perceive and identify upcoming problems and new trends and research questions earlier than others do.

The principal aim of this chapter is to show that the generation, evaluation, and adoption of creative ideas and products vary spatially and that the spatiality and spatial distribution of creative processes should not be ignored. With the study of creativity becoming increasingly multifaceted, my second aim is to explain why various lines of research and insights from multiple disciplines should be brought into a common framework. The purpose is to encourage the cross-fertilization of ideas and to avoid the trap of disciplinary (and dogmatic) insularity. My third concern is to discuss results and methodological problems of creativity research from a geographical perspective and to clarify misunderstandings that might complicate transdisciplinary discourse between human geography and other social and behavioral sciences.

Definitions, Types, Domains, and Degrees of Creativity

Definitions of Creativity

The aspects of creativity, the scales and units of research, and the approaches that are of interest in the study of creativity differ from one group of scholars to the next. Be they philosophers, psychologists, historians, art critics, geographers, sociologists, economists, architects, urban planners, or scholars from some other discipline, they all bring their own questions, approaches, concepts, and methodologies to bear on the subject. This variety has a number of advantages, for it enables each field of inquiry to provide new insights not obtainable in the others. But it also entails problems and misunderstandings. With regard to indicators, theoretical concepts, and empirical methods, research on creative individuals diverges as much from the study of creative organizations as the latter does from inquiry into creative environments. The more the term *creativity* has acquired buzzword status in public debate (e.g. creative industries and creative class) and the more remote its use has become from what it is understood to mean in psychology, the greater the need has become to clarify the word. I do not intend to comment on each of the more than one hundred different definitions of creativity. Instead, I wish to discuss some problems of delimitation and specification.

Most definitions of creativity contain the key adjectives (or synonyms) that appear in the definition by Boden (2004): "creativity is the ability to come up with ideas or artifacts that are *new*, *surprising* and *valuable*" (p. 1; italics added). Creative people typically address topics that are unnoticed, underrated, or not understood by others. New, original, and valuable ideas or products are inevitably scarce in their initial stage (start-up period) and are therefore confined to a small proportion of people and places. As soon as a creative idea or product has been accepted by much of the population, as soon as it disseminates to a large number of places, it ceases to be considered novel or surprising. The definition of creativity therefore implies scarcity, which is the opposite of ubiquity.

Many authors, such as the psychologists Simonton (2000) and Funke (2000, p. 284), have argued that creativity is an all-pervasive phenomenon of human nature. All people constantly need and apply creativity to solve their everyday problems. However, creativity is a matter of degree, and its meaning shifts according to discipline and spatial scale. One should therefore distinguish between various types and fields of creativity. Boden (1994, pp. 76–77, 2004, p. 2), for instance, recommends drawing a line between psychological creativity and historical creativity. To her, psychological creativity involves coming up with a surprising, valuable idea that is new to the person to whom it occurs, regardless of how many other people have had that idea before. Historical creativity means that no one else has had the idea or made the artifact before (as far as is known), that it has arisen for the first time in human history. Both types of creativity interact with their environment, and both can result from environmental stimuli, cues, or prompts. In this sense, a distinction between psychological and historical creativity or between everyday problem-solving and outstanding achievements is necessary in order to avoid misunderstandings.

Geographically speaking, it is an important fact that most definitions of creativity include a relation to a context, environment, organization, group, or field. Briskman (1980) claims that one of the most striking features about creative products "is their appropriateness, the 'internal connection' which exists between these products and the background against which they emerge" (p. 98). Stein (1953) suggests that "creative work is a novel work that is accepted as tenable or useful or satisfying by a group in some point in time" (p. 311). Oldham and Cummings (1996) define creative performance "as products, ideas, or procedures that satisfy two conditions: (1) they are novel or original and (2) they are potentially relevant for, or useful to, an organization" (p. 608).

In the field of management, Woodman, Sawyer, and Griffin (1993, p. 293) define creativity as the generation of a valuable, useful new product, service, idea, procedure, or process by individuals working together in a complex social system. According to Ford (1996) "creativity is a subjective judgment made by members of the field about the novelty and value of a product; it is not an inherent quality that can be measured independent of social-construction processes within a field" (p. 1115). He defines creativity "as a domain-specific, subjective judgment of the novelty and value of an outcome of a particular action" (p. 1115). Similarly, Shalley, Gilson, and Blum (2000, p. 215) stated that "creativity involves the production, conceptualization, or development of novel and useful ideas, processes, or procedures by an individual or by a group of individuals working together." According to D'Agostino (1984, pp. 88–102), true human creativity involves novelty, value, appropriateness to context, and unpredictability in terms of antecedent knowledge, available recipes, existing rules, and environmental stimuli. I note, however, that values are a result of evaluations and vary over time and space. Quality in itself has no meaning in the absence of a domain in which it is realized and a field by which it is judged (Gardner, 1995, p. 38).

A work or idea is not necessarily novel merely by being different from what preceded it. There must be some merit or value in being different. As the philosopher Hausman (1979) states, eccentricities lack the criterion of value. Briskman (1980) claims that one of the most striking features about creative products "is their appropriateness, the 'internal connection' which exists between these products and the background against which they emerge" (p. 98).

To avoid misunderstandings, creativity as a trait or input variable should be differentiated from creativity as a process and creativity as an achievement or output variable (see Wierenga & van Bruggen, 1998, p. 84). The psychologists True (1966, p. 34) and Klausmeier (1961, p. 4) distinguish between creative ability and creative capacity, with ability meaning the power to perform an act now and capacity referring to what the person might be able to do given maturation, education, and interaction with other people.

Categories of Problems and Modes of Thinking

When focusing on the interaction between perceived problem, problem-solver, and the environment or when differentiating between innovation and creativity, between creativity and intelligence, or between types and degrees of creativity, I find it helpful to categorize problems according to the modes of thinking they call for. The first distinction is between convergent and divergent thinking. Guilford (1967) considered convergent thought to be a logical process that leads to an exact solution. In contrast, divergent thought describes atypical conceptual associations, a change of perspective, a deviation from the mainstream, and a broadening of the horizon. Convergent thinking is oriented to one correct or conventional answer that is deemed best. It is tied to existing knowledge, it emphasizes logic and accuracy, and it leaves no room for ambiguity. Divergent thinking means thinking in different directions and searching for new paths. It involves the production of multiple or alternative answers from available information and requires unexpected combinations, links between remote associations, and transformations of information into unexpected forms (Cropley, 2006; Runco & Okuda, 1988). "Convergent thinking usually generates *orthodoxy*, whereas divergent thinking always generates *variability*" (Cropley, 2006, p. 392).

DeBono (1968) distinguishes between vertical and lateral thought processes. The vertical thinker, in solving a problem, digs a preexisting hole deeper; the lateral thinker digs a new hole. The work of the vertical thinker can be monitored by management more easily than the work of the lateral or creative thinker can. Creative thinkers are self-directed; it is almost impossible to wedge them into a uniform scheme, especially if it involves detailed supervision of all aspects of work (Suojanen & Brooke, 1971, p. 19).

Koestler (1964) discriminates between associative and bisociative thinking. Associative thinking is based on habit; set routines; adherence to rules, disciplinary paradigms, and boundaries; and the use of rationality and logic. Bisociative thinking is characterized by overlapping separate domains of thought, a lack of attention to existing rules and disciplinary boundaries, and an emphasis on imagery and intuition. According to Scott and Bruce (1994, p. 587), associative thinking represents the systematic problem-solving style working within established methods or procedures, whereas bisociate thinking stands for the intuitive problem-solving style. Similarly, the Gestaltists (K. Duncker, W. Köhler, and M. Wertheimer) discriminated between *productive* and *reproductive* thought (see Funke, 2000, p. 290). Reproductive thought describes cognitive processes that need only to be recalled in order to solve a problem or task. An example is the recalling of a mathematical operation, a physics equation, or a cooking recipe. Even if the cake has never before been baked and the equation never solved—that is, even if the result is new—the calculation method or the recipe is known. In the case of productive thought, the path to the solution must first be discovered or construed.

Another way of demarcating various categories of creativity is to discriminate between well-defined and ill-defined problems (Unsworth, 2001), open and closed problems (Jaušovec, 2000), and analytical and creative problems. A well-defined problem is a means–end analysis, the most frequent process that humans use when they solve everyday problems. It is precisely what rational behavior or rational problem-solving is about. "The information necessary to solve a well-defined problem is usually specified precisely in the statement of the problem itself. In the case of ill-defined problems, it is often unclear what kind of information exactly is relevant to the problem at hand" (Jaušovec, 2000, p. 214). Means–end analysis is not suitable for studying open goals, dynamic evolution, or ill-defined problems. In an analytical problem, all necessary conditions are stated and only one solution is possible. A mathematical equation can be solved by logic alone. Success at solving it depends primarily on whether the solver is familiar with the logic and rules. A creative problem is one that is open to a variety of solutions. It takes flexibility, imagination, and interaction with the environment to solve that kind of problem. The artist or poet is not praised as creative for following rules known before producing his or her picture or poem but rather for bringing forth something that did not previously exist. The creative process in art, music, and many fields of basic research is open ended. Artists or scientists in basic research normally do not know from the beginning what they are about. If they knew completely where their work was heading, they could not be engaged in creative work (see Maitland, 1976, p. 397; Tomas, 1958, pp. 1–3).

In early studies, convergent thinking and divergent thinking were often presented as conflicting or competing processes. Convergent thinking was regarded as detrimental to creativity, and divergent thinking was almost equated with creativity. Opinion in this regard seems to have shifted somewhat, however. Cropley (2006) argues that a creative process requires a combination of divergent and convergent thinking. "Divergent thinking and convergent thinking seem to add something to each other or even to compensate for defects in each other" (p. 401). He suggests a distinction between generating novelty and evaluating novelty's risk. The generation of novelty stems mainly from divergent thinking. But the risks of introducing novelty have to be explored by convergent thinking and logic.

From a geographer's point of view, it seems important to distinguish between knowledge based on cases, that is, knowledge acquired in *places* (e.g., in the field, archives, museums, and laboratories) and knowledge based on rules and logic (mathematics). The former kind of knowledge depends more on interactions with the environment than the latter does.

Intelligence, Knowledge, Creativity, and Innovation: Their Interplay, Interrelationship, and Delimitation

High levels of intelligence or knowledge do not guarantee creativity. Intelligence and knowledge are necessary, but not sufficient, conditions for creativity. Intelligence and creativity are separate, albeit interdependent, variables (see Chapter 7 by Kaufman in this book; Sternberg & O'Hara, 1999). According to Shekerjian (1990), creativity is not the direct result of intelligence, talent, or skills. It comes instead from having an open "beginner's mind," being curious, practicing divergent thinking, seeing relationships between apparently unrelated factors, drawing on intuition, and tolerating the "long dance of uncertainty" that precedes most breakthroughs (Saaty, 1998, p. 10). The interplay between creativity and intelligence varies according to the problems to be solved and the different phases of a creative process. Intelligence is needed for in-depth thinking and for the development of techniques to solve *defined* problems. Creativity is needed in order to conceive new ideas and new alternatives with which to solve problems. "To analyze problems in detail, we need intelligence. But we need creativity to synthesize and create structure to obtain higher level abstraction of problems" (Saaty, 1998, pp. 9–10). Robinson (1970) states that intelligence is not the initiator or driving force of creativity, that it becomes important only at a later stage, when the new ideas already produced by the mind must be critically evaluated and their implications worked out.

Couger (1995) shows that creativity and intelligence are only moderately related. They are strongly related up to an IQ of 120, after which point the relation disappears. In a relaxed and unconstrained environment, intelligence and creativity do not seem to be related (see also Saaty, 1998, p. 10). In other words, intelligence tests are not very useful in measuring potential for creativity. Another nonlinear relationship is that between knowledge and creativity. One must have knowledge to produce something new, but creativity goes beyond knowledge (Weisberg, 1999). "Knowledge may provide the basic elements, the building blocks out of which are constructed new ideas, but in order for these building blocks to be available, the mortar holding the old ideas together must not be too strong" (p. 226).

When a creative process enters the phase of elaboration and verification (see below, "Stages of the Creative Process"), then creativity joins the stage of innovation. Of course, it may be futile to draw a line between creativity and innovation when analyzing the spatial distribution of creative or innovative products on the spatial macroscale (e.g., patent intensity in provinces or nation-states). But the distinction should be made at least in microscale analysis and in theoretical discussions. Creativity is related to the generation of new and valuable ideas, whereas innovation is more about the *implementation* of those ideas. Most innovations begin with creative ideas, but many highly creative ideas are never implemented or adopted. Many creative individuals fail to act on their ideas because they lack the resources or interest to continue developing them. In many cases the innovator applying a creative idea did not generate it.

It is therefore hardly surprising that the skills, personal traits, organizational structures, and styles of leadership needed for creativity are not the same as those needed for innovativeness. Creativity is linked to an intuitive problem-solving style; innovation, more to a systematic problem-solving style. Successful innovation depends not only on creative ideas but also on the ability to attract venture capital, design new organizational processes, communicate the value of a new idea, persuade people, and "manage impressions" (Kasof, 1995; Magyari-Beck, 1998). In the innovative process, leaders have to set goals, manage attention, coordinate and control actions, raise capital, promote the cohesiveness of their team, and study the market. In the creative process, leaders have to arrange for incentives, supportive environments, new interactions, and exchanges between knowledgeable people. They have to encourage the autonomy and self-esteem of the group members and their willingness to take risks.

Drawing on Amabile (1988) and Staw (1990), Oldham and Cummings (1996) apply these distinctions between creativity and innovation specifically to the realms of performance and organization: "Creative performance refers to products, ideas, and so forth produced at the individual level, whereas innovation refers to the successful implementation of these products at the organizational level" (Oldham & Cummings, 1996, p. 608). Woodman et al. (1993, p. 293) understand creativity as a subset of the broader domain of innovation. They characterize innovation as part of an even broader construct of organizational change. Organizational change can include innovations and creativity, but a good deal of organizational change takes place without innovation.

Categories and Domains of Creativity

To avoid misunderstandings, one should distinguish between various levels and forms of creativity. Many creative processes include some kind of problem-solving, but not all kinds of creativity can be reduced to problem-solving. In the performing arts, visual arts, music, and similar fields, creativity is expressed through performance, self-expression, or self-actualization. Finding and identifying new problems and raising new research questions may be much more creative than solving the problem itself is—and may contribute more to the advance of science (see Sadler & Green, 1977, p. 157).

Differentiating between open and closed problems (see Jaušovec, 2000), Unsworth (2001) proposes four categories of creativity: expected and proactive creativity for open problems and responsive and contributory creativity for closed problems (see Fig. 6.1). Responsive and expected creativity are externally driven; contributory creativity and proactive creativity are self-determined or driven by internal motivation. Responsive creativity can be planned and organized (e.g. the contributions of a think tank, or the Manhattan project). Expected and proactive creativity involve scanning, categorizing, and interpreting the environment to find a problem, evaluating a perceived situation, and then defining the problem in such a way that it can be solved (Unsworth, 2001, p. 294). Expected and responsive creativity have external drivers for engagement; proactive and contributory creativity have internal drivers.

Saaty (1998) distinguishes between deductive and inductive creativity. Deductive creativity is the ability "to face a new application instance to which we might bring to bear past knowledge of similar situations" (p. 10). Inductive creativity "looks at all that experience and attempts to induce from it a description of the larger system from which the problem instances flow" (p. 10). In Chapter 12 of this book, Boden distinguishes combinatorial, exploratory, and transformational. Abel (Chapter 4) suggests a distinction between strong and weak creativity. Most authors agree that a distinction between various types and levels of creativity is necessary.

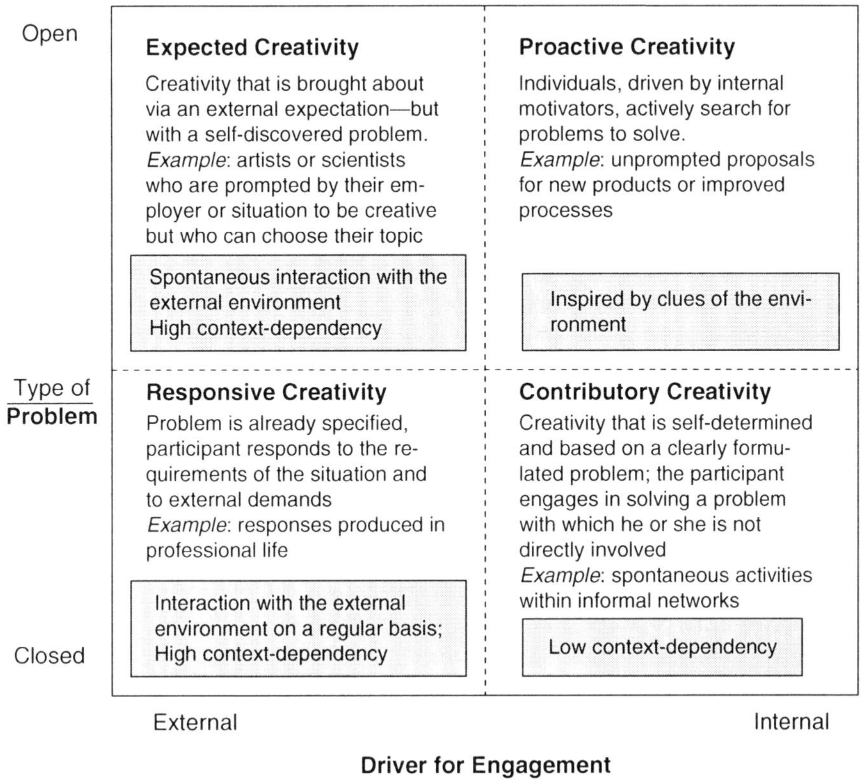

Fig. 6.1 Types of creativity (Unsworth, 2001, p. 291, modified by P. Meusburger)

Judgment of creativity needs a frame of reference. In science and the arts alike, certain rules, paradigms, conventions, expectations, and evaluation procedures develop and come to be regarded as binding in the relevant community, discipline or area, at least for a particular period of time (see Kroeber, 1944). Potential geniuses who cannot fit in or who are unwilling to abide by the given rules or conventions will be either doomed or frustrated. If they are successful, they might launch a new paradigm. Some authors use the term *domain* to refer to these knowledge and evaluation structures of discipline and control. Li and Gardner (1993) define "domain as bodies of disciplined knowledge that have been structured culturally and that can be acquired, practiced, and advanced through the act of creating. … A domain can be described as a unified structure that is rooted in culture" (p. 95). According to Csikszentmihalyi (1999, p. 314), an environment has "two salient aspects: a cultural, or symbolic aspect which here is called the domain; and a social aspect called the field. Creativity is a process that can be observed only at the intersection where individuals, domains, and fields interact" (p. 314). A domain consists of a set of rules; a field consists of persons working within the same domain.

Boden (2004) uses the term *conceptual space* instead of domain. She defines conceptual spaces as structured styles of thought (see also Chapter 12 in this book). Conceptual spaces include ways of writing, styles of painting, theories in science, fashions in couture or cooking, that is, any disciplined way of thinking that is familiar and valued by a certain social group. Li and Gardner (1993, pp. 96–97) explain their concept of domains by comparing Chinese and Western painting. The two domains differ in the way objects are represented and in the materials and media that are used. The differences between Chinese and Western paintings have little to do with personal choices or capabilities but with "choices imposed by the different painting domains in which the respective artists practice and create. … The chief distinguishing characteristics of the two domains impose unique constraints on the process of creativity" (Li & Gardner, 1993, p. 94). Similar rules exist in some scientific disciplines or research departments. They stipulate the approaches or methods that should be preferred and where good research should be published. The anticipatory obedience or self-censorship that such spatially divergent expectations induce in young researchers might favor their professional career but not necessarily their creativity.

As long as a product, scientific concept, or piece of art has not been validated by peers, experts, or users, it is not regarded as creative. These frames of reference or domains can comprise a few dozen experts or millions of supporters. The relevance and power of domains grows and shrinks in the course of time and varies in the spatial dimension. Many scientific disciplines and fields of art are fragmented into domains contradicting and opposing each other. Concepts that are highly respected within one domain (e.g., rational agent in neoclassical economics or constructivism in cultural studies) may be heavily criticized or even ridiculed in other domains. Kasimir Malevich's painting *Black Square* on a white ground is considered by some art critiques to be one of the most important artworks of the twentieth century. For them Malevich (1878–1935) is the legendary, radical, and influential representative of abstract modernism in Russia and the founder of Suprematism. However, the constructivists of Moscow and St. Petersburg (e.g., Vladimir Tatlin) fundamentally criticized the spiritual aspects of Suprematism. They represented a radically materialistic position and sponsored "production art," which emerged after the October Revolution. It would be an interesting challenge to map and interpret the development and diffusion of networks clinging to the one or other of these domains.

The Measurement of Creativity

I do not intend to elaborate on the numerous methods of measuring creativity (for details see Eysenck, 1994; Chapter 7 by Kaufman in this book). But because theories are influenced by empirical results and because empirical results are influenced by ways of measuring, some remarks on measurement seem necessary. Most authors probably agree that creativity is not reducible to mere quantitative scientific analysis. Some aspects of creativity are accessible only through qualitative methods.

At least four main approaches to ascertaining and studying creativity exist. The first one involves analyzing the biographies of creative persons or evaluating what eminent scientists or artists reveal about how they produced their creative work. This methodology has been in use since the mid-nineteenth century (Galton, 1869; Ostwald, 1909). Biographies include quantitative data (e.g., about creative output and age), qualitative data (e.g., narratives and reported emotions and frustrations of creative people), and interpretations of the meaning of relationships with other people. Many Nobel Prize winners, artists, and other thinkers have left diaries, letters, records, or autobiographies or have been interviewed about the steps of their creative process, the way in which they achieved their outstanding results, and the manner in which their environment supported or impeded their work. With the autobiography being primarily a self-interpretation and a means of self-creation (see Mayer, 1999, pp. 455–457; Vidal, 2003, p. 76), this procedure does not deliver objective results but rather evidence that "this is how those who are accepted as being creative say they work" (Westland, 1969, p. 128). Analysts of interactions and relations ask which teachers creative people have been predominantly influenced by, which interactions with other scientists creative people have profited from most, and which kind of field work triggered their new insights. Researchers can standardize and objectively score such self-reports by distributing personality inventories to the test persons (see Rees & Goldman, 1961).

The second main approach to creativity centers on the question of whether people who are regarded as creative in certain fields exhibit similar personality traits. Scientists using this approach design and apply tests to measure the creative potential of persons in order to predict their possible creativity. One kind of such creativity test measures performance from which creativity can be judged. Another kind assesses the creative personality. Both must grapple with two issues. The first is whether they identify creative ability and distinguish between intelligence and creativity or rather measure something else (e.g., originality) instead. The second issue is the fact that they focus on just one side of the interaction—the creative potential or disposition of a person—and more or less ignore the role of the environment.

The third approach is experimental and "uses cognitive task analysis to specify the component processes in creative thinking" (Mayer, 1999, p. 454). Its experiments are performed in controlled environments (artificial contexts). The purpose of this psychometric approach is to find out whether people generally produce a greater number of creative ideas under certain social, organizational, or environmental conditions than under others (for an overview see Plucker & Renzulli, 1999). By changing the procedure, situation, or information to which the participants are exposed in an experiment and by then comparing the responses of those persons, psychologists try to discover which environmental factors help or hinder creativity. A similar method is to evaluate the work environment or climate of organizations. The climate for creativity can be measured by several psychometrical instruments, such as KEYS, which is designed to provide reliable and valid assessments of factors in the work environment that are likely to influence innovation and the generation of creative ideas (Amabile, 1995; Amabile et al., 1996, p. 1155).

The fourth approach is preferably applied on the spatial meso- and macroscales. It aids investigations into the question of why certain time periods, places, cities, or contexts have produced more eminent creative artists and scientists or more creative products (as measured by the number of patents, scientific awards and inventions, for example) than others have (see Brix, 1993, 2003).

The question of whether researchers should tend to concentrate on creative persons, creative products, creative processes, or rather on preconditions for creativity touches not only on disciplinary traditions and scientific interests but also on the matter of data availability. Process models of creativity are not superior to outcome models. Both are needed, for they have complementary functions and fulfill different purposes.

Stages of the Creative Process

Creativity is not a sudden insight but a lengthy process. Since Wallas (1926) and Patrick (1937), it has been generally accepted that a creative process has four or more stages (Funke, 2000; Runco, 1993; Weisberg, 1999; see also Funke's Chapter 1 in this book). Insight is only one of them. Equally important are the stages of preparation; incubation; verification; and acceptance within an organization, discipline, or market. The classical model has four stages. Funke (2000) distinguishes between five stages. Cropley (2006) described six stages: preparation, incubation, illumination, verification, communication, and validation (p. 402). The following description follows Funke (2000, pp. 288–289, see chapter 1 in this book).

Stage 1: Preparation. The preparation phase includes problem-finding, information gathering, and formulation of preliminary ideas. It involves conscious work and draws on a person's education, analytical skills, and problem-relevant knowledge (Lubart, 2000–2001, p. 296). In order to be creative, a person has to have been involved in a specific problem or to have studied the foundations of a discipline and the works of his or her predecessors over a substantial period. The institutions and places of early learning and scientific or artistic socialization, fortuitous events in life, and path dependencies in the early career of a person can have a crucial influence on that individual's intellectual development and future creative processes.

Stage 2: Incubation. A number of authors (e.g., Dreistadt, 1969; Koestler, 1964; Miller, 2000; Niederland, 1967) have studied the importance of latent stages of creativity and have underlined the role of incubation. Incubation refers to the temporary abandonment of conscious, rational, problem-solving endeavors, which are often pursued when a person has reached an apparent dead end in his or her work. It often makes sense to temporarily set aside a problem for which a creative solution is being sought. Throughout phases of inactivity, the human brain evidently continues working. During this incubation stage, existing associative ideas diminish in the memory and are altered by new information from the environment that is superimposed.

Being unaware of the processes that take place during the incubation stage, the creative individual cannot actively influence them (Funke, 2000, p. 288). Mandler

(1992) has extensively studied processes in which solutions to a problem seem to appear suddenly after a period of incubation. Miller (2000) describes how information held in long-term memory can be processed in the unconscious and then find its way into conscious thought:

> Activation is maintained in the unconscious as the result of a previous intense conscious desire to solve the problem at hand. This activation can spread in the unconscious in ways that might not have been possible within the confines of conscious thought. (p. 337)
>
> Psychological tests indicate that people do not experience sudden illuminations without previous conscious or unconscious reasoning. (p. 332)

According to Westfall (1983, pp. 41–42) and Holmes (1986, p. 22), Newton's law of universal gravitation was not a sudden moment of insight but the result of incredibly intense concentration that Newton sustained over some 30 months.

Stage 3: Insight. During incubation, various psychological processes culminate in the moment of insight (Bloomberg, 1967, p. 130; Götz, 1981, p. 300). At a certain point in time, a recombined conceptual association penetrates the threshold of consciousness and delivers a moment of inspiration. Such an illumination occurs when the individual becomes aware of the creative moment, which follows the relevant preparation and subsequent incubation (Funke, 2000, p. 289).

Stage 4: Evaluation. The creative insight gained from the moment of inspiration must be evaluated for its usefulness or appropriateness by the creator and his or her audience. In the history of scientific discoveries, many new phenomena (e.g., those found by Copernicus, Kepler, Galileo, and Freud) were noticed long before they were accepted, but their significance was not realized by the relevant gatekeepers or audience (see Koestler, 1964; Symington, 1987). Kepler knew for 3 years that the planets did not move in perfect circles, but he would not believe it. Freud took up the earlier observations of Breuer, Charcot, and Chrobak, all of whom knew about the sexual aetiology of hysteria, but was less tenaciously attached to the medical manner of practice and was therefore able to countenance the bisociation of two matrices (Symington, 1987, p. 284). Because guidelines of domains or hegemonic science cultures and paradigms vary spatially, the same idea or product will be evaluated differently at various locations. If an evaluation takes place too early, many creative ideas will be killed by critics before they reach the stage of elaboration. At some places scientists have more freedom and time to develop and elaborate their ideas and experiments than in other areas.

Stage 5: Elaboration and verification. After an idea or work has been evaluated, factors such as the availability of resources (venture capital), alliances, public opinion, and power structures become significant in its verification and implementation. Individuals sense problems, develop ideas, present them to the group, learn from the group, work out issues in solitude, and then return to the group to keep modifying and enhancing their ideas (Drazin et al., 1999, p. 290). If the team embraces the creative idea, it has to be approved by the organization's decision-makers and then financed and transformed into a creative product. It is then up to the market to accept that product. Creative processes at the organizational level may emerge from a process of negotiating multiple and potentially competing interests both between different groups within an organization and with external institutions (Drazin et al., 1999).

Stages 1, 4, and 5 of this lengthy process are thought to be much more concatenated with environments, spatial contexts, and spatial relations than phases 2 and 3 are. But future contributions from perceptual psychology and brain research may show that incubation and insight are more influenced by environmental prompts than is presently known. Lubart (2000–2001) discusses some limitations of stage-based models of creative processes and suggests that these models need to be revised or replaced. He and others (e.g., Treffinger, 1995) suggest moving away from the idea of a fixed sequence of activities and toward three sets of processes—understanding the problem, generating ideas, and planning for action (see p. 300). These processes take place in all stages, and their sequence is not fixed.

Explaining Spatial Disparities of Creativity and Conceptualizing the Impact of Places, Environments, and Contexts on Creativity

From the Creative Personality to Multidimensional Models of Creativity

Since the early 1980s, an increasing number of authors have sought to explain creativity by introducing multidimensional models that include contexts and environments (Amabile, 1983a, b, 1988, 1996; Amabile et al., 1990, 1996; Clitheroe et al., 1998; Csikszentmihalyi, 1988, 1990; Ford, 1996; Hennessey & Amabile, 1988; Isaksen et al., 2000–2001; Oldham & Cummings, 1996; Shalley, 1995; Shalley et al., 2000; Woodman et al., 1993). They have dealt primarily with the question of how personal traits, group characteristics, work environments, organizational structures, cultures, and political conditions interact and how they thereby foster or inhibit creativity or innovative activities. The units of research, number of dimensions, and levels of aggregation taken into account by these models vary according to the research interests, disciplines, and methodologies in question.

Frensch and Funke (1995, p. 7) stated that a problem is defined not only by task features but also by the interaction between task requirement, environmental context, and the personal goals and traits of the person attempting to solve it. Scholars emphasizing task features rather than environment–solver interaction seem to take it for granted that persons do not typically differ in terms of their knowledge and absorptive capacity. These researchers therefore tend to believe that all actors perceive situations in a similar (rational) way. The interactionist approach suggests that the prior knowledge, expertise, experience, and "positionality"[1] of persons determine whether and how soon they are able to perceive a problem, risk,

[1] The term comes from gender studies and means that the position in a social system (gender, for example) influences the goals and perception of a person.

or opportunity and how they interpret and react to environmental prompts. Place, space, milieu, network, and spatial context become important as soon as the issue of interaction between person, organization, and environment is broached and as soon as existing models of creativity are expanded to include external influences and intraorganizational factors.

> A contextual approach to organizational creativity focuses on the prompts that initiate creative behaviour, including the role of prompts in suggesting appropriate goals and potential outcomes, and prompts as the basis for judgments about the creativeness of outcomes of the creative process. (Clitheroe et al., 1998, p. 108)

There is abundant empirical evidence of the close relationship between social environment and creativity. In one study, for example, projects rated high in creativity had significantly different work environments from those rated low in creativity (Amabile & Conti, 1999, p. 631). Much less is known about how an environment influences or triggers creative processes. Scholars disagree on how to conceptualize the term *environment*. Is an environment a social or a spatial phenomenon? Is an environment the sum of socioeconomic variables or a social macrophenomenon? What motives are there for including spatial concepts in creativity research?

One reason to turn to spatial contexts and environments as factors of creativity is that creativity research centering predominantly on isolated variables has come to very inconsistent or contradictory results. Some of these inconsistencies have resulted from taking individual variables out of context and from failing to take into account that correlations between variables differ in the spatial dimension from one context to the next. A second reason is that humans are highly contextual beings and that context-dependencies rather increase than diminish in highly specialized societies. On pages 122–126 I specify in more detail why working in the presence of others or moving from one environment to the other may affect the creativity of individuals and groups. Single variables or models ignoring spatial contexts apparently have less power to explain creativity than some spatial categories do. The various personal, organizational, material, cultural, and political factors affecting creativity interact, merge, and modify each other in specific places or areas and lead to spatially rooted macrophenomena called milieu, environment, action setting, context, or "knowledgescapes." According to Matthiesen (2006, 2007), knowledgescapes focus on the interplay of formal and informal interaction networks and milieus in knowledge-based spatial dynamics. Knowledgescapes are contextualized by different knowledge cultures and constituting distinctive knowledge-based forms of habitus of a specific city region. The knowledgescapes heuristics are focused on comparing particular and distinctive developmental pathways within knowledge-based city-regional developments.

Explaining how an environment or context can have an impact on creative processes makes it necessary to clarify concepts. In a book intended to bridge gaps between disciplines and address nongeographers, it does well to remember that geographers do not entirely agree on the exact definitions of the categories presented in the next section. Some authors use the terms *environment* and *spatial context* synonymously.

Spatial Categories and Their Possible Relations to Creativity

Spatial categories have various characteristics in common but differ from each other in several ways. All spatial categories have the capacity to facilitate or impede interactions. All spatial categories comprise institutions, infrastructure, resources, job opportunities, challenges, and risks. All categories can function as a projection screen of symbolic capital and are used in everyday, reductionist language to absorb the reputation of individuals and organizations working in that spatial unit.

Place. The smallest spatial category is the location (site or spot) where a person performs an activity, faces a challenge, or perceives stimuli and clues. Locations are transformed to culturally determined places by a process that Graumann (1996, 2002) calls appropriation (*Aneignung*). Appropriation literally means making something one's own and taking it for one's own use. (Fischer-Kowalski & Erb, 2003, use the term *colonization* instead of appropriation.) Appropriation of space occurs:

> by marking, naming, defining, categorizing, and evaluating space as appropriate or inappropriate, owned or free, by signs, words, rules, regulations, and laws; but also by regular locomotion resulting in paths and roads; by the cultivation of nature as subsistence of supply of resources; ... by building, constructing, and settling; but also by the artistic and scientific representation of space; and finally by the overcoming of distance by developing means of communication. (Graumann, 2002, p. 104)

According to Canter (1977), "a place is the result of relationships between actions, conceptions, and physical attributes" (p. 159). Places are "the major building blocks for understanding human actions in their naturally occurring context" (Canter, 1985, p. 215). A place has an address, materiality, image, reputation, and an *individual* history and identity. Places are part of shared experience—"they cannot be specified independently of the people experiencing them" (Canter, 1986, p.8). The experience of place has been characterized by Canter (1985) "as having three integrated components: activities, evaluations, and physical form" (p. 231). Unlike situations, places have a distinct, enduring existence and are inevitably intertwined with the physical properties and history of their location (Canter, 1985, p. 216). The physical form of places and their functions have *Aufforderungscharakter* (valence, or the capacity to unite, react, or interact with something else). They summon people who have particular intentions and who command specific knowledge and experience, calling upon them to engage in or refrain from certain activities. Whether the term *place* stands for a room, location, neighborhood, or city depends on the scale of analysis.

Action setting. The terms *behavior setting* (Barker, 1968) or *action setting* (Weichhart, 2003) indicate consistent and organized people–environment interaction that result in extraindividual behavior patterns. Action settings have a physical structure intentionally designed for a specific cultural purpose (e.g., church, office, motorway, or laboratory). They have a clearly defined function maintained and organized by certain material settings, programs, and rules that structure behavior and define certain kinds of actions as inappropriate. The action setting consists of the interdependence between a physical milieu and standing patterns of behavior that are unique to the setting. The expectations elicited by the purpose, rules, and restrictions of an action setting induce in people a collective behavior based on anticipatory obedience or self-censorship in the sense that individuals evaluate their own performance, actions, and artifacts through the eyes of their audience, supervisors, or critics. By complying with the rules of an action setting, people try to avoid sanctions or to achieve approval. At every place, an individual must be aware of the pattern of activities that might be expected in that location, but action settings have more stable and more controlled place rules.

Administrative area and region. Administratively demarcated areas may gain relevance for creativity research when certain rules, regulatory frameworks (abolition of censorship, freedom of expression), practices, resources or other factors influencing creativity are valid in or restricted to clearly defined spaces. Mild forms of such influences are research policies, research funding, and evaluation procedures that authorities prescribe for certain administrative areas (states). Harsher versions are censorship, interdiction or restriction of research, sanctions on certain research topics, and prohibition of study abroad. The ideological background of Stalinism for research in biology and the Allied ban of a large variety of research in occupied Germany from 1945 and valid through the early 1950s (Gimbel, 1990) were effective in precisely defined administrative areas.

The concept of region shares many characteristics with the term *area*, but most authors would agree that regions are defined by former or present *functional* relations. Because functional relations are influenced by available transport lines, communication technologies, politics, language barriers, power relations, and so on, their porous and ill-defined boundaries constantly change. Demarcating functional relations (regions) is a scientific task; demarcating administrative areas is a political act. However, long-standing functional regions can become administrative areas by political decision. For pragmatic reasons, administrative areas are also important for creativity research. Most statistics and indicators describing the socioeconomic preconditions for creativity or its output are related to administrative units (e.g., census wards, municipalities, metropolitan areas, provinces, states). The interpretation of spatial patterns of indicators related to administrative areas is a highly efficient heuristic method of evaluating the preconditions for creative processes and discovering underlying factors of influence.

Environment. The term *environment* refers to both subjectively perceived and objective, relatively stable qualities of an individual's or group's physical and social surroundings (see Clitheroe et al., 1998, p. 105). Unlike a place that is locally fixed in absolute or relative space, an environment is actor or system centered. A subject of intentionality is physical; it occupies a place and has a viewpoint from which environmental objects are perceived and remembered (Graumann, 2002, p. 98) and from which relations are maintained.

With regard to social systems (e.g., organizations and institutions), an *internal environment* differs from the *external environment*. A social system's external environment comprises all those elements (persons, material objects, and places) to which a social system has established relatively stable and regular relations and by which a person or social system can be affected in its goals, motivation, learning processes, and capacities. It consists of institutions, competitors, customers, infrastructure, job opportunities, and social structures as well as of attitudes, values, policies, hegemonic ideologies, and pressures of public opinion prevailing in the spaces of a person's or an organization's activities. Organizations live in symbiosis with their external environment and shape their environment. They communicate with it; make transactions; obtain and exchange energy, goods, and services; and supply the environment with their products and services. Individuals and social systems must make timely, effective responses to environmental changes; profit (to varying degrees) from the environment's potential; and are endangered by its risks and uncertainty. Both the definition of important environmental elements and the dependency of a person or social system on the external environment vary with the type of task, the available resources and capacities, the autonomy of a social system, and the degree of uncertainty confronting the social system. This is the main reason why interactions between actor (organization) and environment have an ideographic character that cannot be put into general rules.

Spatial context. In this chapter the term *spatial context* is conceptualized as a social macrophenomenon that represents a totality of interdependent factors of influence. It is an intersection of various social, cultural, economic, and material spheres or matrices. It represents a kind of superordinate concept that includes objective and subjectively perceived factors, the psychological aspects of interpersonal relationships, and the rules and programs of action settings. A spatial context it is not represented by an accumulation but rather a bundling—in the sense meant by Schumpeter (1912/1934)—and an interdependence of various factors. It is not equivalent to a container but to a catalyst in which various objectively and subjectively perceived facts, individuals, institutions, resources, infrastructure, opportunities, restrictions, norms, rules, and cultures interact, intermingle in their mutual dependencies, and modify each other in a defined area. A slight change in one variable, such as the trust in institutions or personal relationships, the availability of resources, or the competence of leaders, will also affect other variables, such as the openness of information exchange, the style of supervision, or intrinsic motivation. The totality of these interacting items offers potential that may stimulate or hamper creative processes.

Atomistic or holistic perspective? The conventional approach to perception assumes that a stimulus activates a specific receptor in the nervous system and that the pattern of receptor stimulation is interpreted with the memory of past experiences to glean information about the environment. In this approach a person must interpret disconnected stimuli in order to construct something meaningful about the environment (for details see Bell et al., 2001, p. 65). Gibson (1979) believed that perception is more holistic, meaning that properties of the environment are perceived as meaningful entities, not as distinct points. Gibson (1960), Canter (1977), and other psychologists have explained the need to have environmental psychologists orient their research to molar units of the environment, that is, to "wholes endowed with significance" (Canter, 1977, p. 1; for details about molar and molecular approaches, see also Bonnes & Secchiaroli, 1995, pp. 68—71, 134–135, 170–171). Gibson (1960) proposes that stimuli be considered in a "molar" rather than a "molecular" sense. Molar stimuli are represented by "what an organism is responding to, and not by what excites all the little receptors" (p. 700). To Gibson and many other psychologists, cognitive perception is a holistic phenomenon, for it deals with the perception of meanings and not, as with visual perception, to the perception of simple stimuli or patterns.

A similar turn is apparent in other social sciences. Schluchter (2005, p. 24) explains why sociology should not restrict itself to methodological individualism but should include social macrophenomena. However, some human geographers, aware that terms such as *landscape* have been applied unscientifically in the past, still hesitate to use them to refer to a spatial totality. Social and behavioral sciences still see-saw between the two poles of atomistic/mechanistic and holistic or system-oriented perspectives.

From the viewpoint of human geography, one can describe the nature of a spatial context by using the metaphor of a seedbed. The seedbed's outcome depends on many variables, including the type and quality of the soil, the characteristics and quality of the seeds, the amount of precipitation and evaporation, the average annual temperature, the duration of the vegetation period, the availability of fertilizers, and competition between the crops. Though perhaps trivial, it is worth stating that soil, seeds, water, temperature, and other necessary factors must interact at a clearly defined place. Proximity without interaction will not lead to the desired results. If one important factor is absent or does not contribute to the processes in the usual way, the crops will not develop as expected. Some deserts have very fertile soils and are full of seeds, but as long there is not enough water the seeds cannot develop. The seedbed metaphor makes it plain that the study of discrete variables has little power to explain the outcome of the seedbed. What counts is the interaction of these variables at a certain location. And that interaction does not occur automatically; it must be initiated. Hot spots of brain power are like seedbeds, too; they are potential, not independent, factors of influence. It would be a mistake to assume that nearby agents (neighbors) automatically interact and that each of them perceives and uses the available potential or that they depend to a similar extent on exchanging information with the environment. Most observers of the seedbed will not be aware of the chemical, physical, and biological processes occurring between the elements of the seedbed. Neither do they have the knowledge to understand the processes occurring within the plants. Most agents, in their life worlds, have a reductionist view focused on the lot, for the details simply are not known.

The Impact of the Macroscale on Creativity: Socioeconomic Structures and Value Systems

The historical persistence of large-scale spatial disparities of socioeconomic structures. Highly developed, wealthy societies provide better or more expensive educational institutions; attract more eminent scientists and artists; have greater rates of specialization, experimentation, and risk-taking; provide more venture capital; have more complex economic relations; and can invest more in expensive basic research than subsistence or illiterate societies. However, even very wealthy societies and hegemonic states have extremely large internal spatial disparities of creativity. When asking why creative acts concentrate at certain places or in specific environments and regions, one must study both the process of creativity itself and the preconditions or antecedents of creativity.

The antecedents enabling or promoting creativity operate on various spatial scales. On the spatial macroscale they comprise the hierarchy of urban systems and the established capacity of places and regions to attract, keep, support, and inspire talented, knowledgeable, and potentially creative people. Such large-scale disparities in economic and scientific attractiveness are the result of long historical processes governing the spatial division of labor, uneven economic development, educational policies, spatial power relations, migration patterns of talented people, and various path dependencies.

Places differ in the complexity of their economic structure, the quality of the jobs they offer, the location of power and decision-making, the spatial extension of their economic relations, and their scientific reputation. Some places and areas act like magnets attracting the most ambitious artists, scientists, and other talented persons. They offer these people outstanding scientific equipment and resources (venture capital), give them the freedom to experiment and break with traditions (as illustrated in Vienna by Schönberg's 12-tone music, Freud's psychoanalysis, and the artists of the "Secession" movement). These places offer them platforms from which to exhibit their creative ideas and products, grant them access to high level international networks, and enable them to gain reputation. The reputation of a place is interrelated with its degree of international connectivity and its potential for face-to-face contact with top-ranking experts of various domains.

Other places and regions constantly lose most of their talented young generation, intellectuals, scientists, and artists through persistent brain drain or brain overflow caused by lack of opportunities and demand, underdevelopment, or political restrictions. These spatial disparities and hierarchies of social, cultural, and economic attractiveness are constantly in flux. Though disparities between the centers and the peripheries have remarkable historical continuity and are perpetually reproduced by the asymmetry of power relations, some magnets eventually lose their attractiveness for various reasons, and the artists, intellectuals, and scientists leave for more engaging places. These processes on the spatial macroscale create certain hot spots of brainpower. They are characterized by a high density and large spatial concentration of workplaces for scientists, intellectuals, artists, innovative entrepreneurs, highly skilled experts, top-level decision-makers of various sectors, and sophisticated research infrastructure (Meusburger, 1998, 2000). In 1980, Budapest and the capitals of the 19 Hungarian provinces (0.6% of all Hungarian cities and villages) accounted for 35% of the country's total population, but offered 96.1% of all jobs for university graduates in personal and business services, 88.7% of all jobs for university graduates in science and research, 88.6% of all jobs for university graduates in cultural services, and the vast majority of highly skilled jobs in other economic sectors (Meusburger, 1997, pp. 132–133). The more than 3,000 remaining Hungarian villages and towns almost entirely lacked the most important precondition for creative milieus, a highly skilled labor force. Consequently, these peripheral areas had virtually no foreign capital in after the introduction of market economy, an imbalance that only reinforced the disparities that already existed between the center and the periphery.

The impact of cultures, world-views, and research policies. Both the generation and evaluation of creative ideas are affected by prevailing cultures, worldviews, political systems, academic evaluation systems, policies on the recruitment of scholars, and research funding. Ruling classes, powerful bureaucracies, or religious leaders who are afraid that novelties potentially threaten their status or the power structure of their system will be reluctant to introduce the relevant innovations. In many cases it is epistemic power that determines what is to be regarded as a creative product, art, or useful technology. In some places creative ideas are recognized, fostered, and implemented very early; in others, they are not regarded as useful or appropriate, so their implementation may be delayed or rejected.

> Ancient Chinese society is a good example of a central authority supported by a powerful bureaucracy that was able to resist for centuries the spread of new ideas. Despite enormous early cultural advances and a great number of creative individuals, Chinese society believed the use of gunpowder for weapons and that of movable type for the printing of books were bad ideas. (Csikszentmihalyi, 1999, p. 323)

> A new idea will face difficulties in being recognized as creative if the field is defensive, rigid, or embedded in a social system that discourages novelty. For instance, the aridity of Soviet genetics in the thirties was not … a fault of the scientists who made up the field, but of the peculiar agenda of the broader social system of which the field was a part. (Csikszentmihalyi, 1988, p. 331)

Research policies, ways of research funding, and evaluation systems define what good scholarship or good publishing is. However, the criteria and methods that are used to evaluate scientific quality differ from one group of disciplines to the next, and applying the criteria of one discipline to evaluate all other disciplines interferes with scientific creativity. Drawing on Torrance's (1995) claim that incompleteness and disharmony can present openings for creativity, Ambrose (2006) and Schorske (1997) distinguish between two types of academic disciplines: the fractured-porous (e.g., political science, English studies, or human geography) and unified-insular (e.g., analytic philosophy, large parts of psychology, economics).

The fractured-porous disciplines lack internal consensus about methods and theories and are less policed by gatekeepers than the fields of the other category are. The fractured-porous disciplines "tend to be internally contested, inclusive of diverse ideas, and in the process of re-conceiving their fundamental conceptual frameworks" (Ambrose, 2006, p. 77). They lack intradisciplinary interdependence because of their fragmented conceptual frameworks, which reflect warring camps of conflicting groups that vie for attention and supremacy. Because they show strong *interdisciplinary* interdependence and because their porous borders allow the importation of constructs from a wide variety of neighboring disciplines, they seem to be more open to new topics and theoretical perspectives (Ambrose, 2006, p. 82).

The tightly unified-insular disciplines are firmly bounded and heavily policed by journals and career promotions and reflect confidence in their conceptual foundations (see Ambrose, 2006, p. 77). Scholars in unified-insular disciplines "face much less incompleteness and disharmony because they receive reminders about the certainty of investigative methods and the apparent solidity of the conceptual foundations that underpin those methods" (Ambrose, 2006, p. 80). According to Schorske (1997) the unified disciplines achieved their unity by tenaciously preserving their core assumptions and methods, purging themselves of diverse inquiry methods, narrowing the scope of investigation, and marginalizing investigative trajectories ill suited to its entrenched core assumptions (see also Ambrose, 2006, pp. 77–78). "Members of unified-insular disciplines are highly interdependent *within* their fields because they strongly agree on the fundamental conceptual frameworks and research methods that define and constrain their work. However, they strongly reject interdependence when it comes to *interdisciplinary* sharing and collaboration because of their impervious disciplinary borders" (Ambrose, 2006, p. 82).

Unified-insular disciplines are more likely to suppress divergent thinking because they enforce consensus about fundamental assumptions and research methods through their policing of publishing and academic recruitment. In contrast, the fractured-porous disciplines appear to encourage divergent thought. The "porous border that surrounds a fragmented field enables the importation of multidisciplinary constructs that stir up additional divergence of thought" (Ambrose, 2006, p. 81). Centrifugal impulses, lack of consensus, and fragmentation and splintering of topics seem to entail more dynamics and creativity than is the case in disciplines that tenaciously preserve their core assumptions and methods for long periods (Ambrose, 2006, p. 78). These two types of disciplinary outlook also shape the scientific policies and evaluation systems of universities, research foundations, and nations. Some evaluation systems allow more fragmentation of scientific cultures, more splintering of topics, and more diverse criteria of scientific excellence than others, with the result that they can respond more readily to societal and political changes.

The Impact of the Mesoscale on Creativity: Organizational Structures and Climates

Organizations are the spatial mesoscale on which creative processes are greatly influenced. An organization is a goal-oriented and information-processing social system that perceives, scans, interprets, and diagnoses information from the environment and from its own elements in order to remain competitive and adapt its goals and structures to new challenges. A primary task of organizations is to reduce the degree of uncertainty and complexity confronting them, both internally and in relation to the external environment.

Through their vertical division of labor, organizations contribute to internal hierarchies of decision-making and spatial disparities of professional skills. These structural features and characteristics of the organizational climate bear significantly on employees' creative performance at work (Amabile, 1988; Oldham & Cummings, 1996; Redmond & Mumford, 1993; Staw, 1990), either supporting or impairing it. Along with organizational culture and spatial relations, they affect whether organizations can attract, recruit, and keep creative persons and whether such individuals are positively stimulated by the work environment and able to develop their creativity. All these factors determine whether creative people are hindered in realizing their ideas, whether the organization accepts and promotes creative ideas in time, and whether incompetent supervisors or mediocre teams block them.

To conceptualize the faculties and dynamics affecting creativity at the mesolevel, it is justified in organization theory and system theory to speak of organizational intelligence, organizational learning, and organizational memories. "Organizational intelligence is an organization's capability to process, interpret, encode, manipulate, and access information in a purposeful, goal-directed manner, so it can increase its adaptive potential in the environment in which it operates" (Glynn, 1996, p. 1088). One may also speak of creative organizations. Indeed, organizations themselves *have* to be creative if they are to attain their goals and survive competition, for it is within organizations that most creative ideas are evaluated, executed, or verified.

This kind of influence extends beyond organizations, too. Even isolated artists or poets must deal with some organization or other if they want to exhibit or publish their work. With the assistance of organizations, individual agents can shape socioeconomic structures on the *macro*scale.

Organizational structure. Many structural features of an organization are important for creative processes. Some of them are:

- The job complexity and skill variety within the organization
- The design of work settings
- The architecture or arrangement of formal communication, decision-making, and authority within the organization
- The degree of centralization or decentralization of decision-making
- The professional and social competence of supervisors and decision-makers
- The style of supervision, evaluation, and conflict-resolution
- Principles of staff recruitment
- The span of control at various hierarchical levels
- The degree of autonomy that different levels of the organization enjoy
- The availability of resources
- The organization's ability to respond to opportunities and risks of the environment; and
- The formalized structure of communication with the environment

Effective obstacles to creativity are rigid adherence to rules, strong hierarchies, and incompetence of supervisors unfamiliar with the subordinates' areas of specialization (for details see Williams & Yang, 1999, p. 375).

Structures of decision-making and competence distribution within an organization cannot be deliberately altered. They depend on the goals, the external environment, and internal resources of the social system. Disruption of formerly stable relations to the environment, rapid changes of the environment, and growing uncertainty imply the need to shift power to those who have the competence necessary for solving the new challenges and averting crises. In threat-rigidity theory, this kind of response is referred to as the "mechanistic shift" (Staw et al., 1981,

p. 516) that organizations go through under threatening conditions. In those times, they centralize control, conserve resources, restrict information flow, and rely on familiar, well-practiced routines (see also Amabile & Conti, 1999, p. 631).

Group composition affects group performance and plays an important role in creating a potentially stimulating environment (Smith, 1971, p. 495). Research on the relationship between team composition and team performance (Hoffman, 1959; Hoffman & Maier, 1961; Hoffman & Smith, 1960; Smith, 1971) has found that groups in which the members' personalities are heterogeneous tend to produce high-quality solutions to problems. The reason given is that the members apparently stimulate one another constructively and that they mutually correct their errors. Some large companies assess cultural diversity so important that they introduced diversity management.

Organizational culture and organizational climate. Some authors use the terms *organizational culture* and *organizational climate* synonymously; others do not. Schein (1985) defines culture as "a pattern of shared basic assumptions that the group learned as it solved its problems of external adaptation and internal integration, that has worked well enough to be considered valid and, therefore, to be taught to new members as the correct way to perceive, think, and feel in relation to those problems" (p. 19). To Forehand and Gilmer (1964), organizational climate means "the set of characteristics that describe an organization and that (a) distinguish the organization from other organizations, (b) are relatively enduring over time, and (c) influence the behavior of people in the organization" (p. 362). Tagiuri (1968) considers this definition deficient because it takes no account of individual perceptions. He notes that members of an organization interpret its climate in ways that affect their attitudes and motivation. He defines climate as "the relatively enduring quality of the total [organizational] environment that (a) is experienced by the occupants, (b) influences their behavior, and (c) can be described in terms of the values of a particular set of characteristics (or attributes) of the environment" (p. 25). With similar emphasis on the individual, Isaksen et al. (2000–2001) break the term climate down into two aspects. They define organizational climate "as the recurring patterns of behavior, attitudes, and feelings that characterize life in the organization" (p. 172) and regard psychological climate as referring to the individual perceptions of the patterns of behavior.

In an additional consideration of organizational culture and climate, Denison (1996) writes that:

> [I]nteraction reproduces a symbolic world that gives culture both a great stability and a certain precarious and fragile nature rooted in the dependence of the system on individual cognition and action. *Climate*, in contrast, portrays organizational environments as being rooted in the organization's value system, but tends to present these social environments in relatively static terms, describing them in terms of a fixed (and broadly applicable) set of dimensions. (p. 624)

> [C]limate refers to a *situation* and its link to thoughts, feelings, and behaviors of organizational members. Thus, it is temporal, subjective, and often subject to direct manipulation by people with power and influence. Culture, in contrast, refers to an evolved context (within which a situation may be embedded). Thus, it is rooted in history, collectively held, and sufficiently complex to resist many attempts at direct manipulation. (p. 644)

Another difference between culture and climate is that the climate of an organization can be altered relatively quickly by external pressure—such as the downsizing of a firm—or internal changes such as the replacement of group leaders. Culture refers more to the deep and enduring structure of organizations, which is rooted in values, beliefs, traditions, and assumptions held by organizational members (Denison, 1996; Gagliardi, 1986; Isaksen et al., 2000–2001).

With regard to creativity, the important questions are whether the culture of an organization tends to breed conformity and risk aversion (see Cummings, 1965, p. 224), whether certain paradigms should be followed, or whether the culture tolerates deviation from what is traditional. Are new members of an organization confronted with ready-made questions, solutions, and answers, or is a diversity of viewpoints, novelty, and innovation encouraged? Phenomenologically, *climate* is external to the individual, but cognitively the climate is internal to the extent that it is affected by individual perceptions (Woodman & King, 1978, p. 818). Following James, Hater, Gent, and Bruni (1978, p. 786), Scott and Bruce (1994) define the climate of an organization "as individual cognitive representations of the organizational setting expressed in terms that reflect psychologically meaningful interpretations of the situation" (p. 581). The climate of the work environment represents "signals [that] individuals receive concerning organizational expectations for behavior and potential outcomes of behavior" (Scott & Bruce, 1994, p. 582). According to Campbell, Dunnette, Lawler, and Weick (1970):

> [We] might define climate as a set of attributes specific to a particular organization that may be induced from the way that organization deals with its members and its environment. For the individual member within the organization, climate takes the form of a set of attributes and expectancies which describe the organization in terms of both static characteristics (such as degree of autonomy) and behavior-outcome and outcome-outcome contingencies. (p. 390)

The measurement of organizational climate poses a difficult methodological problem (for details see Hellriegel & Slocum, 1974; Isaksen et al., 1999; James & Jones, 1974). Allegedly objective attributes will be perceived and evaluated differently by the members of an organization, depending on their motivation, status, and personal experience. A senior manager, for instance, will perceive the climate of his or her organization differently than a blue-collar worker at the bottom of the hierarchy, so comparison of the two perspectives does not contribute much to research knowledge. Nor does an attempt to average people's job satisfaction or perception of organizational climate. Variance in perceptually measured climate scores might be related to differences in individuals rather than to differences in situations (James & Jones, 1974, p. 1103). Because most large and complex organizations are spread over many locations, the category "climate of an organization" should be expressed in the plural. Each large organization has many different climates that vary from one hierarchic level to the other and from one work group to the next.

Dimensions of social relations in the structure, climate, and culture of organizations. The structure, climate, and culture of organizations are shaped by social relations and can be fully understood only if one takes the following three issues into account.

1. The role of group work, brainstorming, solitude, and working in the presence of others

A vast amount of literature on idea generation and creativity is influenced by the brainstorming paradigm (Osborn, 1953; for an overview see Paulus & Yang, 2000). It rests on two assumptions: that group work and group discussion offer a chance to combine formerly unrelated ideas and that a group's chances of generating a creative response to a challenge improve with the number of different views and knowledge bases there are in a discussion. Under certain conditions, group discussion may contribute to "combining known but previously unrelated facts and ideas in such a way that new ones emerge" (Shalley, 1995, p. 484), generating new ideas and boosting overall productivity. The cognitive theory of group creativity (Brown et al., 1998; Paulus, 2000; Paulus et al., 2000) posits conditions under which effects of cognitive stimulation can be observed in groups. However, the underlying assumption seems to be that all participants in a brainstorming group (a) have the same prior knowledge and the same intrinsic motivations and goals, (b) are all interested in sharing their knowledge with others, and (c) will automatically consider the arguments of others. Recent research is inconclusive about the effects of brainstorming and group work on creativity (Mullen et al., 1991; Paulus, 2000; Paulus & Dzindolet, 1993; Paulus et al., 1995, 2000; Paulus & Yang, 2000). A certain skepticism stems from the argument that brainstorming and group work can act as a stimulus by offering new experiences, ideas, and learning opportunities but that it can also exert pressure on the individual to conform to the mediocre ideas of the majority. Collaborative projects tend to suppress individual initiative and independence. It is thus asserted that brainstorming and group work apparently are not major factors in scientific creative achievement. "There is little doubt that in the realm of highest creativity there is only one creative instrument: the individual mind and spirit of the creator. The landmarks of scientific invention have been established by a handful of *lone* investigators" (Raudsepp, 1958, p. 71).

Empirical results of many studies have supported the hypothesis that group discussion can lead group members to copy each other's responses. In a study on the effects of prior group discussion on individual creativity, for example, Andre, Schumer, and Whitaker (1979) found that group discussion led to a lower total number of different responses being produced by the group than by a comparable number of individuals working alone (p. 111). It appears as though "conformity within small groups inhibits individual creativity. Group discussion acts not as a catalyst to individual divergent thinking but merely allows individuals to adopt the different ideas of other individuals" (p. 119). Similarly, an experiment on group influence on creativity in mathematics found that "those pupils working as individuals have a higher creativity score than those working in groups" (Banghart & Spraker, 1963, p. 260). These observations may explain why tight networks or scientific schools eventually lead to a certain fixation on specific topics, methods, or paradigms. In short, group work or strong integration in teams may keep individuals from following their own inclinations and interests.

Many authors seem to understate the difficulties of and barriers to knowledge management, knowledge exchange, and idea-sharing in groups or organizations.

This tendency is especially noticeable among researchers who do not distinguish between knowledge and information or who apply a naïve model of knowledge transfer built on the neoclassical concept of knowledge diffusion (for a critique of this approach, see Meusburger, 2008, pp. 70–74). Andre et al. (1979) have shown that groups using the brainstorming rules generate substantially fewer ideas than the same number of individuals brainstorming in isolation. Similarly, Lindgren and Lindgren (1965, p. 23) and Dunnette, Campbell, and Jaastad (1963) have shown that individuals brainstorming alone are more creative than when brainstorming in a group. As Andre et al. (1979) suggest, "in a practical situation where many alternative solutions to a problem are desired, it would be better to set individuals working alone and then pool their contributions" (p. 122). Separateness and solitude can serve as a protective shield for the scientist or artist.

There are a number of reasons why brainstorming does not always lead to the desired effect and why creativity may be blocked by group work (the following overview draws mainly on Paulus & Yang, 2000). Explaining why group discussion can inhibit creativity, Andre et al. (1979) write:

1. The social nature of the group makes individual performance public, and individuals may be reluctant to express unusual or "far out" ideas because of anxiety over how they will be received. Under this rationale conformity within the group will inhibit individual productivity[,] leading to lower group performance.
2. An individual within the group, perhaps the most creative, may dominate discussion[,] with the remaining individuals simply following his lead (p. 112).

Other scholars concur that creativity can be undermined by such unwillingness to state one's ideas and by the social loafing or free-riding that may ensue when individuals do not feel accountable or feel their efforts are not needed by the group (Karau & Williams, 1993; Kerr & Bruun, 1983). Another factor that reduces creativity is that some individuals cannot adequately express their ideas when someone else is talking (Diehl & Stroebe, 1991; Lamm & Trommsdorff, 1973). Instead, certain members use such meetings to establish their own reputation rather than listen to others. That kind of encounter frequently ends in competition for reputational capital, not in the exchange of ideas (Drazin et al., 1999, p. 295; Sutton & Hargadon, 1996, p. 706). Antipathy between group members and fear of losing status are additional problems that often thwart attempts to generate and exchange ideas within groups. Lack of support in a group quickly extinguishes the spark of originality, and a hostile group climate often wipes out minority views. Moreover, some team members may be disinclined to take responsibility for their actions, and others may not be prepared to share the benefits of their creative ideas with the group. Some participants may be preoccupied with their own ideas and not pay much attention to those generated by others (Paulus & Yang, 2000, p. 86). Glynn (1996) and Nemeth (1986) demonstrated that the positions expressed by a majority of individuals tended to foster convergent thinking, limit the number of presented alternatives, and enforce conformity to the prevailing views.

Hearing others generate ideas does not necessarily enhance one's performance, either. Individuals cannot be forced to participate in the interaction process of other group members, for some members of the group may lack the prior knowledge necessary for understanding the possible benefits for their own work. Still others may be preoccupied with different topics and lack the attention or interest required for the transfer of knowledge. Time, too, partly explains why brainstorming and its attendant levels of creativity may be disappointing. The incubation hypothesis suggests that "individuals do carefully process the shared information but may lack sufficient occasion to demonstrate the stimulation value of this information during the sharing session. ... It may take some time to reflect on the shared information and to integrate this with one's own ideas" (Paulus & Yang, 2000, p. 79; see also Paulus et al., 2000).

However, Kurtzberg and Amabile (2000–2001) point out that researchers have focused predominantly on brainstorming in groups and have paid little attention to team-level creativity. They note that the vast majority of studies on group brainstorming has occurred in a laboratory setting, with groups of participants who had no knowledge of each other's strengths and weaknesses and no strong incentive to create mutual understanding. Effective group problem-solving can occur with the right combination of personalities, the right amount of diversity, sufficient resources, and cooperative process behaviors (see Kurtzberg & Amabile, 2000–2001, p. 288).

Working in the presence of others is not the same as brainstorming. Various authors (Amabile et al., 1990; Shalley, 1995; Zajonc, 1965) have examined the effects that co-action, audience surveillance, expected evaluation, and goal-setting have on creativity. Shalley (1995) reported that the highest levels of creativity occurred when individuals had a creativity goal while working alone and expecting to be evaluated. Matlin and Zajonc (1968) found that surveillance had a significant negative effect on originality, which is an important component of creativity. According to Zajonc (1965), audiences and co-actors intensify "drive/arousal," which facilitates simple, well-learned, dominant responses but impairs complex, counterinstitutional, subordinate processes (quoted in Shalley, 1995, p. 486).

Working in the presence of others can have ambiguous effects. It can be distracting or inspiring. It can either energize people or disrupt their performance, "depending upon whether their attention is focused on dominant or subordinate responses" (Shalley, 1995, p. 486). Large office spaces or the presence of many co-actors lead to an increase in the speed and accuracy with which simple, routine tasks are performed and to a decline in performance on complex tasks (Bond & Titus, 1983; Shalley, 1995). "Individuals working alone on an open-ended, ill-structured task will have higher levels of creativity than individuals working in the presence of co-actors" (Shalley, 1995, p. 486).

Several studies support the hypothesis that solitude, sensory deprivation, and restricted environmental stimulation techniques promote cognitive processes, problem-solving creativity, and scientific thinking and reinforce aspects of imagery (Andre et al., 1979; Arieti, 1976; Forgays & Forgays, 1992; Norlander et al., 1998; Shore, 1971; Suedfeld, 1968, 1980; Suedfeld & Landon, 1970; Suedfeld et al., 1987). According to Shalley (1995), attentional overload may cause individuals to rely on preexisting schemata and routines and may lead to a restriction of cognitive focusing. In contrast, it is assumed that solitude may aid complex cognition (Forgays & Forgays, 1992; Shore, 1971). Arieti (1976) and Suedfeld (1974) consider aloneness conducive to creativity. They mention a long list of eminently creative philosophers, scientists, and religious leaders whose lives are marked by solitude and remoteness (Suedfeld et al., 1987, p. 220).

Arieti (1976) regards inactivity, daydreaming and the lack of external distraction as cultivating creative thought. Suedfeld et al. (1987) hypothesized that low stimulation will enhance creative behavior. They tested their hypothesis by using the Restricted Environmental Stimulation Technique (REST) in an experiment chamber or flotation tank that induces relaxation, lowers arousal, and allows the individual to concentrate on thoughts, feelings, and memories. Suedfeld and Landon (1970) reported evidence that divergent thinking is enhanced under conditions of sensory deprivation. Suedfeld et al. (1987), Forgays and Forgays (1992), and Norlander et al. (1998) explored whether flotation facilitated creative problem-solving ability and originality and whether it positively affected creative performance. All three studies arrived at the same conclusion, which is summed up by Forgays and Forgays (1992):

> Floaters increased their post- over their pre-scores on creativity and vigor, and maintain their curiosity level while they decreased anxiety, tension, depression, and fatigue scores, as compared with controls. It appears, then, that float subjects are alert but relaxed and that these conditions may conduce to the creativity benefits obtained. (p. 333)

Forgays and Forgays also found that "higher creative persons respond to isolation more positively and that such a response pattern may aid their creative process" (p. 333). Floating is associated with increased vigor and a reduction in depression, confusion, hostility, and fatigue (p. 334).

Many studies confirm the important role of solitude, interpersonal distance, and reduced ambient stimulation in enhancing creative thought, but others underline the importance of environmental stimuli as well. This apparent contradiction is explicable in several ways. First, it indicates that creative individuals need periods of intensive stimulation as well as periods of solitude. People seek solitude in order to avoid overstimulation and various stressors associated with excessive proximity. A sequence of group work (brainstorming, stimulation) and individual work (solitude) may produce the best results. Second, group experience does inhibit creativity for the items discussed by the participants, but it may serve as a trigger or catalyst for individual creativity for subsequent items (Andre et al., 1979, p. 122). Third, many past psychological studies on creativity used ad-hoc groups. The relationship found between individual and group performance may not hold for preexisting groups whose members trust each other and have a long history of successful cooperation (Andre et al., 1979, p. 119). Fourth, the composition of the group, the age of group members, and the duration of their successful cooperation may have an important bearing.

> Young groups might be expected to benefit more from the stimulation of heterogeneously composed groups than will old groups. (Smith, 1971, p. 491)

> Findings generally support the expectation that a heterogeneously composed team will be superior in group performance. This is particularly the case for young groups, where there may be greater need for a higher level of stimulation and energetic interaction. (Smith, 1971, p. 493)

2. Personality of leaders and leadership style

The effect and perception of an organizational structure and climate with regard to creativity is modified by the personality of leaders. It shapes their style of goal-setting, decision-making, supervising, and evaluating, which, in turn, influence the effect of other factors. Leaders express organizational norms and values, structure the nature of group interaction, condition subordinates' perceptions of the work environment, and affect in many other ways the performance and creativity of organizations. They can enhance or diminish the subordinate's feeling of self-efficacy and self-esteem and the individual's willingness to pursue risky or original ideas (see Redmond & Mumford, 1993). Creative achievement is enhanced by superiors who show consideration for the feelings and interests of employees; encourage their subordinates to voice their own concerns; give positive, mainly informative feedback; facilitate employee skill development; and promote personal initiative and risk-taking at work. Creative performance in a group or organization is likely to languish under supervisors, who rigidly control their subordinates, closely monitor employee behavior, introduce rigid operating procedures, make decisions without employee input, prove less competent than their subordinates, and suffer from a sense of inferiority (Amabile, 1988; Amabile & Conti, 1999; Amabile et al., 1996; Amabile & Gryskiewicz, 1989; Andrews & Farris, 1967; Oldham & Cummings, 1996; Shalley et al., 2000; Stahl & Koser, 1978; West, 1989).

Collectively, these authors have reported that teams of scientists produced their most creative results when they were:

- Allowed substantial freedom in the way the work was done
- Encouraged rather than controlled by their supervisors
- Given opportunities to influence important decisions and choose the processes of evaluation
- Permitted to ask novel or disturbing questions, and
- Invited to come up with unusual solutions

Other prerequisites of creativity that supervisors can influence are:

- The degree of open information flow across departments
- Mutual receptiveness to other domains or disciplines
- Shared commitment to a project
- Fair and supportive assessment of new ideas
- Reward and recognition of creativity
- Participatory management and decision-making
- Open interaction between supervisor and subordinates, and
- Diversity in team members' knowledge bases (disciplinary background)

Hage and Dewar (1973) found that the values of organizational leaders explained more variance in organizational innovation rates than any single structural dimension. To sum up, research has shown that inducing positive affects can enhance creative problem-solving, whereas generating negative emotions, particularly anxiety, has been associated with lowered creative performance (for an overview see Clapham, 2000–2001).

3. Job satisfaction

Many authors assume that a high level of job satisfaction contributes positively to organizational effectiveness and enhances creative performance and that job dissatisfaction is detrimental to organizations (for an overview see O'Reilly, 1991). They argue that dissatisfied members will quit their organizations and that other employees will remain in the organization but respond either passively by accepting their situation or exhibit passive withdrawal behavior. However, March and Simon (1958), Simonton (2000), van Gundy (1987), Zhou and George (2001) and others question this idea and suggest that job dissatisfaction can also lead to creativity when individuals have "an active and constructive response to their dissatisfaction" (Zhou & George, 2001, p. 684). They argue that discontentment with the status quo can spur organizational change when dissatisfied members of the organization come up with new ideas to do things better or gain new intrinsic motivation to improve their current work situation. Under certain conditions, employees' job dissatisfaction may actually lead to a more creative performance because some employees dissatisfied with the situation will actively try to improve conditions. Exceptional creativity seldom emerges from the most supportive or harmonious environment.

Indeed, some studies maintain that creative potential seems to require exposure to diversifying experiences or cultural diversity that help weaken the constraints imposed by conventional socialization (Simonton, 2000, p. 153). This view is too simplistic, however. Its accuracy depends on the type and circumstances of cultural diversity. The consequences of cultural diversity that stems from cooperating elites differ from those of cultural diversity arising from segregated, disadvantaged and unskilled minorities being in conflict with each other.

Surprisingly, more research has been done on environmental factors enhancing creativity than on those that undermine it. This preoccupation belies the fact that impediments to creativity, such as incompetent leaders, lack of resources, and internal strife, are experienced on a daily basis much more explicitly than factors enhancing creativity, which may not be recognized at all because they are regarded as normal or self-evident.

The Spatial Microscale: Personal Traits of Creative Individuals and Their Relations to the Environment

For many decades, creativity research in psychology, management studies, career studies, art criticism, and philosophy focused mainly on the relationships between creativity and personality, on the search for personal characteristics relating positively or negatively to creativity, and on the identification of traits predictive of creative performance. Some of the early studies related creativity to pathology or neurotic conflicts (for an overview see Becker, 2000–2001; Burchard, 1952; Eissler,

1967; Rees & Goldman, 1961; Schuldberg, 2000–2001; Symington, 1987). Some psychoanalysts believe that artistic creation arises from the unconscious and is motivated by an unfulfilled ambition, an erotic wish, or oral regression (Bergler, 1950; Rees & Goldman, 1961, p. 145). Eissler (1967), Symington (1987), and others have shown that many "great men" were suicidal or suffered from severe injury to infantile narcissism. Other authors (e.g., Slochower, 1967, pp. 4–5) regard crises, periods of sharp transition, and conflicts as a precondition for the highest forms of creativity. Dante had to go through the Inferno to find Eden, and through Purgatory to reach Paradise.

Other authors criticize this psychopathological approach, charging that too much emphasis has been placed on the neurotic affiliations of artistic creation or scientific excellence. Symington (1987) claimed that "psycho-analytic theory is too pathology-oriented to be of much help to us in understanding creative genius" (p. 286). According to Scott (1965), there is "no support for the common-sense belief that creativity stems from neurosis or that creative individuals are inevitably psychologically sick, physically frail, and socially irresponsible" (p. 219). However, there is ample evidence that large numbers of highly creative people had life histories marked by severe frustration, deprivations, and traumatic experiences.

Attributes of creative people. Most authors agree that certain personality characteristics relate to, enhance, or correlate with creativity. In the literature of various disciplines, the most frequently mentioned attributes of this kind are motivation; mental energy; a high level of multiple intelligence; ideational fluency; wide-ranging interests; a disposition to the integration of diverse stimuli; attraction to complexity; intuition; aesthetic sensitivity; general effectiveness of performance; existential security that permits a high degree of nonconformity and allows time for reflective thought; an urge to seek self-fulfillment in a unique manner of the individual's own choosing; a heightened desire for attention, reputation, and social appreciation; above-average adaptive flexibility; originality; curiosity; self-confidence; a marked degree of independence in thought and action (field independence); tolerance of uncertainty and ambiguity; openness to experiencing the inner self and the outer world; a rich and vivid fantasy life; above-average memory for imaginary information; spontaneous flexibility; high level of self-assertion, self-esteem, and self-confidence; a willingness to take risks; nonconformity; an ability to synthesize and rearrange existing facts in order to come up with a new and useful answer to a problem; persistence in the development of ideas; a strong sense of destiny with a marked degree of resoluteness; and a tendency to cling tenaciously to a project in the face of repeated or perceived failure or disagreement and critique of others (Amabile, 1988; Banghart & Spraker, 1963; Barron, 1955; Barron & Harrington, 1981; Carlozzi et al., 1995; Earl, 1987; Eulie, 1984; Funke, 2000; Gardner, 1995; Givens, 1962; Goldsmith & Matherley, 1988; Gough, 1979; Hocevar, 1980; Martindale, 1989; Niederland, 1967; Norlander et al., 1998; Oldham & Cummings, 1996; Rokeach, 1954, 1960; Runco, 1988, 1994; Shalley, 1995; Shaw, 1987; Simonton, 1999; Thomas, 1955).

Rather than discuss or evaluate the pertinence of these variables to creativity studies and the inconsistency of some of the research results, I assert that most of these personal traits are neither innate nor isolated from the social environment. They are learned, activated, affected, or developed through relations and interactions between actor and environment. Some of the characteristics or cognitive frameworks do influence a person's goals and motivation others impair the perception and interpretation of the environment; another group of traits influences social relations. Also associated with the social environment are personal constraints and personal traits *inimical* to creativity—such as depression, learned helplessness, lack of self-esteem, inability to concentrate, or a sense of being controlled by others.

Scholars disagree on the existence of innate ideas or innate knowledge. Most researchers prefer the terms *innate capacities* or *propensities* instead of innate knowledge. Rescher (1966) uses the term *innate* (p. 210) to refer to all intellectual capacities that enable humans to develop other capacities. He regards innate mental propensities as a range of tendencies, inclinations, and dispositions relating to the functioning of the human mind (p. 206). The capacity to *learn* calculus or attain fluency in a foreign language are innate; the ability to *solve* calculus problems or speak a foreign language are unquestionably acquired (Rescher, 1966, p. 206). The capacities to recognize and discern structural patterns, regularities, and similarities or to use analogies are also innate.

Motivation. The creative performance of individuals is influenced both by intrinsic and extrinsic motivation. Intrinsic motivation is one of the most important determinants of creativity. Extrinsic motivation, according to a number of studies, does not enhance creativity (Amabile & Gryskiewicz, 1989; Amabile, 1988; Oldham & Cummings, 1996; Runco, 1993; Shalley, 1995; Shalley & Perry-Smith, 2001; Woodman & Schoenfeldt, 1990). Other researchers (e.g., Eisenberger & Rhoades, 2001; Eisenberger et al., 1999) find that extrinsic motivation, too, can increase creativity when the reward is contingent on creativity. They have shown that extrinsic motivation can do so by enhancing self-determination and intrinsic task interest. Extrinsic motives might enhance self-determination and influence choice of field, type of task, or implementation strategy rather than a person's work on the task itself (Mumford, 2003, p. 112).

Intrinsic motivation not only constitutes tenacity and a strong inner drive for achievement. Together with prior knowledge, sensitivity, and curiosity, it also moderates the relationship between external stimuli, attention, perception, and behavior. Intrinsic motivation and prior knowledge direct attention to certain environmental signs and patterns and help avoid attentional overload, which is regarded as an obstacle to creativity. Ample empirical evidence shows that people's creativity peaks when their motivation is primarily intrinsic and fed by the challenge of their work, their interest in and curiosity about problem-solving, and their satisfaction and enjoyment (Amabile, 1983a, 1988; Amabile et al., 1996, p. 1158).

Intrinsic motivation can be encouraged or eroded by external factors. Organizational impasses, internal strife, rigid supervision, lack of resources, corruption, and incompetent leaders may decrease intrinsic motivation. According to the Cognitive Evaluation Theory (Deci & Ryan, 1980, 1985), situational factors can affect intrinsic motivation by two means: control (e.g., pressure to achieve a certain outcome) and information (e.g., feedback that people receive about their self-determination and task competency) (Shalley & Perry-Smith, 2001, p. 3). Some authors (Amabile, 1979; Amabile et al., 1990; Shalley, 1995; Shalley & Oldham, 1985) suggest that expecting evaluation can have dysfunctional consequences for intrinsic motivation. Individuals, they argue, may no longer dare to take risks or play with ideas, for any failure of a risky experiment may prompt a negative assessment. However, the direction of the response to expected evaluation seems to depend at least partly on the *kind* of evaluation involved. The research of Shalley and Perry-Smith (2001), for instance, indicates that "individuals who expected an informational evaluation had higher intrinsic motivation and creativity than those who expected a controlling evaluation" (p. 15).

Cooper and Jayatilaka (2006) propose a third type of motivation important for creativity: motivation that comes from an individual's feelings of obligation.

> This motivation has characteristics in common with extrinsic motivation in that it is linked to extrinsic rewards. However, in contrast to extrinsic motivation, rewards that may result in obligation motivation are not contingent on task performance. (p. 154)

> There appears to be a powerful rule of reciprocation that is pervasive in all human societies, and which results in our feeling obligated to the future repayment of help, favors, and gifts. (p. 156)

Openness of cognitive frameworks. Another focus of research on the spatial micro-scale of creativity is the relative openness of a person's cognitive framework that has an impact on how a person receives, understands, and evaluates information and acts on stimuli of the environment. Rokeach (1954, 1960) suggested that highly dogmatic people exhibit a closed way of thinking, a tendency to distort incoming information and meanings, and an intolerant attitude toward those with dissimilar values or beliefs. Creativity and dogmatism seem to be inversely related to each other. According to Ohnmacht and McMorris (1971), for example, the problem-solving ability of dogmatic individuals has been found to decline when the problems to be solved require the ability to synthesize rather than analyze. Highly creative experimental participants exhibit greater flexibility and tolerance for novelty, ambiguity, and incongruity than less creative ones do (Martindale et al., 1974, p. 317) and are therefore able to attend to a wider range of environmental stimuli and to absorb or activate a variety of knowledge bases.

Openness and flexibility are also related to empathy and emotions. Several studies support the hypothesis that empathy (affective sensitivity) and creativity are related processes (Alligood, 1991; Gallo, 1989; Kalliopuska, 1992). Carlozzi et al. (1995) hypothesized that empathy is positively related to creativity and expressiveness and inversely related to dogmatism. An affective concern for the relationship between self and others seems to heighten one's sensitivity to the environment as well. Research shows that there is a strong connection between cognitive and affective processes and that emotion can contribute to creativity in several ways (Lubart & Getz, 1997; Russ, 1993). Lubart and Getz (1997, p. 286) argue that people in an emotional state may notice stimuli in the environment that they would usually overlook and that they may interpret stimuli in novel ways because of their emotional perspective.

Ego development, autonomy, and field independence. Many individual traits of creativity and the quality of interaction between two or more individuals are related to self-esteem and ego development. Investigations by Workman and Stillion (1974) seem to confirm "a positive relationship between creativity and ego development, since all creativity sub-scores, as well as total creativity, were significantly related to ego development" (p. 193). A creative person needs self-esteem for various reasons. Abandoning common frames of reference entails uncertainty, isolation, criticism, opposition, and self-doubt. A person needs perseverance not only in order to concentrate on a certain topic for lengthy period but also in order to overcome the various hurdles encountered during the processes of acceptance and implementation. Creativity requires one to take risks, and self-trust makes one immune to rejection by evaluators, journal editors, and peer groups. Creative people seem to be less threatened by failure and criticism than noncreative individuals are. If the self-trust feedback loop is strong enough, individuals will be intensely concerned with their initial vision and will be able to carry on despite apparent rejection (Earl, 1987, p. 423). Funke (2000, p. 295) points out that individuals must be entirely convinced of the value and significance of their creative action and not allow themselves to be negatively affected by others' criticism and disparagement. Self-perception and self-confidence develop in the framework of social relations and spatial settings. They are shaped by past success and failure and derived from the image and status of one's place of work, the reputation of the organization one belongs to, and places and spatial contexts with which people identify themselves.

Witkin, Dyke, Faterson, Goodenough, and Karp (1962) introduced the concept of field independence, which means autonomy, freedom from a scientific discipline's strict rules or shared beliefs, and freedom from the restrictive effects of dependency on others. Creative persons must free themselves from orthodox means of solving problems, from given structures and rules. Field independence is a necessary, but not sufficient, condition for creativity. Creative individuals are likely to be field independent, but not all field-independent subjects are creative. Gordon and Marquis (1966) and Bloomberg (1967, 1971) report that field-independent persons were consistently more creative than field-dependent persons. Similar conclusions about the importance of independence are drawn by Hanák (1993), who analyzes the marginality of creative Jews in the Vienna of 1900. A stage of independence and integrity offers marginal men a sort of cognitive privilege (p. 149). However, Ohnmacht and McMorris (1971) warn not to oversimplify these relations. Autonomy and field independence are closely related with other characteristics, and it may well be that only the cumulative effect of these related variables has an impact on creativity.

Research on self-esteem bears out the theory of cognitive tuning, which rests on the assumption that a relationship exists between affect and creativity (Friedman & Förster 2002). Positive affective states facilitate creative problem-solving (Isen et al., 1987) and bolster cognitive flexibility. It is posited that individuals in a positive affective state become more inclined to take risks and adopt a relatively heuristic processing style in which the likelihood of generating novel alternatives rises (Friedman & Förster 2002; Isen, 1987). In contrast, negative affective states entail risk aversion, increase adherence to established plans of action, and lead to diminished originality.

Self-esteem and self-confidence are important prerequisites of creativity, but ego defenses exert a dysfunctional influence in contexts where change and innovation are desirable. They reduce the organization's ability to learn and adapt. The most important ego defenses seem to be denial of reality, rationalization (an attempt to justify needs, feelings, and behavior), idealization (overevaluation), fantasy (unconscious endeavor to fulfill or gratify difficult or impossible goals and aspirations), and symbolization (Brown & Starkey, 2000; Laughlin, 1970).

Sensitivity to the environment, acuteness of perception and observation. People see what they have become sensitive to, what has made an impression on them. Being aware of and sensitive to the environment, not falling into routine, taking nothing in the external environment for granted, and retaining the capacity to be surprised all belong to the realm of creative activities in both the arts and science (see Robinson, 1970, p. 9). Perceiving the environment in predetermined categories hinders creativity. Artists and other creative individuals often show an "unusually intensive sensitivity to reality" (Kleinschmidt, 1978, p. 52). Creative individuals (especially artists) are said to have greater sensory responsiveness and heightened sensitivity to their inner and outer world (Niederland, 1967, p. 12).

> Only by really looking at the world, and by constantly perceiving it afresh, can we hope to break down our stereotyped sets of responses and open up the opportunity to discover different structures in the reality that surrounds us. ... We must learn to see ... with our eyes and not with our brains, to perceive the world as it appears to us *before* memory with all its habitual associations and interpretations and the intellect with all its categories and conditioned reflexes have time to step in. (Robinson, 1970, p. 5)

Creativity is not merely a matter of inventing solutions, it also involves identifying the important problems of the future (see Patton, 2002, p. 125). Many of these problems are identified by dealing with challenges and risks of the environment. Many traits of creative persons are learned through interaction with the environment, or they shape the perception of the environment or influence the interaction with the environment (see Fig. 6.2).

Theoretical Issues Concerning the Spatiality of Creative Behavior

Chance favors only the prepared mind.

Louis Pasteur (as quoted in Cropley, 2006, p. 394)

Relations between the physical environment and creative processes. As long as scholars focus on the social environment, they can avoid a host of philosophical and scientific controversies. However, a growing number of researchers demand that social and behavioral scientists no longer ignore the physical world, especially the ecological environment. The relationships between characteristics of the physical environment and psychological processes are still hotly contested. Is the physical environment only what appears through perceptual experience or does there exist a "real world" outside perception? How much do the "perceptual world" and the "real world" correspond? To what extent are individual differences in perceiving the environment due to biological or psychological dispositions and unconscious

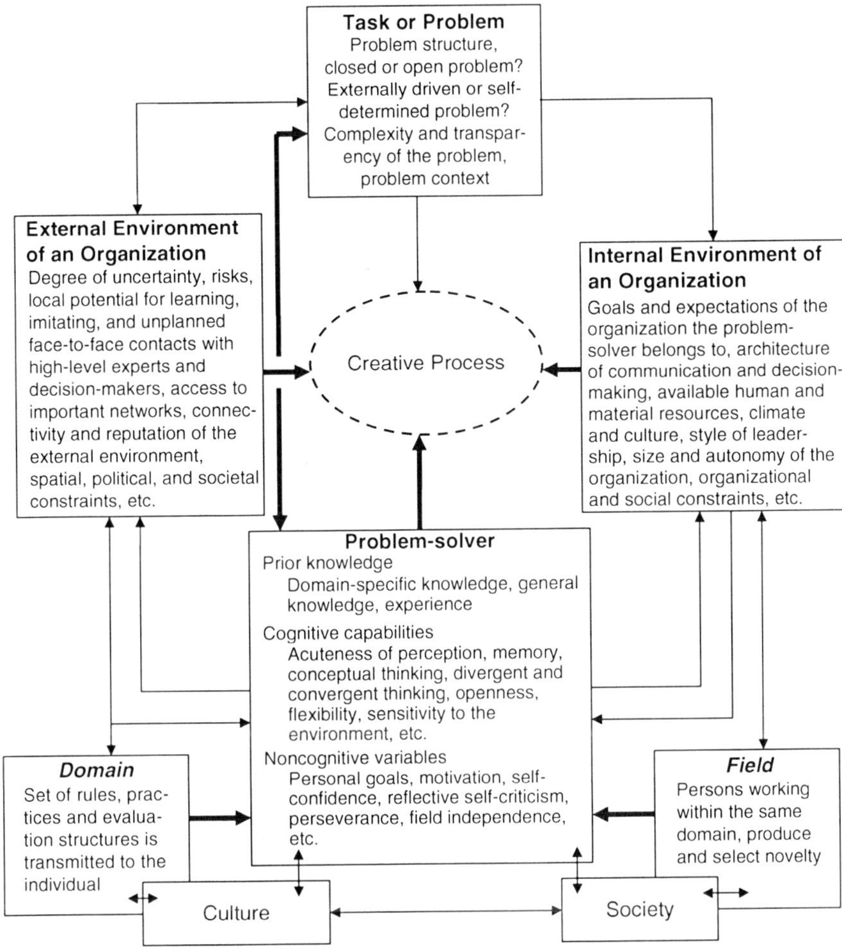

Fig. 6.2 Relations between task, environment, and problem-solver (design P. Meusburger)

processes? Can the social and cultural meaning of an environment be separated from that environment's materiality? Does the Cartesian distinction between the material and symbolic world still make sense? Is perception a holistic or a mechanistic phenomenon? What do we know about subconscious processes of perception and information-processing? Philosophers and psychologists have been concerned with some of these questions since the beginnings of their disciplines. Disputes continue to pit the phenomenological orientation (Gestalt theory) against the "New Look" school of perception, Brunswik's (1956, 1957) theory of the lens model against Gibson's (1979) ecological perception (see Bonnes & Secchiaroli, 1995, pp. 20–58), and holistic approaches against mechanistic ones.

In the conventional approach to perception, perception of the external environment is a function of several different interpretive psychological processes. It is assumed that a stimulus activates a specific nervous system receptor and that the individual interprets the pattern of receptor stimulation by drawing on memories of past experiences in order to gather information about the environment. According to this approach, humans must interpret unconnected stimuli in order to construct something meaningful about the environment (for details see Bell et al., 2001, p. 65). Gibson (1979) believes that perception is more holistic than that. He holds that properties of the environment are perceived not as distinct points but rather as meaningful entities.

Until the 1990s, most social and behavioral scientists argued that it was mainly the psychological, symbolic, and functional meaning of places, environmental structures, and spatial arrangement of objects that influenced behavior and creativity. In recent years, however, researchers have argued against neglecting the role of the materiality of objects and of the corporeality of persons (Funke, 2007; Gieryn, 2000, 2002a, b; Graumann, 2002; Jöns, 2003, 2006; Meusburger, 2008). A Cartesian distinction between the material and symbolic worlds no longer makes sense in an explanation of the actions of individuals in space. Therefore, environmental psychology (for an overview see Bell et. al., 2001; Bonnes & Secchiaroli, 1995; Funke, 2007; Graumann, 2002; Graumann & Kruse, 2003), environmental anthropology (Berkes, 1999; Biersack, 1999; Neves-Graça, 2003, 2007), human geography (Jöns, 2003, 2006; Klüter, 2003; Koch, 2003; Salbaum, 2008; Weichhart, 2003), and sociology (Fischer-Kowalski & Erb, 2003; Fischer-Kowalski & Weisz, 1998) have developed concepts to bridge the old dichotomy between the material and social. Given Bateson's epistemology (Bateson, 1972; Bateson & Bateson, 1987), environmental anthropologists no longer accept views that reduce ecosystems and environments to culturally constructed categories. A main premise of an approach derived from Bateson is that human–environmental relations are characterized by their mutual causality, that they co-construct one another (Berkes, 1999; Neves-Graça, 2007; Rappaport, 1979). According to Neves-Graça (2007), "human-environmental relations are recursively linked, … such that knowledge, forms of knowing, and experiences of past human-environmental practices are constantly being re-embedded in new contexts that are themselves simultaneously sociocultural, historical and ecological. Hence, the ontological condition of human-environmental relations is irreducibly interactive and dynamic" (pp. 149–150). The subject is integrated into the context and socialized by the context. It adapts to the environment but also actively adjusts and shapes the environment according to its capabilities, resources, and position in the context. Places and environments have a physical reality that can be structured by humans. This spatial positioning of objects is decoded, synthesized, and interpreted by other agents who give the spatial patterns a social meaning.

In many cases the symbolic meaning and function of an object cannot be separated from its materiality. It is not only the symbolic meaning of a banknote that enables its bearer to buy a good; it is also the note's materiality, which is carefully checked by the banker or shop owner in order to distinguish legal tender from counterfeits. It is the materiality of a painting that enables experts to decide whether it is an authentic work of a certain artist or a well-made copy. It is the materiality of shelters that constitutes their function and meaning. The physical environment provides a wide variety of experiences and learning opportunities and can enhance creative processes (McCoy & Evans, 2002, p. 409). After traumatic events the physical environment assumes the role of a conditioned stimulus that elicits a conditioned response. Just the sight, sound, or smell of a traumatic environmental experience may trigger a negative reaction such as a phobia, anxiety disorder, or panic attack (Anthony & Watkins, 2002, p. 131). Furthermore, some material objects can direct or stop actions long before individuals have had a chance to attach a meaning to them. An unknown danger, too, can have devastating effects.

In studying the relations between environment and psychological processes or the role of environments as archives and external memories, one might find it helpful to follow Neisser (1987), who distinguishes between perceptual-visual processes (seeing) and perceptual-cognitive processes (thinking, categorizing) and between direct perception and theory-dependent categorizations. Perceptual-visual processes are based on the direct intake of objectively existing information, whereas perceptual-cognitive processes go beyond that information on the basis of beliefs about the world. Thinking and categorizing rely on inferential processes of cognitive functioning and are "anchored to the sociocultural characteristics of the context" (Bonnes & Secchiaroli, 1995, p. 32). A similar point is made by Funke (2007), who argues that the perception of space takes place at two levels, the "surface level" and the "deep level" (pp. 245–250). The perception of space at the surface level is based on the physiology of the sensory system, which analyzes depth cues visible to the human eye, interprets smells and sounds, and reconstructs the environment in a three-dimensional model of reality in the perceiver's head. The perception of space at the deep level has to do with functional aspects, that is, with the meaning and significance of space as communicated between persons. At the deep level, places, regions, and landscapes have the function of an external memory or archive (see Funke, 2007, p. 245). A detailed description of the role of landscapes as external memory is presented by Wassmann (2003) and Wassmann & Keck (2007). Because space perception "requires an analysis of personal memories for certain places" (Funke, 2007, p. 251), perceptual processes are inseparable from human memory. A similar argument is used by the transactional school of perception, whose adherents focus on perceptual phenomena but tend to consider the perceptual issue to be part of the more general one of the individual's relationship to the surrounding world.

Some ecologically relevant affordances, or "invariant functional properties of objects as they are encountered in the course of an organism's active exploration" (Bell et. al., 2001, p. 65; see also Gibson, 1979), can be perceived directly with little or no complex interpretation. They convey a great deal of information without elaborate processing by higher brain centers. Other environmental signals, prompts, and juxtapositions send ambiguous or polyvalent messages, which are perceived and interpreted very differently by actors with varying prior knowledge, experience, memory, absorptive capacities, or ambitions. If a stimulus is ambiguous, then perceivers must draw more heavily on their own knowledge base, their ability to recognize and interpret patterns, their memory, and the social-psychological group dynamics than if the stimulus is unambiguous. Only motivated, experienced, and capable persons are able to perceive and interpret ambiguous environmental clues and stimuli in a way that they can make timely use of the locally offered potential or avoid imminent risks. The ability to recognize the meaning and value of ambivalent cues and patterns early is one of the most important attributes of creative persons and successful decision-makers.

Many orientation skills, habits, and routines, such as driving a car or skiing downhill, are directed through subconscious perception of material objects and codes of the environment. The model by Clitheroe et al. (1998) assumes that "prompts are the starting point of an intentional or unintentional psychological and/or behavioral process. Prompts may come from social or physical features of the contexts; from individuals participating in the context; or from a wide array of extracontextual sources such as published research results, news media, or the internet" (p. 106). Whatever an individual's talent, the conditions under which he or she works can significantly raise or lower the level of creativity (Amabile, 1996, p. 17).

The effects that a place, context, or environment has on creative processes cannot be precisely forecast. A context or environment means potential; it is an offer or risk that some agents will perceive and take into account and that others will disregard. Contexts and environments indirectly "influence creative performance by shaping critical psychological processes rather than directly influencing creative performance" (Choi, 2004, p. 197). A context can affect individual creativity in a number of ways. It can structure problem-solving efforts (Mumford & Gustafson, 1988) or shape the nature and conditions under which individual capacities are developed and applied (Redmond & Mumford, 1993). It can sharpen the individual's intrinsic and extrinsic motivation to pursue new ideas and take risks, and it can furnish or withhold the needed resources. It can raise or lower the likelihood of meeting prominent or inspiring people and can make for solitude or overstimulation and stressors. However, the predictive power of social macrophenomena such as spatial context, environment, or place is much greater than that of any discrete variable studied in laboratory experiments.

Place dependence and emotional attachment to places. Comparatively little controversy exists about which relations between environment (context) and behavior can be subsumed by the terms *place dependence*, and *place attachment*. Individuals depend on places. They are emotionally attached to places, are satisfied with places or gather regularly at specific places for various functional, symbolic, and emotional reasons. Places act as a meeting point or catalyst bringing people, ideas, and resources together. Place dependence exists when scientists need expensive scientific infrastructure for their daily work, when they need frequent and regular face-to-face contact with other experts, or when they want to avail themselves of an institution's reputation or when they seek the liberty of uncontrolled discussions. Even the most talented and best prepared individual cannot be creative if he or she has no chance to work or perform. A surgeon needs a patient, a conductor needs an orchestra, and a chemist needs a laboratory in order to develop and demonstrate his or her creativity. An institution's quality of scientific infrastructure (e.g., libraries, laboratories, and computers), amount of resources, community of experienced researchers, and scientific standing affect the range of possible research questions, the quality of research output, and the institution's capacity to attract and retain outstanding scholars. Being prepared to tackle a challenge is not enough; one must also meet the right people and be at the right place at the right time.

Places differ according to their functions, connectivity, and the range and quality of their interactions with other places. Some professions derive part of their identity, reputation, authority, and authenticity from the places they "belong to." The formation of identity is largely shaped by relations to others, both by identification with others and differentiation from others. The probability of spontaneously meeting the knowledgeable others who are needed for interaction and inspiration as well as for critique, judgment, and evaluation varies from place to place. The dependency on face-to-face communication and infrastructure, and thus the importance of place dependence, varies from one profession and discipline to the next. An artist needs a different environment to develop his or her creativity than a banker; a high-energy physicist, a different environment than a theoretical physicist. Place dependency varies according to stages of a creative process as well. Latent or preliminary phases of inspiration, incubation, and preparatory processes of creativity need different types of environments than do periods of elaboration, which likewise require contexts different from those needed in phases of evaluating, showing and marketing. Each phase has its own demands for stimulation, distraction, solitude, organizational support, publicity, and frequency of face-to-face contact.

Individuals can be emotionally attached to certain places and environments. Places are embedded in contexts of feeling, emotion, and memory and can have an atmospheric quality

> It is above all the symbolic effects and the atmospheric qualities of places that motivate those who have the necessary material, social and cultural resources (wealth, power) to turn places into "scenes," to establish spatial orders that tell "stories" (narrative spaces) and in their symbolic effect convey an atmosphere of amazement, of fascination, and of cultural and social support, of belonging and identity while, at the same time, the existing power and rule relationships can be reproduced and legitimized. (Hennings, 2007, p. 129)

If agents repeatedly have negative experiences at certain places, they may project their negative emotions and fears onto these places with the effect that places evoke in them negative feelings or anxiety when they return to them. A specific environment or spatial context can destroy self-esteem and arouse anxiety, whereas another environment can foster the feeling of security and enhance capability beliefs, motivation, ideational fluency, or the willingness to take risks. With trauma victims a particular place or space may later repeatedly unleash the memory of the trauma (what one has tried to bury in the subconscious). Crowded public spaces often set off panic attacks or agoraphobia (Anthony & Watkins, 2002). "The physical environment assumes the role of a conditioned stimulus that elicits a conditioned response. Just the sight, sound, or smell of a traumatic environmental experience may trigger a negative reaction" (p. 131).

Affective memories can be activated through pictures, objects, names, and scenes. Places and environments can serve as triggers for the episodic memory, which makes the recollection of an episode possible. Lubart and Getz (1997) use the metaphor of a tuning fork to describe the relation between object, memory, and emotion. Depending on a person's emotional state and experience, external stimuli can start a particular tuning fork vibrating.

The environment: Opportunities to learn and experiment. Humans acquire much of their knowledge by interacting with the social and material environment, gathering experience, and developing skills when performing tasks and solving problems in a specific context. Problems to be solved vary spatially, as do the opportunities to imitate successful individuals. The localized necessity of coping repeatedly with specific challenges and solving particular problems steers attention to certain topics, sharpens the perception of particular clues and patterns, builds specialized knowledge bases, and sparks motivational processes that can give rise to new ideas and actions. A stimulating and enriched early environment is an important factor in developing intelligence and creativity and may have an even stronger impact than the environments in later career phases. Children and students learn a great deal through their identification with parents, teachers, mentors, and other role models. If students adopt the attitudes of their teachers, it is not necessarily imitation but rather "the integration of attitudes about the self, about how one finds self-satisfaction and self-fulfillment" (Berlin, 1960, p. 99). In order to meet the expectations or gain the recognition of the mentors or peers, students must internalize the rules of the domain and the context. The context tells the informed actor what kind of behavior will be regarded as appropriate within certain limits.

Learning to survive in a given environment for an extended period equips or adapts agents for particular challenges and situations. Social systems operating for an extended period in specific physical environments (e.g., rain forests, deserts, or polar regions) have developed certain techniques, strategies, competencies, knowledge bases, and value systems to help them adjust to their environment and adapt it. Similar opportunities and necessities to learn derive from social and cultural environments as well. Artists, scientists, and engineers are not the only people who evaluate their work before they complete it. Friends, adversaries, teachers, students, colleagues, reviewers, critics, and the general audience often participate in the creative process by joining in the evaluation of emerging works or previous ones. "The evaluation itself contributes to the value of the work: that is, the evaluative acts that go into the initial creation of the work and its later actualization, form part of an organic whole' " (Leddy, 1994, p. 173). The criticisms and encouragement of others enrich or diminish the scholar's or artist's self-concept and self-esteem and influence that person's intrinsic motivation.

The evaluation process—another important part of the scientist's or artist's creative experience—also varies spatially. An idea, theory, or piece of work that is praised and admired in one location may be misunderstood, heavily criticized, or rejected in others. University departments, research institutions, and art schools differ in their learning environments, conditions of professional socialization, chances to communicate with experienced and knowledgeable experts, material and personal resources, attitudes, expectations, value systems, credentials, reputation, and socioemotional relationships that support or impede creative processes.

Places and spatial contexts as potential for serendipitous interactions. In some special fields of painting, music, or science, only a few select critics and discussion partners serve as a valid reference point (see Funke, 2000, p. 292). Colonies of artists (Murnau in Bavaria, or Worpswede northeast of Bremen), architects (Bauhaus in Dessau), or scientists (development of the nuclear bomb at Los Alamos) were able to achieve their goals successfully in remote rural locations primarily because their projects were self-sustaining in their evaluation processes. Other projects depend on transgressing social and scientific systems (Acham, 2003, p. 292) and crossing disciplinary boundaries; drawing analogies from completely different domains that previously had nothing to do with one another; consulting a broad variety of knowledge bases; and face-to-face contact with experts, evaluators, and critics of many different disciplines. Only major, functionally complex cities offering a large variety of professional skills, scientific disciplines, artistic styles, cultural experiences, and economic and cultural diversity provide such potential for unplanned contact. Díaz de Chumaceiro (1998, 1999, 2004) and others have shown that chance events, unplanned contact, or serendipity have led many scientists (e.g., Pasteur, Fleming, Röntgen, Becquerel, Edison, Galvani, Nobel, and Freud) to breakthroughs or have decisively shaped and affected their choice and pursuit of a career. Serendipity is the unexpected finding of valuable ideas, persons, and things. In this context *true serendipity* is defined as the accidental finding of something valuable but unsought, *pseudoserendipity* is discovery by accidental means of things sought (for details see Díaz de Chumaceiro, 2004, pp. 346–347).

What makes a location attractive is the *possible*, not the actual, contact with other highly creative persons. It is the place's *imagined*, not the real, advantages, that make a location attractive. Whether and how often this potential is activated through communication and interaction is another question. The fact that millions of people play the lottery indicates that possibility is a strong driving force that influences behavior. In most lines of work, the probability of profiting from serendipitous opportunities or interactions is much higher in some spatial contexts than in others.

Members of prestigious university departments, editors of international journals, and researchers integrated into powerful research clusters and long-range research programs demonstrate that long-term proximity to the epistemic centers of disciplinary power affords a scientist prestige, reputation, and strategic advantages and facilitates his or her access to resources. Over time, though, it can have detrimental effects on that person's creativity. There is much empirical evidence that a position near the "key persons" may give scientists the illusion of moving automatically in the right direction or being at the frontier of research. The impact of leading journals makes it difficult for scientists to abandon dominating paradigms or to come up with divergent and provocative thinking. Large research programs and clusters keep scholars on tracks favored or supported by politics or project leaders. Some creative scientists and artists are very proficient at alternating between proximity and distance to epistemological power or between publicity and solitude. They need proximity to experienced, knowledgeable, and challenging people for inspiration, ideas, motivation and emotional support. They then seek distance and solitude in order to avoid strict paradigms and elaborate their still poorly conceived ideas or products without the interference of early critique. In some disciplines, periods of data-gathering, experimentation, and elaboration are place dependent because they depend on expensive instruments or field work. In other fields, this period may also be spent in solitude. However, when it comes to showing and marketing results, proximity to the important decision-makers of a discipline and platforms of attention is helpful.

The environment of a single city or area seldom offers all the incentives, inspiration, and learning opportunities that an artist, artisan, or scholar needs to develop his or her creativity. Mobility increases professional experience and stimulates "new patterns of thinking" (Törnqvist 2004, p. 236). Creative persons are therefore supposed to be mobile and circulate from one place (court, university) to the next so that they experience different work environments; learn from different cultures, professional practices, and knowledge bases; and focus attention on new issues, techniques, and methods. Spatial mobility and—ultimately—location in one of the epistemic centers of one's profession has become a status symbol or sign of excellence in certain professions.

Open Debates and Suggestions for Further Research

Additional research is needed on the role of the time dimension, unconscious information-processing during the incubation phase, the neurological principals of creative thought, the relations between knowledge and action, cultural impacts on creativity, and various other questions. There is much empirical evidence that creativity does not proceed in linear or hierarchical paths but rather in uneven, chaotic bursts in response to problems that erupt over time (Drazin et al., 1999; Kazanjian, 1988; Peterson, 1998). The generation and introduction of creative ideas is easier in some periods than in others. The chance to make certain organizational changes or transform systems is relatively brief. If a creative idea comes too early or too late, it will not be accepted or successfully implemented. Histories of science, inventions, and the arts contain many examples of lengthy lags between the date of an idea's conception or a product's creation and the moment when that idea or product was regarded as valuable and creative. These histories brim with disappointing accounts of highly creative ideas and products that have been ignored by informed, well-educated people (see Magyari-Beck, 1998). Many artists whose paintings are sold today for millions of dollars were not considered as creative in their own lifetimes. Cropley (2006, p. 396) describes how the mathematical propositions of Evariste Galois (e.g., the Galois Theory), who is now regarded as one of history's most original mathematicians, were judged to have no foundation in mathematical knowledge and to lack effectiveness. "His divergent thinking could not gain recognition until convergent thinking had advanced sufficiently to make the effective novelty of his ideas apparent" (pp. 396–397).

The time dimension, too, plays an important role in understanding the interplay of agent and structure. Individuals react differently to the influence of macrolevel factors across multiple periods of time (see Drazin et al., 1999, p. 290). A number of papers (Gray, 1958, 1961, 1966; Munro, 1962; Simonton, 1975, 1979, 1981) suggest that the ups and downs in creativity are the effects of underlying cycles in the political, economic, and social milieu. When the peaks of these cycles converge, the result is a period of great florescence in the arts and sciences (Simonton, 1981, p. 628).

Another problem related to time is the "nemesis of creativity" (Anspach, 1952). All creative actions become habitual after a time. Styles and patterns developed in creative periods may eventually become selective and controlling. Creative insights or new revolutionary paradigms introduced by geniuses are worshipped and finally end in a new orthodoxy constraining the creativity of the next generation. "Leading thinkers often bring forth a visionary new thought framework only to have it concretized by well-meaning but shortsighted followers" (Ambrose, 2006, p. 83). This tendency is described by Koestler (1964):

> The new territory opened up by the impetuous advance of a few geniuses, acting as a spearhead, is subsequently occupied by the solid phalanxes of mediocrity; and soon the revolution turns into a new orthodoxy with its unavoidable symptoms of one-sidedness, over-specialization, loss of contact with other provinces of knowledge, and, ultimately, estrangement from reality. We see this happening—unavoidably it seems—at various times in the history of various sciences. The emergent orthodoxy hardens into a 'closed system' of thought, unwilling or unable to assimilate new empirical data or adjust itself to significant changes in other fields of knowledge. (p. 225)

Answers to some of the following questions would greatly help explain some of the still obscure relationships between environment, learning and action: How are colors, smells, sounds, and spatial patterns perceived and how are they processed in different emotional dispositions? What is the evolutionary perspective of the cognitive unconscious (Reber, 1992)? To which extent can intuition or inspiration be triggered by images of the environment? What are the basic neurological principles of creative thought, inspiration, and memory? How does the mind forget and remember (Schacter, 1992)? Is there a neuroscience of creativity? How far can neural structures be epigenetically influenced by variations in external stimulation or by learning? To what extent is the creative process recursive? What provokes recursion? Which contextual variables lead to modifications in the creative process?

Certain theoretical terms, such as *creative industries*, *creative classes*, and *creative cities* (Department of Culture, Media and Sport, 1998; Florida, 2002, 2005), need further debate and clarification. Suggesting that creative cities can "be made," they fascinate politicians and administrators alike. Such expressions have spawned a new policy area and have become buzzwords in recent years. But they and the empirical methodologies used to define creative cities and creative industries are roundly criticized in academia, for they widely fail to reflect decades of creativity research and have little to do with the definition of creativity developed in psychology. According to the British government's Department of Culture, Media and Sport (1998), for instance, creative industries include advertising, architecture, the art and antiques market, crafts, design, designer fashion, film, interactive leisure software, music, the performing arts, publishing, software, and television and radio (p. 1). Jeffcutt (2005) states that creative industries

> are shaped by interconnection between the media/information industries and the cultural/arts sector. ... The creative industries are shaped by interconnection between diverse domains (or forms) of creative endeavor (i.e. visual art, craft, print, video, music, etc.) that are brought together through new opportunities for the use of digital media technologies. (p. 104)

Do these interconnections really entail creativity? Is any TV production and any yellow press journal creative? Do plagiarism in music or routine products in advertising conform to the definition of creativity (new, original, and valuable)? What about the industries *not* mentioned in the British government's list? Do they lack creative people and products? The attempt to label certain industries or professions as creative and the rest, by omission, as noncreative without evaluating their ideas and products contradicts any definition of creativity accepted in the core disciplines of creativity research.

Many psychologists and human geographers regard the term *creative industry* as inappropriate. For if a person accepts the concept of psychological creativity and the assumption that anybody is creative in his or her daily problem-solving, then it makes no sense to distinguish between creative and noncreative industries. If one adopts the concept of historical creativity, then one must accept that creativity can be neither predicted nor administered, that a given profession has only a tiny proportion of historically creative people, and that members of a given profession are not invariably creative. If these propositions are accepted, then it makes no sense to label a whole industry or whole groups of professions as creative before evaluating their ideas and products. Neither the advertising nor the media industry perpetually engage in creative processes and continuously invent and fabricate creative products.

With regard to creative milieus (environments, contexts, and so forth), it may have become evident that the transdisciplinary discourse between human geography and other social and behavioral sciences has not yet come to an endpoint. Indeed, the promising start of this discourse suggests that this long journey will bring many new insights for all who participate.

References

Acham, K. (2003). Zu den Bedingungen der Entstehung von Neuem in der Wissenschaft [Conditions of the emergence of novelty in science]. In W. Berka, E. Brix, & C. Smekal (Eds.), *Woher kommt das Neue? Kreativität in Wissenschaft und Kunst* (pp. 1–28). Vienna: Böhlau.

Alligood, M. R. (1991). Testing Roger's theory of accelerating change: The relationship among creativity, actualization, and empathy in persons 18 to 92 years of age. *Western Journal of Nursing Research, 13*, 84–96.

Amabile, T. M. (1979). Effects of external evaluation on artistic creativity. *Journal of Personality and Social Psychology, 37*, 221–233.

Amabile, T. M. (1983a). *The social psychology of creativity*. New York: Springer.

Amabile, T. M. (1983b). Social psychology of creativity: A componential conceptualization. *Journal of Personality and Social Psychology, 45*, 357–377.

Amabile, T. M. (1988). A model of creativity and innovation in organizations. In B. M. Staw & L. L. Cummings (Eds.), *Research in organizational behavior*, Vol. 10, (pp. 123–167). Greenwich, CT: JAI Press.

Amabile, T. M. (1995). *KEYS: Assessing the climate for creativity*. Greensboro, NC: Center for Creative Leadership.

Amabile, T. M. (1996). Creativity in context: Update to the social psychology of creativity. New York: Westview Press.

Amabile, T. M. & Conti R. (1999). Changes in the work environment for creativity during downsizing. *The Academy of Management Journal, 42*, 630–640.

Amabile, T. M. & Gryskiewicz, N. (1989). The creative environment scales: The work environment inventory. *Creative Research Journal, 2*, 231–254.

Amabile, T. M., Goldfarb, P., & Brackfield, S. (1990). Social influences on creativity: Evaluation, co-action, and surveillance. *Creativity Research Journal, 3*, 6–21.

Amabile, T. M., Conti R., Coon, H., Lazenby, J., & Herron, M. (1996). Assessing the work environment for creativity. *The Academy of Management Journal, 39*, 1154–1184.

Ambrose, D. (2006). Large-scale contextual influences on creativity: Evolving academic disciplines and global value systems. *Creativity Research Journal, 18*, 75–85.

Andre, T., Schumer, H., & Whitaker, P. (1979). Group discussion and individual creativity. *Journal of General Psychology, 100*, 111–123.

Andrews, F. M. & Farris, G. F. (1967). Supervisory practices and innovation in scientific teams. *Personnel Psychology, 20*, 497–515.

Anspach, E. (1952). The nemesis of creativity: Observations on our occupation of Germany. *Social Research, 19*, 403–429.

Anthony, K. & Watkins, N. (2002). Exploring pathology: Relationships between clinical and environmental psychology. In R. B. Bechtel & A. Churchman (Eds.), *Handbook of environmental psychology* (pp. 129–145). New York: Wiley.

Arieti, S. (1976). *Creativity: The magic synthesis*. New York: Basic Books.

Banghart, F. D. & Spraker, H. S. (1963). Group influence on creativity in mathematics. *The Journal of Experimental Education, 31*, 257–263.

Barker, R. G. (1968). *Ecological psychology: Concepts and methods for studying the environment of human behavior*. Stanford, CA: Stanford University Press.

Barron, F. (1955). The disposition toward originality. *Journal of Abnormal and Social Psychology, 51*, 478–485.

Barron, F. (1969). *Creative person and creative process*. New York: Holt, Rinehardt & Winston.

Barron, F. B. & Harrington, D. M. (1981). Creativity, intelligence, and personality. *Annual Review of Psychology, 32*, 439–476.

Bateson, G. (1972). *Steps to an ecology of mind*. New York: Ballantine Books.

Bateson, G. & Bateson, M. (1987). *Angels fear: Towards an epistemology of the sacred*. Toronto: Bantam Books.

Becker, G. (2000–2001). The association of creativity and psychopathology: Its cultural-historical origins. *Creativity Research Journal, 13*, 45–53.

Bell, P. A., Greene, T. C., Fisher, J. D., & Baum, A. (2001). *Environmental psychology*. 5th edition. Forth Worth, TX: Harcourt College Publishers.

Bergler, E. (1950). *The writer and psychoanalysis*. Garden City, NY: Doubleday.

Berkes, F. (1999). *Sacred ecology: Traditional ecological knowledge and resource management*. Philadelphia, PA: Taylor & Francis.

Berlin, I. N. (1960). Aspects of creativity and the learning process. *American Imago—A Psychoanalytic Journal for the Arts and Sciences, 17*, 83–99.

Biersack, A. (1999). Introduction: From the "new ecology" to the new ecologies. *American Anthropologist*, New Series, *101*, 5–18.

Bloomberg, M. (1967). An inquiry into the relationship between field independence-dependence and creativity. *Journal of Psychology, 67*, 127–140.

Bloomberg, M. (1971). Creativity as related to field independence and mobility. *Journal of Genetic Psychology, 118*, 3–12.

Boden, M. (1994). What is creativity? In M. Boden (Ed.), *Dimensions of creativity* (pp. 75–117). Cambridge, MA: MIT Press.

Boden, M. (2004). In a nutshell. In M. Boden (Ed.), *The creative mind: Myths and mechanisms* (revised and expanded 2nd ed.). London: Routledge (pp. 1–24).

Bond, C. F., Jr. & Titus, L. J. (1983). Social facilitation: A meta-analysis of 241 studies. *Psychological Bulletin, 94*, 265–292.

Bonnes, M. & Secchiaroli, G. (1995). *Environmental psychology*. London: Sage.

Briskman, L. (1980). Creative product and creative process in science and the arts. *Inquiry: An interdisciplinary journal of philosophy 23*, 83–106.

Brown, A. D. & Starkey, K. (2000). Organizational identity and learning: A psychodynamic perspective. *The Academy of Management Review, 25*, 102–120.

Brown, V., Tumeo, M., Larey, T. S., & Paulus, P. B. (1998). Modeling cognitive interactions during group brainstorming. *Small Group Research, 29*, 495–526.

Brunswik, E. (1956). *Perception and the representative design of psychological experiments.* Berkeley, CA: University of California Press.

Brunswik, E. (1957). Scope and aspects of cognitive problems. In J. S. Bruner (Ed.), *Contemporary approaches to cognition* (pp. 5–31). Cambridge, MA: Harvard University Press.

Burchard, E. M. L. (1952). The use of projective techniques in the analysis of creativity. *Journal of Projective Techniques, 16*, 412–427.

Campbell, J., Dunnette, M. D., Lawler, E. E., & Weick, K. E. (1970). *Managerial behavior, performance, and effectiveness.* New York: McGraw-Hill.

Candolle, A. de (1873). *Histoire des sciences et des savants depuis deux siecles* [History of science and scholarship over two centuries]. Geneva, Switzerland: Georg.

Canter, D. (1977). *The psychology of places.* London: Architectural Press.

Canter, D. (1985). Putting situations in their place: Foundations for a bridge between social and environmental psychology. In A. Furnham (Ed.), *Social behavior in context* (pp. 208–239). London: Allyn & Bacon.

Carlozzi, A., Bull, K. S., Eells, G. T., & Hurlburt, J. D. (1995). Empathy as related to creativity, dogmatism, and expressiveness. *Journal of Psychology, 129*, 365–373.

Choi, J. N. (2004). Individual and contextual predictors of creative performance: The mediating role of psychological processes. *Creativity Research Journal, 16*, 187–199.

Clapham, M. M. (2000–2001). The effects of affect manipulation and information exposure on divergent thinking. *Creativity Research Journal, 13*, 335–350.

Clitheroe, H. C., Stokols, D., & Zmuidzinas, M. (1998). Conceptualizing the context of environment and behavior. *Journal of Environmental Psychology, 18*, 103–112.

Cooper, R. B. & Jayatilaka, B. (2006). Group creativity: The effects of extrinsic, intrinsic, and obligation motivations. *Creativity Research Journal, 18*, 153–172.

Couger, J. D. (1995). *Creative problem solving and opportunity finding.* Danvers, MA: Boyd & Fraser.

Cropley, A. (2006). In praise of convergent thinking. *Creativity Research Journal*, 18, 391–404.

Csikszentmihalyi, M. (1990). The domain of creativity. In M. A. Runco & R. S. Albert (Eds.), *Theories of creativity* (pp. 190–212). Newbury Park, CA: Sage.

Csikszentmihalyi, M. (1988). Society, culture, and person: A systems view of creativity. In R. J. Sternberg (Ed.), *The nature of creativity* (pp. 325–339). Cambridge, England: Cambridge University Press.

Csikszentmihalyi, M. (1999). Implications of a systems perspective for the study of creativity. In R. J. Sternberg (Ed.), *Handbook of creativity* (pp. 313–335). New York: Cambridge University Press.

Cummings, L. L. (1965). Organizational climates for creativity. *The Academy of Management Journal, 8*, 220–227.

D'Agostino, F. (1984). Chomsky on creativity. *Synthese, 58*, 85–117.

DeBono, E. (1968). *New Think.* New York: Basic Books.

Deci, E. L. & Ryan, R. M. (1980). The empirical exploration of intrinsic motivational processes. In L. Berkowitz (Ed.), *Advances in experimental social psychology* (Vol. 13, pp. 39–80). New York: Academic Press.

Deci, E. L. & Ryan, R. M. (1985). *Intrinsic motivation and self-determination in human behavior.* New York: Plenum.

Denison, D. R. (1996). What is the difference between organizational culture and organizational climate? A native's point of view on a decade of paradigm wars. *The Academy of Management Review, 21*, 619–654.

Department of Culture, Media and Sport (DCMS) (1998). *The creative industries mapping report.* London: HMSO.

Díaz de Chumaceiro, C. L. (1998). Serendipity in Freud's career: Before Paris. *Creativity Research Journal, 11*, 79–81.

Díaz de Chumaceiro, C. L. (1999). Research on career paths: Serendipity and its analog. *Creativity Research Journal, 12*, 227–229.

Díaz de Chumaceiro, C. L. (2004). Serendipity and pseudoserendipity in career paths of successful women: Orchestra conductors. *Creativity Research Journal, 16*, 345–356.

Diehl, M. & Stroebe, W. (1991). Productivity loss in idea-generating groups: Tracking down the blocking effect. *Journal of Personality and Social Psychology, 61*, 392–403.

Drazin, R., Glynn, M. A., & Kazanjian, R. K. (1999). Multilevel theorizing about creativity in organizations: A sensemaking perspective. *The Academy of Management Review, 24*, 286–307.

Dreistadt, R. (1969). The use of analogies and incubation in obtaining insight in creative problem solving. *Journal of Psychology, 71*, 159–175.

Dunnette, M. D., Campbell, J., & Jaastad, K. (1963). The effect of group participation on brainstorming effectiveness for two industrial samples. *Journal of Applied Psychology, 47*, 30–37.

Earl, W. L. (1987). Creativity and self-trust: A field study. *Adolescence, 22*, 419–432.

Eisenberger, R. & Rhoades, L. (2001). Incremental effects of reward on creativity. *Journal of Personality and Social Psychology, 81*, 728–741.

Eisenberger, R., Rhoades, L., & Cameron, J. (1999). Does pay for performance increase or decrease perceived self-determination and intrinsic motivation? *Journal of Personality and Social Psychology, 77*, 1026–1040.

Eissler, K. R. (1967). Psychopathology and creativity. *American Imago—A Psychoanalytic Journal for the Arts and Sciences, 24*, 35–81.

Ellis, H. (1926). *A study of British genius* (rev. ed.). Boston, MA: Houghton Mifflin.

Ericsson, K. A. (Ed.) (1996). *The road to excellence: The acquisition of expert performance in the arts and sciences, sports, and games*. Mahwah, NJ: Erlbaum.

Eulie, J. (1984). Creativity: Its implications for social studies. *Social Studies, 75*, 28–31.

Eysenck, H. J. (1994). The measurement of creativity. In M. A. Boden (Ed.), *Dimensions of creativity* (pp. 199–242). Cambridge, MA: MIT Press.

Feldman, D. H., Csikszentmihalyi, M., & Gardner, H. (1994). *Changing the world: A framework for the study of creativity*. New York: Praeger.

Fischer-Kowalski, M. & Erb, K. (2003). Gesellschaftlicher Stoffwechsel im Raum. Auf der Suche nach einem sozialwissenschaftlichen Zugang zur biophysischen Realität [Societal metabolism in space: In search of a social science approach to biophysical reality]. In G. Kohlhepp, A. Leidlmair, & F. Scholz (Series Eds.) & P. Meusburger & T. Schwan (Vol. Eds.), *Erdkundliches Wissen: Vol. 135. Humanökologie: Ansätze zur Überwindung der Natur-Kultur-Dichotomie* (pp. 257–285). Stuttgart, Germany: Steiner.

Fischer-Kowalski, M. & Weisz, H. (1998). Society as a hybrid between material and symbolic realms: Towards a theoretical framework of society–nature interaction. *Advances in Human Ecology, 8*, 215–251.

Florida, R. (2002). *The rise of the creative class: And how it's transforming work, leisure, community and everyday life*. New York: Basic Books.

Florida, R. L. (2005). *Cities and the creative class*. New York: Routledge.

Ford, C. M. (1996). A theory of individual creative action in multiple social domains. *The Academy of Management Review, 21*, 1112–1142.

Forehand, G. A. & Gilmer, B. (1964). Environmental variation in studies of organizational behavior. *Psychological Bulletin, 62*, 361–382.

Forgays, D. G. & Forgays, D. K. (1992). Creativity enhancement through flotation isolation. *Journal of Environmental Psychology, 12*, 329–335.

Frensch, P. A. & Funke, J. (1995). Definitions, traditions, and a general framework for understanding complex problem solving. In P. A. Frensch & J. Funke (Eds.), *Complex problem solving: The European perspective* (pp. 3–25). Hillsdale, NJ: Erlbaum.

Friedman, R. S. & Förster, J. (2002). The influence of approach and avoidance motor actions on creative cognition. *Journal of Experimental Social Psychology, 38*, 41–55.

Funke, J. (2000). Psychologie der Kreativität (Psychology of creativity). In R. M. Holm-Hadulla (Ed.), *Kreativität*. Heidelberger Jahrbücher 44 (pp. 283–300). Heidelberg, Germany: Springer.

Funke, J. (2007). The perception of space from a psychological perspective. In J. Wassmann & K. Stockhaus (Eds.), *Experiencing new worlds* (pp. 245–257). New York: Berghahn Books.
Gagliardi, P. (1986). The creation and change of organizational cultures: A conceptual framework. *Organization Studies*, *7*, 117–134.
Gallo, D. (1989). Educating for empathy, reason and imagination. *Journal of Creative Behavior*, *23*, 98–115.
Galton, F. (1869). *Hereditary genius: An inquiry into its laws and consequences*. London: Macmillan.
Gardner, H. (1988). Creative lives and creative works: A synthetic scientific approach. In R. J. Sternberg (Ed.), *The nature of creativity* (pp. 298–321). New York: Cambridge University Press.
Gardner, H. (1993a). *Frames of mind: The theory of multiple intelligences*. New York: Basic Books, London: HarperCollins.
Gardner, H. (1993b). *Creating minds: An anatomy of creativity as seen through the lives of Freud, Einstein, Picasso, Stravinsky, Eliot, Graham, and Gandhi*. New York: Basic Books.
Gardner, H. (1995). Creativity: New views from psychology and education. *RSA Journal*, 143, 5459, 33–42.
Gibson, J. J. (1960). The concept of stimulus in psychology. *American Psychologist*, *15*, 694–703.
Gibson, J. J. (1979). *An ecological approach to visual perception*. Boston, MA: Houghton Mifflin.
Gieryn, T. F. (2000). A space for place in sociology. *Annual Reviews of Sociology*, *26*, 463–496.
Gieryn, T. F. (2002a). Give place a chance: Reply to Gans. *City & Community*, *1*, 341–343.
Gieryn, T. F. (2002b). What buildings do. *Theory and Society*, *31*, 35–74.
Gimbel, J. (1990). *Science, technology, and reparations: Exploitation and plunder in postwar Germany*. Stanford, CA: Stanford University Press.
Givens, P. R. (1962). Identifying and encouraging creative processes: The characteristics of the creative individual and the environment that fosters them. *Journal of Higher Education*, *33*, 295–301.
Glynn, M. A. (1996). Innovative genius: A framework for relating individual and organizational intelligences to innovation. *The Academy of Management Review*, *21*, 1081–1111.
Goldsmith, R. E. & Matherley, T. A. (1988). Creativity and self-esteem: A multiple operationalization validity study. *Journal of Psychology*, *122*, 47–56.
Gordon, G. & Marquis, S. (1966). Freedom, visibility of consequences, and scientific innovation. *The American Journal of Sociology*, *72*, 195–202.
Götz, I. L. (1981). On defining creativity. *The Journal of Aesthetics and Art Criticism*, *39*, 297–301.
Gough, H. G. (1979). A creative personality scale for the Adjective Check List. *Journal of Personality and Social Psychology*, *37*, 1398–1405.
Graumann, C.-F. (1996). Aneignung [Appropriation]. In L. Kruse, C.-F. Graumann, & E. Lantermann (Eds.), *Ökologische Psychologie. Ein Handbuch in Schlüsselbegriffen* (pp. 124–130). Weinheim, Germany: Psychologie Verlags Union.
Graumann, C.-F. (2002). The phenomenological approach to people–environment studies. In R. B. Bechtel & A. Churchman (Eds.), *Handbook of Environmental Psychology* (pp. 95–113). New York: Wiley.
Graumann, C.-F. & Kruse, L. (2003). Räumliche Umwelt. Die Perspektive der humanökologisch orientierten Umweltpsychologie [Spatial environment: The perspective of environmental psychology oriented to human ecology]. In G. Kohlhepp, A. Leidlmair, & F. Scholz (Series Eds.) & P. Meusburger & T. Schwan (Vol. Eds.), *Erdkundliches Wissen: Vol. 135. Humanökologie: Ansätze zur Überwindung der Natur-Kultur-Dichotomie* (pp. 239–256). Stuttgart, Germany: Steiner.
Gray, C. E. (1958). An analysis of Graeco-Roman development: The epicyclical evolution of Graeco-Roman civilization. *American Anthropologist*, New Series, *60*, 13–31.
Gray, C. E. (1961). An epicyclical model for Western civilization. *American Anthropologist*, New Series, *63*, 1014–1037.

Gray, C. E. (1966). Measurement of Creativity in Western civilization. *American Anthropologist*, New Series, *68*, 1384–1417.

Guilford, J. P. (1967). *The nature of human intelligence*. New York: McGraw Hill.

Hage, J. & Dewar, R. (1973). Elite values versus organizational structure in predicting innovation. *Administrative Science Quarterly*, *18*, 279–290.

Hanák, P. (1993). Social marginality and cultural creativity in Vienna and Budapest (1890–1914). In E. Brix & A. Janik (Eds.), *Kreatives Milieu. Wien um 1900* (pp. 128–161). Vienna: Verlag für Geschichte und Politik.

Hausman, C. R. (1979). Criteria of creativity. *Philosophy and Phenomenological Research*, *40*, 237–249.

Hellriegel, D. & Slocum, J. W. (1974). Organizational climate: Measures, research and contingencies. *The Academy of Management Journal*, *17*, 255–280.

Hennessey, B. A. & Amabile, T. M. (1988). The conditions of creativity. In R. J. Sternberg (Ed.), *The nature of creativity* (pp. 11–38). Cambridge, England: Cambridge University Press.

Hennings, W. (2007). On the constitution of space and the construction of places: Java's magic axis. In J. Wassmann & K. Stockhaus (Eds.), *Experiencing new worlds* (pp. 125–145). New York/Oxford, England: Berghahn Books.

Hocevar, D. (1980). Intelligence, divergent thinking, and creativity. *Intelligence*, *4*, 25–40.

Hoffman, L. R. (1959). Homogeneity of member personality and its effects on group problem-solving. *Journal of Abnormal and Social Psychology*, *58*, 27–32.

Hoffman, L. R. & Maier, N. R. F. (1961). Quality and acceptance of problem solutions by members of homogeneous and heterogeneous groups. *Journal of Abnormal und Social Psychology*, *62*, 401–407.

Hoffman, L. R. & Smith, C. G. (1960). Some factors affecting the behavior of members of problem-solving groups. *Sociometry*, *23*, 273–291.

Holmes, F. L. (1986). Patterns of scientific creativity. *Bulletin of the History of Medicine*, *60*, 19–35.

Isaksen, S. G., Lauer, K. J., & Ekvall, G. (1999). Situational outlook questionnaire: A measure of the climate for creativity and change. *Psychological Reports*, *85*, 665–674.

Isaksen, S. G., Lauer, K. J., Ekvall, G., & Britz, A. (2000–2001). Perceptions of the best and worst climates for creativity: Preliminary validation evidence for the situational outlook questionnaire. *Creativity Research Journal*, *13*, 171–184.

Isen, A. M. (1987). Positive affect, cognitive processes, and social behavior. In L. Berkowitz (Ed.), *Advances in experimental social psychology* (Vol. 20, pp. 203–253). New York: Academic.

Isen, A. M., Daubman, K. A., & Nowicki, G. P. (1987). Positive affect facilitates creative problem solving. *Journal of Personality and Social Psychology*, *52*, 1122–1131.

James, L. R. & Jones, A. P. (1974). Organizational climate: A review of theory and research. *Psychological Bulletin*, *81*, 1096–1112.

James, L., Hater, J., Gent, M., & Bruni, J. (1978). Psychological climate: Implications from cognitive social learning theory and interactional psychology. *Personnel Psychology*, *31*, 783–813.

Jaušovec, N. (2000). Differences in cognitive processes between gifted, intelligent, creative, and average individuals while solving complex problems: An EEG study. *Intelligence*, *28*, 213–237.

Jeffcutt, P. (2005). The organizing of creativity in knowledge economies: Exploring strategic issues. In D. Rooney, G. Hearn, & A. Ninan (Eds.), *Handbook on the knowledge economy* (pp. 102–117). Cheltenham/Northampton/UK: Edward Elgar.

Jöns, H. (2003). *Grenzüberschreitende Mobilität und Kooperation in den Wissenschaften. Deutschlandaufenthalte US-amerikanischer Humboldt-Forschungspreisträger aus einer erweiterten Akteursnetzwerkperspektive* [Cross-boundary mobility and cooperation in the sciences: U.S. Humboldt Research Award winners in Germany from an expanded actor-network perspective] (Heidelberger Geographische Arbeiten No. 116). Heidelberg, Germany: Selbstverlag des Geographischen Instituts.

Jöns, H. (2006). Dynamic hybrids and the geographies of technoscience: Discussing conceptual resources beyond the human/non-human binary. *Social & Cultural Geography*, *7*, 559–580.

Kalliopuska, M. (1992). Creative way of living. *Psychological Reports*, *70*, 11–14.

Karau, S. J. & Williams, K. D. (1993). Social loafing: A meta-analytic review and theoretical integration. *Journal of Personality and Social Psychology, 65*, 681–706.

Kasof, J. (1995). Explaining creativity: The attributional perspective. *Creativity Research Journal, 8*, 311–366.

Kazanjian, R. K. (1988). Relation of dominant problems to stages of growth in technology-based new ventures. *The Academy of Management Journal, 31*, 257–279.

Kerr, N. L. & Bruun, S. E. (1983). Dispensability of member effort and group motivation losses: Free-rider effects. *Journal of Personality and Social Psychology, 44*, 78–94.

Klausmeier, H. J. (1961). *Learning and human abilities: Educational psychology*. New York: Harper & Brothers.

Kleinschmidt, H. (1978). American Imago on psychoanalysis, art, and creativity: 1964–1976. *American Imago—A Psychoanalytic Journal for the Arts and Sciences, 35*, 45–58.

Klüter, H. (2003). Raum als Umgebung (Space as environment). In G. Kohlhepp, A. Leidlmair, & F. Scholz (Series Eds.) & P. Meusburger & T. Schwan (Vol. Eds.), *Erdkundliches Wissen: Vol. 135. Humanökologie: Ansätze zur Überwindung der Natur-Kultur-Dichotomie* (pp. 217–238). Stuttgart, Germany: Steiner.

Koch, A. (2003). Raumkonstruktionen [Space constructs]. In G. Kohlhepp, A. Leidlmair, & F. Scholz (Series Eds.) & P. Meusburger & T. Schwan (Vol. Eds.), *Erdkundliches Wissen: Vol. 135. Humanökologie: Ansätze zur Überwindung der Natur-Kultur-Dichotomie* (pp. 175–196). Stuttgart, Germany: Steiner.

Koestler, A. (1964). *The act of creation*. London: Hutchinson.

Kroeber, A. L. (1944). *Configurations of culture growth*. Berkeley, CA: University of California Press.

Kurtzberg, T. R. & Amabile, T. (2000–2001). From Guilford to creative synergy: Opening the black box of team-level creativity. *Creativity Research Journal, 13*, 285–294.

Lamm, H. & Trommsdorff, G. (1973). Group versus individual performance on tasks requiring ideational proficiency (brainstorming): A review. *European Journal of Social Psychology, 3*, 361–387.

Laughlin, H. P. (1970). *The ego and its defenses*. New York: Appleton-Century-Crofts.

Leddy, T. (1994). A pragmatist theory of artistic creativity. *Journal of Value Inquiry, 28*, 169–180.

Li, J. & Gardner, H. (1993). How domains constrain creativity. *American Behavioral Scientist, 37*, 94–101.

Lindgren, H. C. & Lindgren, F. (1965). Creativity, brainstorming, and orneriness: A cross-cultural study. *The Journal of Social Psychology*, (ISSN-0022-4545) *67*, 23–30.

Lubart, T. I. (2000–2001). Models of the creative process: Past, present and future. *Creativity Research Journal, 13*, 295–308.

Lubart, T. I. & Getz, I. (1997). Emotion, metaphor, and the creative process. *Creativity Research Journal, 10*, 285–301.

Magyari-Beck, I. (1998). Is creativity a real phenomenon? *Creativity Research Journal, 11*, 83–88.

Maitland, J. (1976). Creativity. *The Journal of Aesthetics and Art Criticism, 34*, 397–409.

Mandler, G. (1992). Memory, arousal, and mood: A theoretical integration. In S. A. Christianson (Ed.), *The handbook of emotion and memory: Research and theory* (pp. 93–110). Hillsdale, NJ: Erlbaum.

March, J. & Simon, H. (1958). *Organizations*. New York: Wiley.

Martindale, C. (1989). Personality, situation, and creativity. In J. A. Glover, R. R. Ronning, & C. R. Reynolds (Eds.), *Handbook of creativity* (pp. 211–232). New York: Plenum Press.

Martindale, C., Abrams, L., & Hines D. (1974). Creativity and resistance to cognitive dissonance. *The Journal of Social Psychology*, (ISSN-00224545) *92*, 317–318.

Matlin, M. W. & Zajonc, R. B. (1968). Social facilitation of word associations. *Journal of Personality and Social Psychology, 10*, 455–460.

Matthiesen, U. (2006). Raum und Wissen. Wissensmilieus und KnowledgeScapes als Inkubatoren für zukunftsfähige stadtregionale Entwicklungsdynamiken? [Space and Knowledge: Knowledge milieus and knowledgescapes as incubators for sustainable urban development dynamics]. In D. Tänzler, H. Knoblauch, & H. Soeffner (Eds.), *Zur Kritik der Wissensgesellschaft* (pp. 155–188). Konstanz, Germany: UVK Verlagsgesellschaft.

Matthiesen, U. (2007). Wissensmilieus und KnowledgeScapes [Knowledge milieus and knowledgescapes]. In R. Schützeichel (Ed.), *Handbuch Wissenssoziologie und Wissensforschung* (pp. 679–693). Konstanz, Germany: UVK Verlagsgesellschaft.

Mayer, R. E. (1999). Fifty years of creativity research. In R. J. Sternberg (Ed.), *Handbook of creativity* (pp. 449–460). New York: Cambridge University Press.

McCoy, J. M. & Evans, G. W. (2002). The potential role of the physical environment in fostering creativity. *Creativity Research Journal, 14*, 409–426.

Meusburger, P. (1997). Spatial and social inequality in communist countries and in the first period of the transformation process to a market economy: The example of Hungary. *Geographical Review of Japan, Series B, 70*, 126–143.

Meusburger, P. (1998). *Bildungsgeographie. Wissen und Ausbildung in der räumlichen Dimension* [Geography of education: Knowledge and education in the spatial dimension]. Heidelberg, Germany: Spektrum Akademischer Verlag.

Meusburger, P. (2000). The spatial concentration of knowledge: Some theoretical considerations. *Erdkunde, 54*, 352–364.

Meusburger, P. (2008). The nexus between knowledge and space. In P. Meusburger (Series Eds.) & P. Meusburger, M. Welker, & E. Wunder (Vol. Eds.), *Knowledge and space: Vol. 1. Clashes of Knowledge. Orthodoxies and heterodoxies in science and religion* (pp. 35–90). Dordrecht, The Netherlands: Springer.

Miller, A. I. (2000). *Insights of genius: Imagery and creativity in science and art*. Cambridge, MA: MIT Press.

Mullen, B., Johnson, C., & Salas, E. (1991). Productivity loss in brainstorming groups: A meta-analytic integration. *Basic and Applied Social Psychology, 12*, 3–23.

Mumford, M. D. (1995). Situational influences on creative achievement: Attributions or interactions? *Creativity Research Journal, 8*, 405–412.

Mumford, M. D. (2003). Where have we been, where are we going? Taking stock in creativity research. *Creativity Research Journal, 15*, 107–120.

Mumford, M. D. & Gustafson, S. B. (1988). Creativity syndrome: Integration, application, and innovation. *Psychological Bulletin, 103*, 27–43.

Munro, T. (1962). What causes creative epochs in the arts? *The Journal of Aesthetics and Art Criticism, 21*, 35–48.

Neisser, U. (1987). From direct perception to conceptual structure. In U. Neisser (Ed.), *Concepts and conceptual development: Ecological and intellectual factors in categorization* (pp. 11–23). Cambridge, England: Cambridge University Press.

Nemeth, C. (1986). Differential contributions of majority and minority influences. *Psychological Review, 93*, 23–32.

Neves-Graça, K. (2003). Investigating ecology: Cognitions in human-environmental relationships. In G. Kohlhepp, A. Leidlmair, & F. Scholz (Series Eds.) & P. Meusburger & T. Schwan (Vol. Eds.), *Erdkundliches Wissen: Vol. 135. Humanökologie: Ansätze zur Überwindung der Natur-Kultur-Dichotomie* (pp. 309–326). Stuttgart, Germany: Steiner.

Neves-Graça, K. (2007). Elementary methodological tools for a recursive approach to human-environmental relations. In J. Wassmann & K. Stockhaus (Eds.), *Experiencing new worlds* (pp. 146–164). New York: Berghahn Books.

Niederland, W. G. (1967). Clinical aspects of creativity. *American Imago, 24*, 6–34.

Norlander, T., Bergman, H., & Archer, T. (1998). Effects of flotation REST on creative problem solving and originality. *Journal of Environmental Psychology, 18*, 399–408.

Ohnmacht, F. W. & McMorris, R. F. (1971). Creativity as a function of field independence and dogmatism. *Journal of Psychology, 79*, 165–168.

Oldham, G. R. & Cummings, A. (1996). Employee creativity: Personal and contextual factors at work. *The Academy of Management Journal, 39*, 607–634.

O'Reilly, C. A. (1991). Organizational behavior: Where we've been, where we are going. In M. R. Rosenzweig & L. W. Porter (Eds.), *Annual review of psychology* (Vol. 42, pp. 427–458). Palo Alto, CA: Annual Reviews.

Osborn, A. F. (1953). Applied imagination. Principles and procedures of creative thinking. New York: Charles Scribner's.

Ostwald, W. (1909). *Grosse Männer* [Great men]. Leipzig: Akademische Verlagsgesellschaft.

Patrick, C. (1937). Creative thought in artists. *Journal of Psychology, 4*, 35–73.
Patton, J. D. (2002). The role of problem pioneers in creative innovation. *Creativity Research Journal, 14*, 111–126.
Paulus, P. B. (2000). Groups, teams, and creativity: The creative potential of idea-generating groups. *Applied Psychology: An International Review, 49*, 237–262.
Paulus, P. B. & Dzindolet, M. T. (1993). Social influence processes in group brainstorming. *Journal of Personality and Social Psychology, 64*, 575–586.
Paulus, P. B. & Yang, H. (2000). Idea generation in groups: A basis for creativity in organizations. *Organizational Behavior and Human Decision Processes, 82*, 76–87.
Paulus, P. B., Larey, T. S., & Ortega, A. H. (1995). Performance and perception of brainstormers in an organizational setting. *Basic and Applied Social Psychology, 17*, 249–265.
Paulus, P. B., Larey, T. S., & Dzindolet, M. T. (2000). Creativity in groups and teams. In M. Turner (Ed.), *Groups at work: Advances in theory and research* (pp. 319–338). Hillsdale, NJ: Hampton.
Peterson, M. F. (1998). Embedded organizational event: The units of process in organizational science. *Organization Science, 9*, 16–33.
Plucker, J. A. & Renzulli, J. S. (1999). Psychometric approaches to the study of human creativity. In R. J. Sternberg (Ed.), *Handbook of creativity* (pp. 35–61). New York: Cambridge University Press.
Rappaport, R. (1979). *Ecology, meaning and religion*. Richmond, CA: North Atlantic Books.
Raudsepp, E. (1958). The industrial climate for creativity: An opinion study of 105 experts. *Management Review, 47*, 4–8 and 70–75.
Reber, A. S. (1992). The cognitive unconscious: An evolutionary perspective. *Consciousness and Cognition, 1*, 93–133.
Redmond, M. R. & Mumford, M. D. (1993). Putting creativity to work: Effects of leader behavior on subordinate creativity. *Organizational Behavior and Human Decision Processes, 55*, 120–151.
Rees, M. & Goldman, M. (1961). Some relationships between creativity and personality. *Journal of General Psychology, 65*, 145–161.
Rescher, N. (1966). A new look at the problem of innate ideas. *The British Journal for the Philosophy of Science, 17*, 205–218.
Robinson, J. (1970). Valéry's view of mental creativity. *Yale French Studies, 44*, 3–18.
Rokeach, M. (1954). The nature and meaning of dogmatism. *Psychological Review, 61*, 194–204.
Rokeach, M. (1960). *The open and closed mind*. New York: Basic Books.
Runco, M. A. (1988). Creativity research: Originality, utility, and integration. *Creativity Research Journal, 1*, 1–7.
Runco, M. A. (1993). Operant theories of insight, originality, and creativity. *American Behavioral Scientist, 37*, 54–67.
Runco, M. A. (1994). *Problem finding, problem solving, and creativity*. Norwood, NJ: Ablex.
Runco, M. A. & Okuda, S. M. (1988). Problem discovery, divergent thinking, and the creative process. *Journal of Youth and Adolescence, 17*, 213–222.
Russ, S. W. (1993). *Affect and creativity: The role of affect and play in the creative process*. Hillsdale, NJ: Erlbaum.
Saaty, T. L. (1998). Reflections and projections on creativity in operations research and management science: A pressing need for a shift in paradigm. *Operations Research, 46*, 9–16.
Sadler, W. A. & Green, S. (1977). Feature Essay. *Contemporary Sociology: A Journal of Reviews, 6*, 156–159.
Salbaum, J. (2008). Zeichen der Mehrdeutigkeit [Signs of polyvalence]. In H. Eigner, B. Ratter, & R. Dikau (Eds.), *Umwelt als System—System als Umwelt? Systemtheorien auf dem Prüfstand* (pp.155–169). Munich, Germany: Oekom.
Schacter, D. L. (1992). Understanding implicit memory: A cognitive neuroscience approach. *American Psychologist, 47*, 559–569.
Schein, E. (1985). *Organizational culture and leadership*. San Francisco, CA: Jossey-Bass.
Schluchter, W (2005). Handlung, Ordnung und Kultur. Studien zu einem Forschungsprogramm im Anschluss an Max Weber [Action, order, and culture: Studies for a research program building on Max Weber]. Tübingen, Germany: Mohr Siebeck.

Schorske, C. E. (1997). The new rigorism in the human sciences: 1940–1960. In T. Bender & C. E. Schorske (Eds.), *American academic culture in transformation: Fifty years, four disciplines* (pp. 309–329). Princeton, NJ: Princeton University Press.
Schuldberg, D. (2000–2001). Creativity and psychopathology: Categories, dimensions, and dynamics. *Creativity Research Journal, 13*, 105–110.
Schumpeter, J. A. (1934). *The theory of economic development: An inquiry into profits, capital, credit, interest, and the business cycle*. Harvard Economic Studies, Vol. 46. Cambridge, MA: Harvard University Press (Original German work published 1912)
Scott, S. G. & Bruce, R. A. (1994). Determinants of innovative behavior: A Path model of individual innovation in the workplace. *The Academy of Management Journal, 37*, 580–607.
Scott, W. E. (1965). The creative individual. *The Academy of Management Journal, 8*, 211–219.
Shalley, C. E. (1995). Effects of coaction, expected evaluation, and goal setting on creativity and productivity. *The Academy of Management Journal, 38*, 483–503.
Shalley, C. E. & Oldham, G. R. (1985). Effects of goal difficulty and expected evaluation on intrinsic motivation: A laboratory study. *The Academy of Management Journal, 28*, 628–640.
Shalley, C. E. & Perry-Smith, J. E. (2001). Effects of social-psychological factors on creative performance: The role of informational and controlling expected evaluation and modeling experience. *Organizational Behavior and Human Decision Processes, 84*, 1–22.
Shalley, C. E., Gilson, L., & Blum, T. C. (2000). Matching creativity requirements and the work environment: Effects on satisfaction and intentions to leave. *The Academy of Management Journal, 43*, 215–223.
Shaw, G. A. (1987). Creativity and Hypermnesia for words and pictures. *Journal of General Psychology, 114*, 167–178.
Shekerjian, F. (1990). *Uncommon genius: How great ideas are born*. New York: Viking.
Shore, E. (1971). Sensory deprivation, preconscious processes and scientific thinking. *American Journal of Orthopsychiatry, 41*, 574–580.
Simonton, D. K. (1975). Sociocultural context of individual creativity: A transhistorical time-series analysis. *Journal of Personality and Social Psychology, 32*, 1119–1133.
Simonton, D. K. (1979). Multiple discovery and invention: Zeitgeist, genius, or chance? *Journal of Personality and Social Psychology, 37*, 1603–1616.
Simonton, D. K. (1981). Creativity in Western civilization: Intrinsic and extrinsic causes. *American Anthropologist, New Series, 83*, 628–630.
Simonton, D. K. (1999). Creativity from a historiometric perspective. In R. J. Sternberg (Ed.), *Handbook of creativity* (pp. 116–133). New York: Cambridge University Press.
Simonton, D. K. (2000). Creativity: Cognitive, personal, developmental, and social aspects. *American Psychologist, 55*, 151–158.
Slochower, H. (1967). Genius, psychopathology and creativity. *American Imago, 24*, 3–5.
Smith, C. G. (1971). Scientific performance and the composition of research teams. *Administrative Science Quarterly, 16*, 486–496.
Stahl, M. J. & Koser, M. C. (1978). Weighted productivity in R & D: Some associated individual and organizational variables. *IEEE Transactions on Engineering Management, EM-25*, 20–24.
Stamm, J. (1967). Creativity and sublimation. *American Imago, 24*, 82–97.
Staw, B. M. (1990). An evolutionary approach to creativity and innovation. In M. A. West & J. L. Farr (Eds.), *Innovation and creativity at work* (pp. 287–308). Chichester, UK: Wiley.
Staw, B. M., Sandelands, L. E., & Dutton, J. E. (1981). Threat-rigidity effects in organizational behavior: A multilevel analysis. *Administrative Science Quarterly, 26*, 501–524.
Stein, M. (1953). Creativity and culture. *Journal of Psychology, 36*, 311–322.
Sternberg, R. E. & Lubart, T. I. (1991). An investment theory of creativity and its development. *Human Development, 34*, 1–31.
Sternberg, R. E. & Lubart T. I. (1999). The concept of creativity: Prospects and paradigms. In R. J. Sternberg (Ed.), *Handbook of creativity* (pp. 3–15). New York: Cambridge University Press.
Sternberg, R. E. & O'Hara, L. A. (1999). Creativity and intelligence. In R. J. Sternberg (Ed.), *Handbook of creativity* (pp. 251–272). New York: Cambridge University Press.

Suedfeld, P. (1968). The cognitive effects of sensory deprivation: The role of task complexity. *Canadian Journal of Psychology, 22*, 302–307.
Suedfeld, P. (1974). Social isolation: A case for interdisciplinary research. *Canadian Psychologist, 15*, 1–15.
Suedfeld, P. (1980). *Restricted environmental stimulation: Research and clinical applications*. New York: Wiley.
Suedfeld, P. & Landon, P. B. (1970). Motivational arousal and task complexity: Support for a model of cognitive changes in sensory deprivation. *Journal of Experimental Psychology, 83*, 329–330.
Suedfeld, P., Metcalfe, J., & Bluck, S. (1987). Enhancement of scientific creativity by flotation REST (restricted environmental stimulation technique). *Journal of Environmental Psychology, 7*, 219–231.
Suojanen, W. & Brooke, S. (1971). The management of creativity. *California Management Review, 14*, 1, 17–23.
Sutton, R. I. & Hargadon, A. (1996). Brainstorming groups in context: Effectiveness in a product design firm. *Administrative Science Quarterly, 41*, 685–718.
Symington, N. (1987). The sources of creativity. *American Imago, 44*, 275–287.
Tagiuri, R. (1968). The concept of organizational climate. In R. Tagiuri & G. H. Litwin (Eds.), *Organizational climate: Explorations of a concept* (pp. 11–32). Boston, MA: Harvard University Press.
Thomas, C. A. (1955). *Creativity in science: The eight annual Arthur Dehon Little memorial lecture*. Cambridge, MA: MIT.
Törnqvist, G. (2004). Creativity in time and space. *Geografiska Annaler, 86 B*, 227–243.
Torrance, E. P. (1995). *Why fly: A philosophy of creativity*. Norwood, NJ: Ablex.
Treffinger, D. J. (1995). Creative problem solving: Overview and educational implications. *Educational Psychology Review*, 7, 301–312.
True, S. (1966). A study of the relation of general semantics and creativity. *Journal of Experimental Education, 34*, 34–40.
Unsworth, K. (2001). Unpacking creativity. *The Academy of Management Review, 26*, 289–297.
Van Gundy, A. (1987). Organizational creativity and innovation. In S. G. Isaksen (Ed.), *Frontiers of creativity research* (pp. 358–379). Buffalo, NY: Bearly.
Vidal, F. (2003). Contextual biography and the evolving systems approach to creativity. *Creativity Research Journal, 15*, 73–82.
Wallas, G. (1926). *The art of thought*. New York: Harcourt, Brace and Company.
Wassmann, J. (2003). Landscape and memory in Papua New Guinea. In H. Gebhardt & H. Kiesel (Eds.), *Weltbilder* (pp. 329–346). Heidelberger Jahrbücher No. 47. Berlin: Springer.
Wassmann, J. & Keck, V. (2007). Introduction. In J. Wassmann & K. Stockhaus (Eds.), *Experiencing new worlds* (pp. 1–18). New York: Berghahn Books.
Weichhart, P. (2003). Gesellschaftlicher Metabolismus und Action Settings. Die Verknüpfung von Sach- und Sozialstrukturen im alltagsweltlichen Handeln [Social metabolism and action settings: The link between technical and social structures in everyday action]. In G. Kohlhepp, A. Leidlmair, & F. Scholz (Series Eds.) & P. Meusburger & T. Schwan (Vol. Eds.), *Erdkundliches Wissen: Vol. 135. Humanökologie: Ansätze zur Überwindung der Natur-Kultur-Dichotomie* (pp. 15–44). Stuttgart, Germany: Steiner.
Weisberg, R. W. (1999). Creativity and knowledge: A challenge to theories. In R. J. Sternberg (Ed.), *Handbook of creativity* (pp. 226–250). New York: Cambridge University Press.
West, M. A. (1989). Innovation amongst health care professionals. *Social Behaviour, 4*, 173–184.
Westfall, R. S. (1983). Newton's development of the *Principia*. In R. Aris, T. Davis, & R. H. Stuewer (Eds.), *Springs of scientific creativity. Essays on founders of modern science* (pp. 21–43). Minneapolis, MN: University of Minnesota Press.
Westland, G. (1969). The investigation of creativity. *The Journal of Aesthetics and Art Criticism, 28*, 127–131.
Wierenga, B. & van Bruggen, G. H. (1998). The dependent variable in research into the effects of creative support systems: Quality and quantity of ideas. *MIS Quarterly, 22*, 81–87.

Williams, W. M. & Yang, L. T. (1999). Organizational creativity. In R. J. Sternberg (Ed.), *Handbook of creativity* (pp. 373–391). New York: Cambridge University Press.
Witkin, H. A., Dyke, R. B., Faterson, H. F. Goodenough, D. R., & Karp, S. A. (1962). *Psychological differentiation*. New York: Wiley.
Woodman, R. W. & King, D. C. (1978). Organizational climate: Science or folklore? *The Academy of Management Review*, *3*, 816–826.
Woodman, R. W. & Schoenfeldt, L. F. (1990). An interactionist model of creative behavior. *Journal of Creative Behavior*, *24*, 279–290.
Woodman, R. W., Sawyer, J. E., & Griffin, R. W. (1993). Toward a theory of organizational creativity. *The Academy of Management Review*, *18*, 293–321.
Workman, E. A. & Stillion, J. M. (1974). The relationship between creativity and ego development. *The Journal of Psychology*, *88*, 191–195.
Zajonc, R. B. (1965). Social facilitation. *Science*, *149*, 269–274.
Zhou, J. & George, J. M. (2001). When job dissatisfaction leads to creativity: Encouraging the expression of voice. *The Academy of Management Journal*, *44*, 682–696.

Chapter 7
Creativity, Intelligence, and Culture: Connections and Possibilities

James C. Kaufman

The topic of intelligence and culture is a widely studied and often highly controversial area, marked by much debate and emotion. Typically, African Americans and Hispanic Americans score lower than European Americans on a variety of measures of intelligence and ability (see Loehlin, 2000, for an overview). Group tests such as the SAT, ACT, Graduate Records Examinations, and Advanced Placement exams produce similar results (Camara & Schmidt, 1999; Morgan & Maneckshana, 1996). Some researchers argue that these measures reflect actual differences (e.g., Herrnstein & Murray, 1994; Jensen, 1998). Others point to the discrepancy between socioeconomic status and opportunities across ethnicities (Rogers, 1996; Sternberg, 1996), whereas still others argue that current ability measures do not incorporate enough aspects of intelligence (Sternberg et al., 2008).

Indeed, the Kaufman Assessment Battery for Children—Second Edition (KABC-II; Kaufman & Kaufman, 2004) is based on two current theories of intelligence (the Cattell–Horn–Carroll (CHC) theory of intelligence (McGrew, 2005) and Luria's (1970) neuropsychological model). The KABC-II, as well as its predecessor (the K-ABC; Kaufman & Kaufman, 1983), has consistently shown the smallest differences in test scores by ethnicity of the major individual intelligence tests (Cole et al., 2009; Kaufman, 2003; Kaufman et al., 2005). Another instrument based on Luria's theory, the Cognitive Assessment System (Naglieri & Das, 1997), also shows small differences by ethnicity (Naglieri et al., 2005).

Although group mean differences are certainly relevant to the discussion, there are other, more psychometric approaches that offer a more sophisticated view of the problem. The argument underlying such approaches is that, even though two groups may perform differently on an ability test, the test itself may not be in error or biased. These approaches evaluate content that may be inappropriate because it unfairly favors one group over another (for instance, sports examples may stereotypically favor males). The question then becomes whether different constructs may be measured across nominal groups by the same test. A test may

J.C. Kaufman (✉)
Learning Research Institute, California State University, Department of Psychology, 5500 University Parkway, San Bernardino, CA 92407, USA
e-mail: jkaufman@csusb.edu

P. Meusburger et al. (eds.), *Milieus of Creativity*, Knowledge and Space 2,

measure verbal ability in European Americans, for example, but may be measuring something quite different in a Hispanic–American population (Reynolds et al., 1999; Reynolds, 2000).

Creativity in Intelligence Theory

Creativity, despite its growing connection with intelligence theory, is not represented on any major IQ test. Guilford (1967, 1988) pioneered this connection by integrating creativity into a larger framework of intelligence. His Structure of Intellect model organized human cognition along three dimensions. The first dimension was called "operations" and simply meant the type of mental effort required for any kind of task. The second dimension, "content," referred to the general subject area. The third dimension, "product," represented the actual products that might result from different kinds of thinking in different kinds of subject matters. With five operations, four contents, and six products, Guilford's (1967) model had 120 different possible mental abilities.

One of Guilford's operations is divergent production—analyzing responses to questions with no obvious, singular answer (such as "What would happen if people no longer needed sleep?"). Guilford (1967) initially described divergent production as consisting of four specific abilities: fluency, flexibility, originality, and elaboration. A common instrument used to measure divergent production is the Unusual Uses Test, where participants are asked to list all the uses of a familiar object, such as a brick. In this context, fluency is quantitative and measured by the number of responses. Flexibility is measured by the variety of different categories or concepts that are evoked. Elaboration is measured by the level of descriptiveness of each use. Originality is measured by uniqueness of a participant's response in comparison to the responses of other participants. Modern researchers use the broader term "divergent thinking" to describe what Guilford referred to as divergent production.

Many subsequent theories of intelligence incorporated creativity-related abilities into their structures. Two examples of this class of theory are the CHC and Luria's (1970) model. CHC theory is a combination of the Cattell–Horn theory of fluid and crystallized intelligence (Horn & Cattell, 1966; Horn & Hofer, 1992; Horn & Noll, 1997) and Carroll's Three-Stratum Theory (1993). The CHC model proposes ten different broad factors of intelligence. One of them is *Glr* (long-term storage and retrieval), which includes creativity/originality as a component. Specific components of *Glr* and their relationship to creativity are discussed in more detail in A.S. Kaufman and N.L. Kaufman (2008) Luria's theory is also at the heart of the PASS model (Planning, Attention, Simultaneous, and Successive; see Das et al., 1994, for an overview). It has been hypothesized that planning abilities are related to creativity (Naglieri & Kaufman, 2001).

Sternberg's theory of successful intelligence includes creative abilities as one of three essential components, along with analytical and practical abilities. Although not currently used in a major IQ test, this theory is the basis for exciting work in college admissions. Sternberg and his colleagues at Tufts University are adding explicit assessment of creativity as a nonrequired component of college admittance. His measures of successful intelligence (including creativity) predict college success more accurately than standard admissions tests do, and differences by ethnicity are significantly reduced (Stemler et al., 2006; Sternberg, 2006; Sternberg & the Rainbow Project Collaborators, 2006).

Creativity and IQ Tests

Regardless of whether creativity plays a role in theoretical conceptions of intelligence, it does not play a role in traditional measures of intelligence. How does creativity relate to traditional measures of intelligence? Most researchers who study creativity and intelligence use tests of divergent thinking (such as the Torrance Tests of Creative Thinking [TTCT; Torrance, 1966, 1974] and other divergent-thinking measures) or other similarly scored paper-and-pencil tests. Generally, such paper-and-pencil measures are significantly associated with psychometric measures of intelligence and are especially associated with verbally oriented measures. This relationship, however, is not a particularly strong one (see Barron & Harrington, 1981; Kim, 2005).

In most of these studies, the correlation between divergent thinking and IQ is maintained up to a certain level of performance on a traditional individual intelligence test. Researchers pursuing this traditional work have argued that there is a "threshold effect," in which creative potential and psychometric intelligence are positively correlated up to an IQ of approximately 120. In people with higher IQs, the two constructs show little relationship (e.g., Fuchs-Beauchamp et al., 1993; Getzels & Jackson, 1962).

More recently, however, the threshold theory has come under fire. Comparing measures of fluid intelligence and creativity (as measured through divergent-thinking tests), Preckel, Holling, and Weise (2006) found modest correlations across all levels of intellectual abilities. Along similar lines, a 21-study meta-analysis by Kim (2005) showed virtually no support for the threshold theory, with very small positive correlations found between measures of ability and measures of creativity and divergent thinking.

Notably, almost none of these studies involved traditional, individually administered intelligence tests. Much of the research covered by Kim (2005) was more than 30 years old and had therefore been conducted with IQ tests that do not reflect current IQ theory. In addition, most of the studies used group IQ tests. Although group IQ tests serve a strong purpose in research, they are not used by most school psychologists for psychoeducational assessment (Kaufman & Lichtenberger, 2006). One of the few studies to use an individually administered, modern IQ test was conducted by Sligh, Conners, and Roskos-Ewoldsen (2005), who used the Kaufman Adolescent and Adult Intelligence Test (Kaufman & Kaufman, 1993). This instrument is based on Horn's (1989) revision and expansion of the Cattell–Horn Gf-Gc theory (Horn & Cattell, 1966). Sligh et al. delved deeper into the intelligence–creativity relationship than traditional threshold theory research had by specifically examining the relationship between a measure of actual creative innovation and *Gf* (fluid intelligence, which measures a person's ability to adapt and be flexible in new situations) and *Gc* (crystallized intelligence, which measures knowledge acquired from formal schooling and acculturation). A measure of creative innovation had participants create and modify inventions. Sligh et al. found that *Gc* showed the same moderate and positive relationship to composite creativity as had past studies (mentioned above). In contrast, *Gf* showed the *opposite* pattern. *Gf* and composite creativity were significantly correlated for the high IQ group, but they were not significantly correlated for people with average IQs.

Sligh et al.'s (2005) results and the mixed findings on the threshold hypothesis indicate that supplementing traditional intelligence tests with measures of creativity may provide unique information about an individual. Creativity is not simply an inherent part of traditional intelligence. Levels and types of cognitive abilities differ in the ways they are associated with measures of creativity.

I now explore findings about how selected ethnicities and cultures compare on measures of creativity.

Creativity Across Cultures

Hispanic–American and Hispanic

Studies of creativity in Hispanic Americans and European Americans tend to arrive at different results depending on whether the measure of creativity is verbal or nonverbal. For example, Argulewicz and Kush (1984) found that European Americans scored higher than Hispanic Americans on three of four TTCT Verbal (English) forms but found no significant differences on the Figural forms. (The TTCT has been translated into Spanish, among many other languages, and has been shown to have construct validity in many Hispanic cultures; see Wechsler, 2006, for example.)

Studies using only nonverbal assessments have typically found no differences (e.g., Argulewicz et al., 1982) between European Americans and Hispanic Americans or have shown that bilingual Hispanic Americans have a slight advantage in the nonverbal domain (Kessler & Quinn, 1987; Price-Williams & Ramirez, 1977). However, low-income Hispanic–American elementary students scored below the norms on the TTCT (Mitchell, 1988), and teachers rated European American students as being more creative than Hispanic–American students, with highly acculturated Hispanic Americans receiving higher marks than less acculturated Hispanic Americans (Masten et al., 1999).

Some researchers found that European–American parents had more favorable perceptions of creativity than Hispanic–American parents did (Strom & Johnson, 1989; Strom et al., 1992). However, they also found that Hispanic–American parents were more likely to engage in play activities with their children and valued play more than European–American parents did (Strom & Johnson, 1989). Make-believe play can be a valuable component of a child's developing imagination (Singer & Singer, 1990).

Eastern vs. Western Cultures

Studies of the TTCT often show Western cultures outperforming Eastern cultures. Jellen and Urban (1989) administered a measure of creative thinking and drawing to children from several different countries and found that, in general, Western countries (such as Germany, England, and the United States) scored higher than Eastern countries (such as China and India). American college students scored higher on the TTCT than Japanese college students in one study (Saeki et al., 2001), and Americans from five different age groups scored higher than similar individuals from Hong Kong (Jaquish & Ripple, 1984). Zha, Walczyk, and Griffith-Ross (2006) found that although Chinese graduate students outperformed their American counterparts on the GRE, American graduate students scored higher on four out of five measures of divergent thinking. School children in Hong Kong scored higher on the Figural form of the TTCT than their counterparts in Taiwan, Singapore, and America, but lower than German children. On the Verbal form, the results were in the opposite order (Rudowicz et al., 1995).

Self-report and self-assessments tend to show fewer differences than are found on psychometric tests Plucker, Runco, and Lim (2006) found no difference in creative potential (as measured by the Runco Ideational Behavior Scale; Runco et al., 2001) between Korean students and American students. Similarly, Lim and Plucker (2001) found that Koreans and Americans hold very similar concepts about the nature of creativity. Malaysian students scored higher than American, Indian, and Hungarian students on one self-report measure of creativity, but American students scored higher than Malaysian students on a different self-report measure (Palaniappan, 1996).

According to both American and Chinese raters in one study, artwork produced by American college students was more creative than art produced by Chinese students (Niu & Sternberg, 2001). Yet a similar study that compared American and Chinese drawings of geometric shapes found that the two groups were rated similarly for creativity by both American and Chinese raters (Chen et al., 2002). In both studies, American and Chinese judges tended to agree on which products were creative and which were not, although Niu and Sternberg (2001) found that the Chinese judges tended to give higher scores than their American counterparts. There were no differences between Chinese and British school children in terms of rated artwork, except for the higher ratings earned by Chinese children who attended a weekend art school (Cox et al., 1998). Another study found that Japanese children produced drawings that received higher rating than drawings by British children did (Cox et al., 2001).

Differences in styles and values in Eastern and Western cultures may explain some of the findings that Western individuals receive higher scores on creativity assessments. Li (1997) proposed a horizontal and vertical tradition of creativity. Horizontal traditions, which are favored by Western cultures, tend toward changing and modifying pre-existing structures. In vertical traditions, however, the nature of the work is much more constrained and consistent with past work. A piece's worth is more dependent on how well the artist is able to capture his or her subject matter (Li, 1997). This theory is consistent with the idea that, whereas both Eastern and Western cultures value the effectiveness of a piece of creativity, the West values the novelty of a piece much more than the East does. Of much more interest to the East is whether a piece is authentic—"a reflection of an individual's own values and beliefs" (Averill et al., 2001, p. 172).

Why does this difference occur between East and West? One answer may lie in the theory of interdependence vs. independence. This theory argues that Northern Americans and Western Europeans see themselves as independent and that their motivations and goals follow accordingly. In contrast, for example, Asian cultures are more interdependent and have a higher sense of group responsibility. These cultures are motivated by different variables, such as group harmony (Markus & Kitayama, 1991).

Whether a person is part of an independent or interdependent culture can affect his or her personality and style. People from interdependent cultures are more likely to see themselves as fundamentally linked to others and to view themselves in the context of their social relationships (Cross & Markus, 1999). This view translates into a cognitive style; Asians were found to be more field dependent and more holistic than Americans, for example (Ji et al., 2000; Nisbett et al., 2001). People who are more field dependent tend to score lower on tests of creativity (e.g., Chadha, 1985; Noppe, 1985).

Asian Americans

There are many studies that compare Asians and Europeans or Americans. Far fewer studies have compared Asian *Americans* to Americans of different ethnicities. Rostan, Pariser, and Gruber (2002) studied student artwork by Chinese–American and European–American students, with two groups in each culture: students with additional art training and classes and students with no such instruction. Each group's artwork (one drawing from life and one drawing from imagination) was assessed by both Chinese and American judges. Neither set of judges found any significant differences between cultures, only between art students versus nonart students. Pornrungroj (1992) gave the Figural form of the TTCT to Thai children and Thai-American children and found Thai children received significantly higher scores than did Thai Americans. Yoon (2005) gave the TTCT to European–American and Asian–American middle-school students (the latter being a mix of Chinese Americans, Korean Americans, Japanese Americans, and Southeastern

Asian Americans). There were no significant differences either between the European Americans and Asian Americans or between the different subgroups of Asian Americans.

African Americans

Creativity researchers have discerned few differences between African Americans and European Americans. Indeed, some of the only creativity-related differences that have surfaced tend to favor African Americans. These findings have been fairly consistent regardless of the type of measurement. The TTCT and other divergent-thinking measures, with both verbal and figural forms, have been used extensively in these studies (e.g., Glover, 1976; Iscoe & Pierce-Jones, 1964; Kaltsounis, 1974; Knox & Glover, 1978; Torrance, 1971, 1973).

Other work, too, has found no differences between African Americans and European Americans. Some of the studies used questionnaires measuring creative accomplishments (Stricker et al., 2001), whereas others have looked at the ability to be trained on creativity tasks (Moreno & Hogan, 1976). Still other research has examined the development of divergent-thinking abilities in adolescents from South Africa and the United States (Ripple & Jaquish, 1982). Kaufman, Baer, and Gentile (2004) studied poems, stories, and personal narratives written by African–American and European–American eighth-grade students. There were no differences in creativity scores assigned by expert judges. Of 13 measures of giftedness, those of creativity showed some of the smallest differences between these two groups (Harty et al., 1984).

Indeed, some of the only significant differences that have emerged tend to favor African Americans. Torrance (1971, 1973) found that African–American children scored higher on the TTCT's Figural tests in fluency, flexibility, and originality than European–American children, whereas the European Americans outscored the African Americans on Figural elaboration and on all Verbal subtests. The initial sample compared African–American children in Georgia with children of higher socioeconomic status in Minnesota. When Torrance's subsequent work (1973) used European Americans also from Georgia, all differences narrowed significantly. Torrance also found that the African Americans in his sample received higher fluency and originality scores on the TTCT than did the European–American participants. Troiano and Bracken (1983) gave measures of creative thinking to three different kindergarten classes, one comprised of Dutch Americans, another of African Americans, and the third of Native Americans. They found that African Americans and Native Americans scored approximately one standard deviation higher on creative thinking, particularly in fluency, than the Dutch Americans. Kaufman (2006) asked 3,553 individuals (mostly high school and college students) to rate themselves in 56 different domains of creativity. African Americans rated themselves significantly higher than at least one other ethnicity on all factors. All ethnicities except for Asian Americans rated themselves higher than another ethnicity on at least one factor.

African Americans, Creativity, and Personality

It has been suggested that creativity does not benefit African Americans on intelligence tests and may even hurt them (Heath, 1983). Some researchers have proposed that differences on certain IQ or achievement subtests, such as those involved in remembering the details of a story, may show larger differences between African Americans and European Americans in part because African Americans approach the task differently (Heath, 1983; see Manly et al., 1998). According to this theory, European Americans go about it as the test-makers intended—by trying to memorize as many appropriate details as possible and stick to the presented story. In contrast, African Americans may put more emphasis on telling the story creatively. Indeed, another possible negative outcome is that African Americans are penalized for creative behavior in the classroom. Baldwin (1985, 2003) asserts that teachers and other authority figures may mistake creativity in African–American students as unruly or disruptive behavior.

Baldwin (2001) also analyzed a list of creative traits and abilities. These abilities included well-researched aspects, such as being open to experience or having high divergent-thinking ability, and rather theoretical aspects, such as being antiauthoritarian and having a "zany" sense of humor and a low tolerance for boredom (Clark, 1988). Many of these abilities, Baldwin (2001) reasoned, are specifically appropriate and often exhibited by African Americans. These ideas are consistent with Shade's (1986) theory of an African–American cognitive style. Her research with cognitive style tests found that African Americans were more likely to be spontaneous, flexible, and open-minded than European Americans. In contrast, European Americans were more regulated and structured.

Such a connection between open-mindedness and creativity seems to be a natural one. Indeed, being open to new experiences has been shown to be highly correlated with creativity, regardless of the measure used. These results have been found with self-reports of creative acts (Griffin & McDermott, 1998), biographical data on creative accomplishments (King et al., 1996), studies of creative professions (Domino, 1974), analysis of participants' daydreams (Zhiyan & Singer, 1996), creativity ratings on stories (Wolfradt & Pretz, 2001), and psychometric tests (Furnham, 1999; McCrae, 1987).

However, large empirical research studies on openness to experience and culture do not seem to bear out the arguments stated by Baldwin (1985, 2001, 2003) and Shade (1986). There generally tend to be no differences on any personality factors across cultures (e.g., Goldberg et al., 1998; Kyllonen et al., 2005; McCrae & Costa, 1997). However, Heuchert, Parker, Stumpf, and Myburgh (2000) found that White South Africans scored higher on openness to experience than Black South Africans. (However, much of this difference was in the openness-to-feelings subcomponent as opposed to the more creativity-related openness to fantasy and aesthetics subcomponents.) Allik and McCrae (2004) found that people from European and European–American cultures tended to be more open to experience than people from Asian and African cultures. Schmitt, Allik, McCrae, and Benet-Martínez (2007), in a massive study of 17,837 people from 56 nations, found that participants from

South American and European countries were the most open to experience (Chile was the highest), with those from South Asian countries generally being less open to experience. Participants from African countries ranged between those two groups. Lastly, Saucier and Goldberg (2001) studied personality labels in 13 languages (including English) and found that openness to experience was the only one of the big-five traits (the others being emotional stability, extraversion, conscientiousness, and agreeableness) *not* to be found in all languages. Openness to experience, therefore, can be considered specific to Anglo cultures (Benet-Martínez & Oishi, 2008).

Conclusions

Regardless of the nature of the specific differences between ethnicities and across cultures, the patterns are *not* the same as for intelligence. I am not arguing that any one culture is more creative or intelligent than another. Rather, I am arguing that if measures of intelligence show some differences and measures of creativity show other differences, then using *both* intelligence and creativity measures would yield more information and present both a more complete and a fairer picture of an individual than either measure alone would.

Creativity is not the only dimension that could have been selected to expand current conceptions of intelligence. The demand for and interest in noncognitive constructs is increasing in general among educators and admissions committees alike (Kyllonen et al., 2005). Similarly, one may argue that emotional intelligence, motivation, practical intelligence, or personality could all supplement measures of intelligence by offering a more varied and informative insight into an individual than is currently provided. In the world of business, these measures are already being used for both hiring and promotion (e.g., Agars & Kaufman, 2005). Many high-level businesses administer a wide battery of measures to ensure that their workers have the highest level of ability needed to succeed. Any of these constructs is likely to show different patterns across different cultures, with these patterns reflecting cultural values and principles (Hofstede, 2001).

The connection between creativity and intelligence, however, goes back to the days of Galton and Binet (Baer & Kaufman, 2006). Past research indicates that creativity and intelligence, despite the similar cognitive abilities they involve, have different patterns across cultures and ethnicities. As discussed in this chapter, the groups of people who receive the highest scores on ability measures are not necessarily the same as those who receive the highest scores on creativity measures. Work has already started on supplementing group admission tests with measures of creativity (Sternberg, 2006). It is time to continue this trend to discover the additional knowledge that can result from supplementing individual tests of ability and achievement with measures of creativity.

Acknowledgement The author would like to thank Sarah A. Burgess, Alan S. Kaufman, and Elizabeth Skrzynecky for editorial help and suggestions.

References

Agars, M. D. & Kaufman, J. C. (2005). Creativity in the workplace: Introduction to the special issue. *Korean Journal of Thinking & Problem Solving*, *15*, 5–6.

Allik, J. & McCrae, R. R. (2004). Toward a geography of personality traits: Patterns of profiles across 36 cultures. *Journal of Cross-Cultural Psychology*, *35*, 13–28.

Argulewicz, E. N. & Kush, J. C. (1984). Concurrent validity of the SRBCSS creativity scale for Anglo-American and Mexican-American gifted students. *Educational and Psychological Research*, *4*, 81–89.

Argulewicz, E. N., Elliott, S. N., & Hall, R. (1982). Comparison of behavioral ratings of Anglo-American and Mexican-American gifted children. *Psychology in the Schools*, *19*, 469–472.

Averill, J. R., Chon, K. K., & Hahn, D. W. (2001). Emotions and creativity, East and West. *Asian Journal of Social Psychology*, *4*, 165–183.

Baer, J. & Kaufman, J. C. (2006). Creativity in English-speaking countries. In J. C. Kaufman & R. J. Sternberg (Eds.), *International handbook of creativity* (pp. 10–38). Cambridge: Cambridge University Press.

Baldwin, A. Y. (1985). Programs for the gifted and talented: Issues concerning minority populations. In F. Horowitz & M. O'Brien (Eds.), *The gifted and talented: Developmental perspectives* (pp. 223–249). Washington, DC: American Psychological Association.

Baldwin, A. Y. (2001). Understanding the challenge of creativity among African Americans. *Journal of Secondary Gifted Education*, *12*, 121–125.

Baldwin, A. Y. (2003). Understanding the challenge of creativity among African Americans. *Inquiry*, *22*, 13–18.

Barron, F. & Harrington, D. (1981). Creativity, intelligence, and personality. *Annual Review of Psychology*, *32*, 439–476.

Benet-Martínez, V. & Oishi, S. (2008). Culture and personality. In O. P. John, R. W. Robins, & L. A. Pervin (Eds.), *Handbook of personality: Theory and research* (pp. 542–567). New York: Guilford.

Camara, W. J. & Schmidt, A. E. (1999). *Group differences in standardized testing and social stratification* (College Board Rep. No. 99–5). New York: College Board.

Carroll, J. B. (1993). *Human cognitive abilities*. Cambridge, England: Cambridge University Press.

Chadha, N. K. (1985). Creativity and cognitive style. *Psycho Lingua*, *15*, 81–88.

Chen, C., Kasof, J., Himsel, A. J., Greenberger, E., Dong, Q., & Xue, G. (2002). Creativity in drawing geometric shapes: A cross-cultural examination with the consensual assessment technique. *Journal of Cross-Cultural Psychology*, *33*, 171–187.

Clark, B. (1988). *Growing up gifted* (3rd ed.) Columbus, OH: Merrill.

Cole, J. C., Kaufman, J. C., & Dang, J. (2009, in press). Ethnic and gender invariance on the KABC-II: A multigroup structural validation for children 7–18 years. *International Journal of Testing*.

Cox, M. V., Perara, J., & Fan, X. (1998). Children's drawing ability in the UK and China. *Psychologia: An International Journal of Psychology in the Orient*, *41*, 171–182.

Cox, M. V., Koyasu, M., Hiranuma, H., & Perara, J. (2001). Children's human figure drawings in the UK and Japan: The effects of age, sex and culture. *British Journal of Development Psychology*, *19*, 275–292.

Cross, S. E. & Markus, H. R. (1999). The cultural constitution of personality. In L. A. Pervin & O. P. John (Eds.), *Handbook of personality: Theory and research* (2nd ed., pp. 378–396). New York: Guilford.

Das, J. P., Naglieri, J. A., & Kirby, J. R. (1994). *Assessment of cognitive process: The PASS theory of intelligence*. Boston, MA: Allyn & Bacon.

Domino, G. (1974). Assessment of cinematographic creativity. *Journal of Personality and Social Psychology*, *30*, 150–154.

Fuchs-Beauchamp, K. D., Karnes, M. B., & Johnson, L. J. (1993). Creativity and intelligence in preschoolers. *Gifted Children Quarterly*, *73*, 113–117.

Furnham, A. (1999). Personality and creativity. *Perceptual and Motor Skills, 88*, 407–408.

Getzels, J. W. & Jackson, P. W. (1962). *Creativity and intelligence: Explorations with gifted students*. New York: Wiley.

Glover, J. A. (1976). Comparative levels of creative ability in Black and White college students. *Journal of Genetic Psychology, 128*, 95–99.

Goldberg, L. R., Sweeney, D., Merenda, P. F., & Hughes, J. E., Jr. (1998). Demographic variables and personality: The effects of gender, age, education, and ethnic/racial status on self-descriptions of personality attributes. *Personality and Individual Differences, 24*, 393–403.

Griffin, M. & McDermott, M. R. (1998). Exploring a tripartite relationship between rebelliousness, openness to experience and creativity. *Journal of Social Behavior and Personality, 26*, 347–356.

Guilford, J. P. (1967). *The nature of human intelligence*. New York: McGraw-Hill.

Guilford, J. P. (1988). Some changes in the structure-of-intellect model. *Educational and Psychological Measurement, 48*, 1–4.

Harty, H., Adkins, D. M., & Sherwood, R. D. (1984). Predictability of giftedness identification indices for two recognized approaches to elementary school gifted education. *Journal of Educational Research, 77*, 337–342.

Heath, S. B. (1983). *Ways with words*. Cambridge: Cambridge University Press.

Herrnstein, R. J. & Murray, C. (1994). *The Bell Curve: Intelligence and class structure in American life*. New York: Free Press.

Heuchert, J. W. P., Parker, W. D., Stumpf, H., & Myburgh, C. P. H. (2000). The Five-Factor Model in South African college students. *American Behavioral Scientist, 44*, 112–125.

Hofstede, G. (2001). *Culture's consequences: Comparing values, behaviors, institutions, and organizations across nations*. Thousand Oaks, CA: Sage.

Horn, J. L. (1989). Cognitive diversity: A framework of learning. In P. L. Ackerman, R. J. Sternberg, & R. Glaser (Eds.), *Learning and individual differences* (pp. 61–116). New York: Freeman.

Horn, J. L. & Cattell, R. B. (1966). Refinement and test of the theory of fluid and crystallized intelligence. *Journal of Educational Psychology, 57*, 253–270.

Horn, J. L. & Hofer, S. M. (1992). Major abilities and development in the adult period. In R. J. Sternberg & C. Berg (Eds.), *Intellectual Development* (pp. 44–99). Cambridge, England: Cambridge University Press.

Horn, J. L. & Noll, J. (1997). Human cognitive capabilities: Gf-Gc theory. In D. P. Flanagan, J. L. Genshaft & P. L Harrison (Eds.), *Beyond traditional intellectual assessment: Contemporary and emerging theories, tests, and issues* (pp. 53–91). New York: Guilford.

Iscoe, I. & Pierce-Jones, J. (1964). Divergent thinking, age, and intelligence in white and Negro children. *Child Development, 35*, 785–797.

Jaquish, G. A. & Ripple, R. E. (1984). A life-span development cross-cultural study of divergent thinking abilities. *International Journal of Aging and Human Development, 20*, 1–11.

Jellen, H. U. & Urban, K. (1989). Assessing creative potential worldwide: The first cross-cultural application of the Test for Creative Thinking-Drawing Production (TCT-DP). *Gifted Educational International, 6*, 78–86.

Jensen, A. R. (1998). *The g factor: The science of mental ability*. Westport, CT: Praeger.

Ji, L., Peng, K., & Nisbett, R. E. (2000). Culture, control, and perception of relationships in the environment. *Journal of Personality and Social Psychology*, 78, 943–955.

Kaltsounis, B. (1974). Race, socioeconomic status and creativity. *Psychological Reports, 35*, 164–166.

Kaufman, A. S. & Kaufman, N. L. (1983). *K-ABC interpretive manual*. Circle Pines, MN: American Guidance Service.

Kaufman, A. S. & Kaufman, N. L. (1993). *Manual for Kaufman Adolescent & Adult Intelligence Test (KAIT)*. Circle Pines, MN: American Guidance Service.

Kaufman, A. S. & Kaufman, N. L. (2004). *Kaufman Assessment Battery for Children—Second Edition (KABC-II) administration and scoring manual*. Circle Pines, MN: American Guidance Service.

Kaufman, A. S. & Lichtenberger, E. O. (2006). *Assessing adult and adolescent intelligence* (3rd ed.). New York: Wiley.
Kaufman, J. C. (2003, August). Gender and ethnic differences on the KABC-II. In M. H. Daniel (Chair), *The KABC-II: Theory, content, and administration*. Symposium conducted at the meeting of the American Psychological Association, Toronto.
Kaufman, J. C. (2006). Self-reported differences in creativity by gender and ethnicity. *Journal of Applied Cognitive Psychology, 20*, 1065–1082.
Kaufman, J. C. & Kaufman, S. B. (2008). *Using intelligent testing to find creativity: Searching for divergent production in IQ tests* (Unpublished manuscript).
Kaufman, J. C., Baer, J., & Gentile, C. A. (2004). Differences in gender and ethnicity as measured by ratings of three writing tasks. *Journal of Creative Behavior, 38*, 56–69.
Kaufman, J. C., Kaufman, A. S., Kaufman-Singer, J. L., & Kaufman, N. L. (2005). The Kaufman Assessment Battery for Children—Second Edition (KABC-II) and the Kaufman Adult and Adolescent Intelligence Test (KAIT). In D. P. Flanagan & P. L. Harrison (Eds.), *Contemporary Intellectual Assessment: Theories, tests, and issues* (2nd ed., pp. 344–370). New York: Guilford.
Kessler, C. & Quinn, M. E. (1987). Language minority children's linguistic and cognitive creativity. *Journal of Multilingual and Multicultural Development, 8*, 173–186.
Kim, K. H. (2005). Can only intelligent people be creative? A meta-analysis. *Journal of Secondary Gifted Education, 16*, 57–66.
King, L. A., McKee-Walker, L., & Broyles, S. J. (1996). Creativity and the five-factor model. *Journal of Research in Personality, 30*, 189–203.
Knox, B. J. & Glover, J. A. (1978). A note on preschool experience effects on achievement, readiness, and creativity. *Journal of Genetic Psychology, 132*, 151–152.
Kyllonen, P. C., Walters, A. M., & Kaufman, J. C. (2005). Noncognitive constructs and their assessment in graduate education. *Educational Assessment, 10*, 153–184.
Li, J. (1997). Creativity in horizontal and vertical domains. *Creativity Research Journal, 10*, 107–132.
Lim, W. & Plucker, J. (2001). Creativity through a lens of social responsibility: Implicit theories of creativity with Korean samples. *Journal of Creative Behavior, 35*, 115–130.
Loehlin, J. C. (2000). Group differences in intelligence. In R. J. Sternberg (Ed.), *Handbook of intelligence* (pp. 176–193). Cambridge: Cambridge University Press.
Luria, A. R. (1970). Thc functional organization of the brain. *Scientific American, 222*, 66–78.
Manly, J. J., Miller, S. W., Heaton, R. K., Byrd, D., Reilly, J., Velasquez, R. J., et al. (1998). The effect of Black acculturation on neuropsychological test performance in normal and HIV positive individuals. *Journal of the International Neuropsychological Society, 4*, 291–302.
Markus, H. & Kitayama, S. (1991). Culture and the self: Implications for cognition, emotion, and motivation. *Psychological Review, 98*, 224–253.
Masten, W. G., Plata, M., Wenglar, K., & Thedford, J. (1999). Acculturation and Teacher Ratings of Hispanic and Anglo-American Students. *Roeper Review, 22*, 64–65.
McCrae, R. R. (1987). Creativity, divergent thinking, and openness to experience. *Journal of Personality and Social Psychology, 52*, 1258–1265.
McCrae, R. R. & Costa, P. T., Jr. (1997). Personality trait structure as a human universal. *American Psychologist, 52*, 509–516.
McGrew, K. S. (2005). The Cattell-Horn-Carroll theory of cognitive abilities: Past, present, and future. In D. P. Flanagan & P. L. Harrison (Eds.), *Contemporary intellectual assessment: Theories, tests, and issues* (2nd ed., pp. 136–181). New York: Guilford.
Mitchell, B. M. (1988). Hemisphericity and creativity: A look at the relationships among elementary-age low-income Hispanic children. *Educational Research Quarterly, 12*, 2–5.
Moreno, J. M. & Hogan, J. D. (1976). The influence of race and social-class level on the training of creative thinking and problem-solving abilities. *Journal of Educational Research, 70*, 91–95.

Morgan, R. & Maneckshana, B. (1996, April). *The psychometric perspective: Lessons learned from 40 years of constructed response testing in the Advanced Placement Program*. Paper presented at the Annual Meeting of the National Council of Measurement in Education in New York.

Naglieri, J. A. & Das, J. P. (1997). *Cognitive Assessment System*. Chicago, IL: Riverside.

Naglieri, J. A. & Kaufman, J. C. (2001). PASS theory, intelligence, and giftedness. *Roeper Review, 23*, 151–156.

Naglieri, J. A., Rojahn, J. R., Aquilino, S. A., & Matto, H. C. (2005). Black European American differences in intelligence: A study of the planning, attention, simultaneous, and successive theory of intelligence. *Journal of Psychoeducational Assessment, 23*, 146–160.

Nisbett, R. E., Peng, K., Choi, I., & Norenzayan, A. (2001). Culture and systems of thought: Holistic vs. analytic cognition. *Psychological Review, 108*, 291–310.

Niu, W.-H. & Sternberg, R. J. (2001). Cultural influences on artistic creativity and its evaluation. *International Journal of Psychology, 36*, 225–241.

Noppe, L. D. (1985). The relationship of formal thought and cognitive styles to creativity. *Journal of Creative Behavior, 19*, 88–96.

Palaniappan, A. K. (1996). A cross-cultural study of creative perceptions. *Perceptual and Motor Skills, 82*, 96–98.

Plucker, J., Runco, M., & Lim, W. (2006). Predicting ideational behavior from divergent thinking and discretionary time on task. *Creativity Research Journal, 18*, 55–63.

Pornrungroj, C. (1992). *A comparison of creativity test scores between Thai children in a Thai culture and Thai-American children who were born and reared in an American culture*. Unpublished doctoral dissertation, Illinois State University.

Preckel, F., Holling, H., & Weise, M. (2006). Relationship of intelligence and creativity in gifted and non-gifted students: An investigation of threshold theory. *Personality and Individual Differences, 40*, 159–170.

Price-Williams, D. R. & Ramirez III, M. (1977). Divergent thinking, cultural differences, and bilingualism. *The Journal of Social Psychology*, (ISSN-0022-4545) *103*, 3–11.

Reynolds, C. R. (2000). Methods for detecting and evaluating cultural bias in neuropsychological tests. In E. Fletcher-Janzen, T. L. Strickland, & C. R. Reynolds (Eds.), *Handbook of cross-cultural neuropsychology* (pp. 249–285). New York: Plenum.

Reynolds, C. R., Lowe, P. A., & Saenz, A. L. (1999). The problem of bias in psychological assessment. In C. R. Reynolds & T. B. Gutkin (Eds.), *The handbook of school psychology* (pp. 549–596). New York: Wiley.

Ripple, R. E. & Jaquish, G. A. (1982). Developmental aspects of ideational fluency, flexibility, and originality: South Africa and the United States. *South African Journal of Psychology, 12*, 95–100.

Rogers, K. B. (1996). What *The Bell Curve* says and doesn't say: Is a balanced view possible? *Roeper Review, 18*, 252–255.

Rostan, S. M., Pariser, D., & Gruber, H. E. (2002). A cross-cultural study of the development of artistic talent, creativity, and giftedness. *High Ability Studies, 13*, 125–156.

Rudowicz, E., Lok, D., & Kitto, J. (1995). Use of the Torrance Test of Creative Thinking in an exploratory study of creativity in Hong Kong primary school children: A cross-cultural comparison. *International Journal of Psychology, 30*, 417–430.

Runco, M. A., Plucker, J. A., & Lim, W. (2001). Development and psychometric integrity of a measure of ideational behavior. *Creativity Research Journal, 13*, 393–400.

Saeki, N., Fan, X., & Van Dusen, L. (2001). A comparative study of creative thinking of American and Japanese college students. *Journal of Creative Behavior, 35*, 24–36.

Saucier, G. & Goldberg, L. R. (2001). Lexical studies of indigenous personality: Premises, products, and prospects. *Journal of Personality, 69*, 847–879.

Schmitt, D. P., Allik, J., McCrae, R.R., & Benet-Martínez, V. (2007). The geographic distribution of Big Five personality traits: Patterns and profiles of human self-description across 56 nations. *Journal of Cross-Cultural Psychology, 38*, 173–212.

Shade, B. J. (1986). Is there an Afro-American cognitive style? An exploratory study. *Journal of Black Psychology, 13*, 13–16.

Singer, D. G. & Singer, J. L. (1990). *The house of make-believe: Children's play and the developing imagination*. Cambridge, MA: Harvard University Press.

Sligh, A. C., Conners, F. A., & Roskos-Ewoldsen, B. (2005). Relation of creativity to fluid and crystallized intelligence. *Journal of Creative Behavior, 39*, 123–136.

Stemler, S. E., Grigorenko, E. L., Jarvin, L., & Sternberg, R. J. (2006). Using the theory of successful intelligence as a basis for augmenting AP exams in psychology and statistics. *Contemporary Educational Psychology, 31*, 344–376.

Sternberg, R. J. (1996). *Successful intelligence: How practical and creative intelligence determine success in life*. New York: Simon & Schuster.

Sternberg, R. J. (2006). How can we simultaneously enhance both academic excellence and diversity? *College and University, 81*, 17–23.

Sternberg, R. J. & The Rainbow Project Collaborators (2006). The Rainbow Project: Enhancing the SAT through assessments of analytical, practical, and creative skills. *Intelligence, 34*, 321–350.

Sternberg, R. J., Kaufman, J. C., & Grigorenko, E. L. (2008). *Applied intelligence*. Cambridge: Cambridge University Press.

Stricker, L. J., Rock, D. A., & Bennett, R. E. (2001). Sex and ethnic-group differences on accomplishment measures. *Applied Measurement in Education, 14*, 205–218.

Strom, R. & Johnson, A. (1989). Hispanic and Anglo families of gifted children. *Journal of Instructional Psychology, 16*, 164–172.

Strom, R., Johnson, A., Strom, S., & Strom, P. (1992). Parental differences in expectations of gifted children. *Journal of Comparative Family Studies, 23*, 69–77.

Torrance, E. P. (1966). *The Torrance Tests of Creative Thinking: Norms-technical manual*. Lexington, MA: Personal Press.

Torrance, E. P. (1971). Are the Torrance Tests of Creative Thinking biased against or in favor of the "disadvantaged" groups? *Gifted Child Quarterly, 15*, 75–80.

Torrance, E. P. (1973). Non-test indicators of creative talent among disadvantaged children. *Gifted Child Quarterly, 17*, 3–9.

Torrance, E. P. (1974). *Torrance Tests of Creative Thinking*. Lexington, MA: Ginn.

Troiano, A. B. & Bracken, B. A. (1983). Creative thinking and movement styles of three culturally homogeneous kindergarten groups. *Journal of Psychoeducational Assessment, 1*, 35–46.

Wechsler, S. (2006). Validity of the Torrance Tests of Creative Thinking to the Brazilian culture. *Creativity Research Journal, 18*, 15–25.

Wolfradt, U. & Pretz, J. E. (2001). Individual differences in creativity: Personality, story writing, and hobbies. *European Journal of Personality, 15*, 297–310.

Yoon, S. N. (2005). Comparing the intelligence and creativity scores of Asian American gifted students with Caucasian gifted students. Unpublished doctoral dissertation, Purdue University, Lafayette, IN.

Zha, P., Walczyk, J. J., & Griffith-Ross, D. A. (2006). The impact of culture and individualism-collectivism on the creative potential and achievement of American and Chinese adults. *Creativity Research Journal, 18*, 355–366.

Zhiyan, T. & Singer, J. L. (1996). Daydreaming styles, emotionality, and the big five personality dimensions. *Imagination, Cognition, and Personality, 16*, 399–414.

Chapter 8
Exploring the Relationships Between Problem-Solving Style and Creative Psychological Climate

Scott G. Isaksen

Richard Florida (2002), a popular U.S. economist, argues that place is the key economic and social organizing unit of modern times and asserts that future models for economic growth need to focus on technology, talent, and tolerance. Technology includes innovation and concentration of high-tech industry. Talent is the number of people in creative occupations—creative capital. Tolerance is about places that are open and accepting and that have an edge in attracting different kinds of people. Implied in these three areas is the interaction of people and place, or person–environment fit.

The purpose of this chapter is to approach the issue of person–environment fit through an operational framework of creativity and innovation. The concept that behavior is a function of both the person and the environment has its roots in the work of Lewin (1936, 1951). A more recent perspective is that taken by Sternberg and Vroom (2002), who discuss the importance of the person–environment issue within the context of leadership.

The person–environment issue has a strong basis in the current ecological approach to creativity research (Isaksen et al., 1993). An ecological approach to creativity research must include consideration of not just the methods and results involved but also the people and context. The practical application of this research program focuses on taking a systemic approach to organizational innovation and transformation (Isaksen & Tidd, 2006).

There is a great variety of approaches to understanding the characteristics of creative people. A current trend is to investigate style and level of creativity. Drawing on research conducted within the Cognitive Styles Project (Isaksen, 2004), Selby, Treffinger, Isaksen, and Lauer (2004) have offered a new way to assess problem-solving style. It is a tool that is used in this study to inquire into people's preferences for how they process information, deal with change, and make decisions. The environment (i.e., context, situation) is examined here with a second assessment tool, one that has grown out of the Creative Climate Project (Isaksen & Ekvall, 2007).

This chapter also presents a review of previous research within both the Cognitive Styles Project and the Creative Climate Project, with a particular focus

S.G. Isaksen (✉)
Creative Problem Solving Group, 6 Grand View Trail, PO Box 648, Orchard Park NY 14127, USA
e-mail: sgiaway@cpsb.com

P. Meusburger et al. (eds.), *Milieus of Creativity*, Knowledge and Space 2,

on previous comparisons of the relation between cognitive style and individual perceptions of the climate for creativity. In addition to sharing new and preliminary results obtained by the two new measurement instruments, this chapter identifies tentative implications for future research and practice.

An Ecological Approach Includes People and Environment

An ecological, or systemic, approach to fathoming creativity offers the greatest likelihood of productive research and practice (Harrington, 1990; Isaksen & Tidd, 2006; Isaksen & Treffinger, 2004). It is concerned with the interaction of several variables within a specific context, much like the ecologist who explores the interactions among living and nonliving components within an ecosystem (Rodgers & Kerstetter, 1974). An explicit goal of the ecological approach to creativity research is to deepen the understanding of and build effectively *on* the multifaceted nature of creativity rather than consider creativity a monolithic construct. The hope is to comprehend the natural interactions among the sources that lead to creative productivity. An ecological approach to creativity research rests simply on the idea that the whole may be greater than the sum of its parts.

Creative productivity and performance can be viewed through at least four major independent lenses. Rhodes (1961) described them as person, product, process, and press. Numerous other scholars have described them similarly (MacKinnon, 1978; Mooney, 1963; Raven, 1984; Stein, 1968). The interactionist approach used in this chapter includes five categories of contingencies that build upon these early frameworks. The categories are personal orientation, situational outlook, task, creative problem-solving process, and outcomes.

Personal orientation includes an emphasis on understanding the style preferences, competencies, motivation, personality traits, characteristics, and knowledge or expertise of the people involved in creative problem-solving. This contingency subsumes all aspects relating to cognitive, affective, motivational, and individual difference. My colleagues and I have focused on the individual differences that affect the learning and application of creative problem-solving.

Situational outlook encompasses many elements surrounding the context within which creative problem-solving will be learned or applied. They are culture, climate, organizational structure, physical environment, resources, constraints, and systems, to name a few. Situational outlook is the contingency that provides the opportunity to improve the understanding of the place or press aspect of the creativity system.

The third category of contingency is what I call task. It encompasses the content domain within which creative problem-solving will be applied. The preliminary analysis of task includes an understanding of the nature of the intended outcomes and the factors influencing success. This category of contingency includes the task's importance, kind, and degree of ownership; the degree of ambiguity, novelty, or complexity; and the vision of the desired future state, among other elements. The task is the immediate focus of attention for the integration of the other contingencies. It is the work that needs to be done.

The contingency category referred to as the creative problem-solving process has to do with the balance required between generating (or divergence) and focusing (or convergence), the tools and techniques to be deployed, and a descriptive framework of problem-solving components. This category of contingency has a 50-year tradition of research and development and now includes a variety of tools and techniques; guidelines for generating and focusing; and an open, descriptive, process framework. It now also includes a metacognitive component to appraise the task and design the specific application of the process (Isaksen & Treffinger, 2004).

The final contingency category, outcomes, focuses on the degree of diffusion required; the blend of novelty, usefulness, and resolution in the outcome itself; and the degree of satisfaction and enjoyment that is necessary. The task contingency is the more local focus of attention, whereas the outcomes contingency addresses the characteristics of what results from the application of creative problem-solving.

The ecological or interactionist approach to creativity research describes a rather broad design to guide inquiry (Puccio & Murdock, 1999; Woodman & Schoenfeldt, 1999). The aim has been to increase the understanding of which creative problem approaches work best for whom and under what circumstances. Within this broad design, attention has concentrated particularly on two of the contingencies: (a) people and personal orientation, and (b) place, or situational outlook.

Problem-Solving Style Is a Way to Understand People

The endeavor to understand creative people has a rather long and substantial history. The terms creativity and imagination appear in writings as early as those of the ancient Greeks and Romans. Modern interest in creativity among educators and psychologists is widely thought to have been kindled in the mid-twentieth century partly by J. P. Guilford's presidential address to the American Psychological Association in 1950 (Guilford, 1987).

The major historical approach to understanding creativity in people has centered on identifying the traits, characteristics, and other personal attributes that distinguish eminently creative people from their less creative counterparts. Much of this work includes an emphasis on the cognitive characteristics or intellectual patterns and mechanisms that guide and direct the person's intellectual processes or activities (Boden, 1992; Dacey, 1989; Guilford, 1967; Runco, 1991; Torrance, 1987). Other work explores personality traits, values, temperament, and motivational dispositions that influence the ends to which people direct their thinking (Amabile, 1983; Barron, 1969; MacKinnon, 1978). A third realm of work comprises the study of biographical events and life experiences leading to creative achievement (Csikszentmihalyi, 1996; Davis, 1998; Gardner, 1993; Simonton, 1987).

Much of the research into the characteristics of creative people has focused on high-level creativity (Albert, 1983; Cox & Terman, 1926; Goertzel et al., 1978). These efforts have included a focus on understanding genius (Galton, 1869) and differentiating between varying levels of creativity. Stein (1983) uses a lower-case "c" to refer to generally distributed creativity and an uppercase "C" to refer to the ex nihilo level. Boden (1994) drew a similar distinction, suggesting *H-creativity* to denote creativity of a historically significant level and *P-creativity* to denote new and meaningful creativity of the more general type.

A more recent line of inquiry has added a complementary, but distinct, perspective on the pursuit of understanding creativity in people. Offering the sharpest distinction yet between level and style of creativity, Kirton (1994) asserts that level focuses on capacity or degree and on the question "How creative are you?" Style focuses on preference or modality and answers the question "How are you creative?" This emerging line of inquiry has its proponents (Isaksen, 2004) and its opponents (Kaufmann, 2004).

The Cognitive Styles Project, major research initiated at the Center for Studies in Creativity in the 1980s, investigated the relationship of psychological type, cognitive styles, and learning styles of those engaged in creative problem-solving. The main thrust was to sharpen insight into individual differences in the ways people learn and apply creative problem-solving. Previous research has already established that the level of creativity can be enhanced through deliberate instruction (Parnes, 1987; Torrance, 1987). Some interesting individual differences were found between participants who stayed with the experimental program and those who dropped out. Various assessments were used, and meaningful and significant differences were found in a range of studies (Isaksen, 2004).

One of the most recent outgrowths of this research project has been the development of a new measurement tool, *VIEW: An Assessment of Problem Solving Style*™ (hereafter referred to as VIEW).[1] For its development Selby, Treffinger, Isaksen, and Lauer (2004) defined problem-solving styles as consistent individual differences in the ways people prefer to plan and carry out generating and focusing activities in order to gain clarity, produce ideas, and prepare for action. VIEW assesses three independent dimensions of problem-solving style.

Orientation to Change

The first dimension in VIEW is the orientation to change (OC). It addresses three questions: "How do I prefer to deal with boundaries and parameters?" "How do I feel about and react to structure?" and "How do I prefer to respond to novel

[1] Conflict of interest: The author derives monetary benefit from the distribution of *View: An Assessment of Problem Solving Style*™ and the *Situational Outlook Questionnaire*® described in this chapter.

challenges?" Scores below the mean in this dimension indicate an explorer style. In ordinary language an explorer is an individual who thrives on venturing in uncharted directions, seeks to break new ground, and follows adventurous or promising new possibilities wherever they may lead. Scores above the mean on the OC scale indicate a developer style. In ordinary language a developer is an individual who brings tasks to fulfillment. It is a person who begins with the basic elements or ingredients and then organizes, synthesizes, refines, and enhances them, forming or shaping them into a more complete, functional, useful condition or outcome.

Manner of Processing

The second dimension in VIEW is called manner of processing (MP). It, too, addresses three questions: "How do I prefer to manage information and its flow when problem-solving?" "When do I share my thinking?" and "Does interacting with others build or spend energy?" Scores below the mean indicate a preference for an external style of processing. Individuals who exhibit a well-developed preference for this style draw their energy from interaction with others, discussing possibilities, and building from the ideas of others. Scores above the mean reflect a preference for an internal style of processing. Those with a well-developed internal style look first reflectively to their own inner resources and draw energy from their reflection.

Ways of Deciding

The third dimension within VIEW is called ways of deciding (WD) and addresses such questions as "What factors get first priority when I focus or decide?" "Where do I start?" and "How do I make trade-offs?" Scores on this scale indicate whether one's primary focus in decision-making is on "people" or "task." Individuals with scores below the mean tend to adopt a people style as their primary emphasis when deciding. They consider first the impact that choices and decisions have on people's feelings and support and on the need for harmony and positive relationships. Scores above the mean indicate a focus on a task style. Individuals with a task style tend to look first at choices and decisions that are logical, sensible, and objectively justifiable. They prefer making impersonal judgments resting on well-reasoned conclusions.

VIEW has strong conceptual foundations, has demonstrated acceptable psychometric properties (Selby et al., 2007), and offers high potential for future research and practical applications. It has already been applied in order to improve understanding of how style preferences of the people involved in creative problem-solving affect the learning and application of creative problem-solving tools, guidelines, and process (Isaksen & Geuens, 2007).

Psychological Climate Is a Way to Understand Context

Context can be taken to mean something as broad as society, zeitgeist, or national culture and something very limited, such as the working climate within a team. When interpreting context in its broad sense of culture, writers have offered various definitions (Hofstede, 2001; Trompenaars & Hampden-Turner, 2004). There are consistent themes within that diversity, however. In general, culture is seen as something that all or most of the members of some social group share and that older members usually try to pass on to younger members. It is usually regarded as something that shapes behavior and structures perceptions of the world.

Organizational culture is a concept different from that of culture in its generic sense. Most people have exercised a choice to join a place of work, whereas people are born into particular societies. People employed in organizations usually have limits on how much time they spend at work and have other discretionary time available. They are generally free to leave an organization and may do so more easily than they can leave a society. As a concept, organizational culture describes the shared mental programming of people within the same organization, particularly if they share the same nationality. Organizational cultures can differ in meaningful ways, and they are formed by founding leaders, learning experiences of members, and new beliefs and values brought into the organization by new members and leaders (Schein, 1992).

Organizational climate is the recurring patterns of behavior, attitudes, and feelings that characterize life in the organization. Climate exists objectively in the organization and can be observed and studied in a number of different ways. Climate is a manifestation of culture and can be viewed as an intervening variable. As such, climate is affected by numerous other variables within the organization such as the people, resources, concepts, and the physical environment. Climate exerts a direct affect on a variety of organizational and psychological processes such as problem-solving and communication, which, in turn, affect the quality, profitability, and productivity of the organization.

As a rather broad and inclusive concept, culture in its generic sense subsumes climate (Denison, 1996). Culture is usually considered within the discipline of anthropology; climate, within the discipline of social psychology. Cultural dimensions have remained relatively descriptive, meaning that one set of assumptions or values is neither better nor worse than another. Climate is usually normative in that people generally look for environments that are not just different but better for certain things than for others. Lastly, culture is such a deep and stable concept that climate is more easily observed and influenced than culture.

Within the ecological research framework described above, the contingency of situational outlook has been approached through use of the *Situational Outlook Questionnaire®* (SOQ). The translation, validation, and development of the SOQ have been major tasks within the Creative Climate Project initiated at the Center for Studies in Creativity and currently being undertaken by the Creative Problem Solving Group.

The SOQ has grown out of more than 50 years of research and development and currently has nine dimensions (Isaksen, 2007b; Isaksen & Ekvall, 2007).

Challenge and Involvement

The dimension of challenge and involvement refers to the degree to which people are involved in daily operations, long-term goals, and visions. High levels of challenge and involvement mean that people are intrinsically motivated and committed to contributing to the success of the organization. People find joy and meaning in their work, and therefore invest a great deal energy. In the opposite situation people are not engaged and feelings of alienation and indifference are present. The common sentiment and attitude is apathy and lack of interest in that work and interaction are both dull and listless.

Freedom

Freedom is defined as the independence of behavior exhibited by the people in the organization. In a climate with a high level of freedom, people are given autonomy to define much of their own work. People are able to exercise discretion in their day-to-day activities. People take the initiative to acquire and share information and to make plans and decisions about their work. In the opposite climate people work within strict guidelines and roles. People carry out their work in prescribed ways with little room to redefine their tasks.

Trust and Openness

The dimension of trust and openness refers to emotional safety in relationships. When there is a high degree of trust, individuals can be genuinely open and frank with one another. People sincerely respect one another and can count on each other for personal support. Where trust is missing, people are suspicious of each other and therefore closely guard themselves and their ideas. People in that situation also find it extremely difficult to communicate openly with each other.

Idea Time

The dimension of idea time is defined as the amount of time people can use (and do use) for elaborating new ideas. In situations with a great amount of idea time, there are possibilities to discuss and test sudden insights and fresh suggestions that are not planned or included in the task assignment. There are opportunities to take the time to explore and develop new ideas. Flexible timelines permit people to explore new avenues and alternatives. In the reverse case every minute is booked and specified. The time pressure makes thinking outside the instructions and planned routines impossible.

Playfulness and Humor

Playfulness and humor refer to the level of spontaneity and ease displayed within the workplace. A relaxed atmosphere where good-natured jokes and laughter often occur is indicative of this dimension. People can be seen having fun at work. The atmosphere is seen as easy-going and light-hearted. The opposite climate is characterized by gravity and seriousness. The atmosphere is stiff, gloomy, and cumbrous. Jokes and laughter are regarded as improper and intolerable.

Conflict

Conflict is defined as the presence of personal and emotional tensions in the organization. When the level of conflict is high, groups and individuals dislike and may even hate each other. The climate can be characterized by interpersonal warfare. Plots, traps, and struggles for power and territory are usual elements in the life of the organization. Personal differences spawn gossip and slander. In the opposite case people behave in a more mature manner. They have psychological insight and control over their impulses. People accept and deal effectively with diversity.

Idea Support

Idea support refers to the ways in which new ideas are treated. In the supportive climate ideas and suggestions are received in an attentive and professional way by bosses, peers, and subordinates. People listen to each other and encourage initiatives. Possibilities for trying out new ideas are created. The atmosphere is constructive and positive when new ideas are considered. When idea support is low, the automatic "no" prevails. Every suggestion is immediately refuted with a destructive counterargument. Fault-finding and obstacle-raising are the usual styles of responding to ideas.

Debate

Debate is the occurrence of encounters and disagreements between viewpoints, ideas, and differing experiences and knowledge. In the debating organization many voices are heard and people are keen on putting forward their ideas for consideration and review. People can often be seen discussing opposing opinions and sharing diverse perspectives. Where debates are missing, people follow authoritarian patterns without question.

Risk-Taking

The dimension of risk-taking is defined as the tolerance of uncertainty and ambiguity exposed in the workplace. In the high risk-taking case bold new initiatives can be taken even when the outcomes are unknown. People feel as though they can take a gamble on some of their ideas. People will often go out on a limb and put an idea forward. In a risk-avoiding climate there is a cautious, hesitant mentality. People try to be on the safe side. They decide to sleep on the matter. They set up committees and cover themselves in many ways before making a decision.

Previous Research on Linkages Between Organizational Climate and Problem-Solving Style

Organizational climate has been the subject of vast research and inquiry, as has cognitive and problem-solving style. There has been much less inquiry that seeks to examine the potential conceptual and empirical linkages between these two domains.

The person–environment fit is a domain that stems from the basic argument that human behavior is attributable either to characteristics of the person or to the environment (Lewin, 1936, 1951; Murray, 1938). The keystone of this domain is the notion that behavior is influenced by both intrapersonal characteristics *and* the environment (Caplan, 1983; Choi, 2004, Edwards et al., 2006; Holland, 1966; Pervin, 1987; Puccio et al., 1995; Schneider, 1987a). The ecological approach to creativity research has strong conceptual linkages to the domain of person–environment fit and suggests the need to explore the relationships between personal orientation and situational outlook.

Until recently, these two contingencies had remained largely independent in the creativity literature. Exceptions are the few sources dealing with cognitive climate (e.g., Kirton & McCarthy, 1988) and a series of articles reporting the results of investigation into the relationship of cognitive style and individual psychological climate (Clapp & Kirton, 1994; Isaksen & Kaufmann, 1990; Isaksen & Lauer, 1999).

The original study (Isaksen & Kaufmann, 1990) used the Kirton Adaption–Innovation Inventory (Kirton, 1976)—KAI—and an early version of the SOQ as the measures of cognitive style and psychological climate, respectively. KAI, a measure of cognitive style, assesses one principal continuum with two styles: an adaptive preference (i.e., the individual stays within boundaries and endeavors to improve performance) and an innovative preference (which implies ease of thinking across paradigms and of doing things differently). The findings arrived at through correlation analysis did not reveal any strong significant relationships between the KAI and the SOQ. Discriminant function analysis performed on findings related to rather extreme scores on the KAI revealed that adaptors perceived more challenge than innovators and that innovators perceived more conflict than adaptors.

Clapp and Kirton (1994), however, challenged the theoretical relationship of the two instruments used in the original study. They called attention to key points, both theoretical and methodological, requiring further explanation and investigation. Their work prompted the second study by Isaksen and Lauer (1999), in which the authors used a sample of 646 subjects. They reported findings similar to those of their original study and were able to clarify the nature of the relationship between cognitive style and individual psychological climate.

Because Isaksen and Lauer had clearly different groups according to KAI theory, they subjected these two groups to discriminant analysis (Hair et al., 1987) to determine whether there were any statistically significant and meaningful differences in their orientation to individual psychological climate. (The discriminant function allows for analysis of both groups across all climate variables.) A difference in how the two groups view climate would mean that they are likely to remain distinctly separate with regard to any particular climate dimension. The results of the discriminant analysis showed that the classification of the participants into their respective groups was 58.8% correct. The dimension of challenge and involvement and that of conflict were found to be optimal predictor variables and were consistent with the original study. These results indicated that adaptors experienced more challenge and involvement and more risk-taking within their individual psychological climates than innovators did in their own. Innovators experienced more conflict within their climates than adaptors did in theirs.

Because previous research utilized only one main dimension of style to assess the personal orientation contingency against the nine dimensions of climate assessed by the SOQ, further research using a more robust measure of style is needed.

The Current Study

Participants

The sample for this exploratory study consisted of 144 participants who had completed both the VIEW and SOQ assessments. It was an aggregated group of samples of convenience including participants from six different organizations. A total of 70 participants came from a leadership program within a global communications and advertising company, 30 from a leadership program within a global electrical engineering company, and the remaining 44 from four other organizations involved in workshops on managing change. The average age of the participants was 33. The sample included 82 men, 43 women and 19 people who declined to indicate gender. These samples of convenience were selected because each of the events included a diversity of participants drawn from North America, Europe, the Middle East, and Africa.

Materials

VIEW. The measure of problem-solving style used in this study was VIEW. VIEW is a 34-item instrument on which respondents are asked to indicate the degree to which two opposing descriptions reflect their answer to the following statement: "When I am solving problems, I am a person who prefers …" The items are scored on a 7-point scale ranging from 1 (*not at all applicable*) to 7 (*highly applicable*). The theoretical range of scores for the OC dimension is 18 for the strongest explorer style and 126 for the strongest developer style. The range for the MP dimension is 8 for the strongest external style and 56 for the internal style. The range for the WD dimension is 8 for the strongest people-oriented style and 56 for the task-oriented style. The theoretical mean for the OC scale is 72. The theoretical mean for the two other dimensions is 32. There is sound evidence of VIEW's test–retest and internal reliability, factor structure, and concurrent validity. For a more complete report regarding VIEW's psychometric properties, see Selby, Treffinger, and Isaksen (2007).

SOQ. The measure of creative climate was the SOQ, which is designed to assess situational conditions related to creativity, innovation, and change in such a way that it does not prescribe the perfect climate for all situations. Rather than providing a simplistic and ubiquitous normative goal toward which everyone should strive, this approach to measurement of climate provides a profile aimed at gauging the current situation for the purpose of developing and implementing organizational improvement initiatives.

The SOQ stems from research by Göran Ekvall, who studied the psychosocial aspects of the work environment and developed the Creative Climate Questionnaire (CCQ). The version of the SOQ used in this study contains 53 close-ended questions designed to measure the nine dimensions of climate described earlier. In addition, the SOQ contains three open-ended narrative questions for which the participants individually describe what is helping and hindering their creativity at work and what actions they would take to ameliorate these conditions. The 53 items are scored on a 4-point scale from 0 (*not at all applicable*) to 3 (*highly applicable*). Scores are reported on a scale ranging from 0 (*not at all applicable*) to 300 (*highly applicable*) for each of the nine dimensions. The narrative comments are subjected to qualitative analysis.

Studies of the SOQ's validity and reliability have been conducted (Isaksen & Ekvall, 2007). Some of them have indicated that organizations described as "innovative" in terms of productivity differ substantially from those described as "stagnant." Higher scores on the eight positive dimensions and a lower score on the negative dimension (conflict) indicate a climate that tends to be conducive to creativity. Scores have consistently discriminated between organizations that are successful at developing new products or services from those that are not. The SOQ is, however, not a direct measure of organizational stagnation or progressiveness.

Isaksen, Lauer, Ekvall, and Britz (2001) found consistent patterns of response across work situations deemed by the respondents to be the "best case" or the "worst case" they have experienced. Across these studies, the researchers found that there was generally no "ideal" score for any of the dimensions of creative climate. A "perfect" score of 300 on the positive dimensions, combined with a perfect score of 0 on the negative dimension (conflict) did not necessarily indicate a best-case scenario.

Therefore, results of the SOQ are not designed to indicate a theoretical or cross-situational ideal. Rather, they act as a barometer, gauging the general perception of how these dimensions are perceived within a given climate. Nor are the results to be treated as though the dimensions all fall on a single continuum. Factor analysis has repeatedly revealed multiple independent dimensions associated with the climate conducive to change and creativity (Isaksen & Ekvall, 2007). The scores on the SOQ are best used as a profile and can help identify strengths and potential weaknesses within any specific working situation (Isaksen, 2007a).

At an organizational level of analysis, the dimensions of the SOQ have significantly discriminated between organizations that were innovative in terms of their ability to develop and commercialize new products and those that were stagnant (Isaksen & Ekvall, 2007; Nyström & Edvardsson, 1980; see Table 8.1).

The SOQ is used primarily to assess an organizational level attribute of climate. For the purposes of this study, the SOQ results were used at an individual level of analysis. The SOQ was thus applied as a measure of individual psychological climate in order to keep the results conceptually parallel with the individual measure of problem-solving style (Brown & Leigh, 1996).

Climate researchers have drawn a distinction between organizational climate and psychological climate. In reviewing organizational climate research and theory, James and Jones (1974) identified the term psychological climate to be consistent with the measurement approach in which individuals are asked to report their perceptions of the work environment and their preferred term when these perceptions

Table 8.1 Mean SOQ[a] scores achieved by innovative and stagnant organizations

Climate	Innovative $N = 10$ (630)[a]	Stagnant $N = 5$ (275)	Difference
Challenge	238 (237)	163 (164)	75 (73***)
Freedom	210 (209)	153 (155)	57 (54**)
Idea support	183 (182)	108 (111)	75 (71***)
Trust	178 (180)	128 (130)	50 (50*)
Dynamism	220 (224)	140 (141)	80 (83***)
Playfulness	230 (233)	140 (139)	90 (94***)
Debates	158 (156)	105 (104)	53 (52**)
Conflicts	78 (79)	140 (140)	–62 (61***)
Risk-taking	195 (194)	53 (55)	142 (139***)

[a] *Situational outlook questionnaire.*

[b] Parentheses enclose means expressed at the level of the individual respondent

*$p = < .05$, **$p < 01$, ***$p < .001$.

are used as the unit of analysis. The researchers suggested that the term organizational climate be used to refer to the organizational attribute and that the term psychological climate be used to refer to the individual characteristic. James and Sells (1981) defined psychological climate as:

> individuals' cognitive representations of relatively proximal situational events, expressed in terms that reflect the psychological meaning and significance of the situation to the individual. A central postulate of psychological-climate theory is that individuals tend to interpret situations in psychological terms; that is, to assign psychological meaning to environmental attributes and events. Psychological climate is regarded as an attribute of the individual. (p. 275)

By applying the SOQ results as indicators of individual psychological climate and by investigating individual differences of problem-solving style, I am considering two concepts at the same level of analysis.

Results and Discussion

This study represents the first attempt to examine problem-solving style and individual psychological climate by using VIEW and the SOQ. The design of the analysis followed a procedure similar to that in previous studies on cognitive style and individual psychological climate. The descriptive statistics for both measures appear in Table 8.2. The sample was very close to the theoretical mean on OC, but was slightly more external on the MP dimension, and slightly more task-oriented on the WD dimension. Correlations were computed to determine the degree of relationship between the two sets of variables. Only two correlations reached the .05

Table 8.2 Descriptive statistics on the dimensions of VIEW[a] and SOQ[b] (N = 144)

	Range of scores			
Dimension	Minimum	Maximum	Mean	SD
		SOQ		
Challenge and involvement	71	300	211.51	48.71
Freedom	17	267	166.09	49.69
Trust and openness	80	280	181.67	47.92
Idea time	0	267	129.63	59.24
Playfulness and humor	17	300	179.05	60.36
Conflict	0	267	90.74	61.32
Idea support	20	300	188.19	55.94
Debate	33	300	200.93	55.01
Risk-taking	20	280	141.53	54.87
		VIEW		
Orientation to change	42	108	72.40	13.91
Manner of processing	12	45	26.09	7.45
Ways of deciding	20	55	36.11	7.83

[a] *View: an assessment of problem solving style.*
[b] *Situational outlook questionnaire.*

level of significance. Explorers had a low correlation with seeing more challenge and involvement in their climates ($r = .21$; $p \leq .01$). Externals had a low correlation with idea support ($r = -.18$; $p \leq .05$).

Because only two significant correlations were found among the 27 possible relationships in the entire sample, further study was conducted to determine whether participants with pronounced problem-solving preferences produce more varied scores on the SOQ. To this end, tests of the equality of group means were administered to those individuals who scored outside the middle standard deviation on each of the three style dimensions (see Table 8.3). The results of 50 participants who scored within one standard deviation were removed from the statistics, so this sample had 48 with a fairly strong explorer style and 46 with a fairly strong developer style.

The next level of analysis was to test the equality of group means by comparing the two stronger preference groups on the OC dimension with the nine dimensions of the SOQ. With 63.8% of the groups correctly classified, no significant differences were found between strong explorers and developers and the nine dimensions of the SOQ.

Table 8.4 contains the descriptive statistics for the MP dimension of VIEW and the scores for the nine dimensions of the SOQ. Forty-six participants were excluded through elimination of those participants with MP scores within the middle standard deviation.

Again, the next level of analysis was to test the equality of group means by using the scores of the participants who were clearly internals and externals. With 67% of the two groups correctly classified, no significant differences were found. The one classification closest to reaching significance was the idea-support dimension of the SOQ, indicating that those with an external preference perceived more idea support in their climates.

Table 8.5 contains the descriptive statistics for the WD dimension of VIEW and the scores on the SOQ. Fifty-four participants were excluded in order to obtain clear preferences on the WD dimension.

Table 8.3 Descriptive statistics for the OC[a] dimension of VIEW[b] and scores on the nine climate dimensions of the SOQ[c]

	Fairly strong explorer (n = 48)		Fairly strong developer (n = 46)	
SOQ dimension	Mean	*SD*	*Mean*	*SD*
Challenge and involvement	215.77	41.49	200.62	51.90
Freedom	170.49	42.00	155.43	60.46
Trust and openness	175.42	49.38	185.22	48.89
Idea time	126.39	62.16	120.65	69.14
Playfulness and humor	182.29	55.27	180.80	67.40
Conflict	94.10	59.40	95.65	68.80
Idea support	192.92	51.78	181.74	57.94
Debate	197.57	52.59	193.12	57.26
Risk-taking	138.75	55.41	128.26	53.26

[a] Orientation to change.
[b] *View: an assessment of problem solving style.*
[c] *Situational outlook questionnaire.*

Table 8.4 Descriptive statistics for the MP[a] dimension of VIEW[b] and scores for the nine climate dimensions of the SOQ[c]

SOQ dimension	Fairly strong external manner of processing (n = 54)		Fairly strong internal manner of processing (n = 44)	
Challenge and involvement	213.49	51.58	201.95	52.16
Freedom	169.14	53.04	158.71	55.51
Trust and openness	181.11	41.56	180.00	52.30
Idea time	127.78	62.53	129.17	56.51
Playfulness and humor	187.04	61.91	175.00	58.79
Conflict	92.59	62.25	82.58	54.63
Idea support	196.67	49.72	176.36	59.65
Debate	203.09	50.89	184.09	59.93
Risk-taking	143.33	58.76	139.55	57.30

[a] Manner of processing.
[b] *View: an assessment of problem solving style.*
[c] *Situational outlook questionnaire.*

Table 8.5 Descriptive statistics for the WD[a] dimension of VIEW[b] and scores on the nine climate dimensions of the SOQ[c]

	Fairly strong people-oriented style (n = 37)		Fairly strong task-oriented style (n = 53)	
SOQ Dimension	Mean	*SD*	Mean	*SD*
Challenge and involvement	224.32	41.63	209.97	56.28
Freedom	172.52	41.61	170.13	54.83
Trust and openness	185.95	54.34	180.75	45.06
Idea time	127.48	60.12	132.08	56.22
Playfulness and humor	178.38	58.12	170.75	60.44
Conflict	89.64	60.90	85.85	63.75
Idea support	195.14	59.52	186.79	50.07
Debate	205.41	60.74	197.80	47.37
Risk-taking	150.27	54.29	140.00	55.61

[a] Ways of deciding.
[b] *View: an assessment of problem solving style.*
[c] *Situational outlook questionnaire.*

A test of the equality of group means was conducted with the participants who scored relatively high as people- or task-oriented deciders. With 60.5% of the two groups correctly classified, no significant differences were found.

Even with individuals with very clear problem-solving preferences, none of the 54 potential differences in the climate dimensions was significant. The first major implication from this exploratory study is that the two constructs of individual psychological climate and problem-solving style are distinct, from a quantitative perspective. Problem-solving style and individual perceptions of the climate may both certainly have an effect on behavior, but the two measures used in this study allow relatively independent treatment of the two concepts.

Access to qualitative data permitted further analysis, including constant comparison and open coding. Most topics cut across style differences, but there were a few instances of divergence. Table 8.6 provides sample quotations from the qualitative analysis, which illustrate some meaningful differences in the way the three narrative questions were answered by participants with pronounced problem-solving style preferences.

Despite the lack of significant quantitative differences, clear and understandable *qualitative* differences emerge in all three of VIEW's dimensions when it comes to individual perceptions of psychological climate. At one level, this finding suggests that the SOQ can pick up on these differences because it is a multimethod assessment. From another point of view, these results suggest that situational outlook and personal orientation remain conceptually and empirically distinct from each other.

It is quite plausible that individuals within the same work group would not assign similar meaning to their work environment. These differences in perception could stem from a variety of variables, including different exposure to tasks, events, or other situational attributes; differences in social roles; or individual differences in personality (James & Sells, 1981). Clearly, further work remains to be done to deepen the understanding of the relationship between a measure of problem-solving style and psychological climate in order to sort out exactly how individuals of different styles may perceive their environments. Further research may also include multivariate assessment of some of the other variables within both situational outlook and personal orientation.

Differences within or between organizations are often attributed to traits and characteristics of the people involved or to the situation and aspects of the climate. However, some researchers (e.g., Schneider, 1987b) believe there has been too much emphasis on either the characteristics of the person or the situation as the fundamental determinant of organizational behavior. These researchers point to the domain of person–environment fit as a more useful approach to widening the understanding of the causes of organizational behavior (Caplan, 1987; Pervin, 1987). In addition, neither of the investigations into just the person or the environment will provide a satisfactory way to understand the origins of meaning within the social situation. The individual and the environment interact with each other, so sharp distinctions and entirely independent lines of inquiry may not be as fruitful as considering a rather holistic relationship between the two concepts (Epstein & O'Brien, 1985; Schneider & Reichers, 1983). In short, the issue of the person–environment fit is a joint function of what the individual has to offer and the situational factors that are relevant.

Although seeking to understand the climate for creativity can be seen as a productive line of inquiry, the situation or environment is clearly only part of the challenge. If researchers are to understand the conditions for creativity, they must go beyond the individual psychological perceptions of the social or work setting and consider the characteristics of the person as well. It is reasonable to assert that the perception of a given social setting or event may differ radically, depending on the orientations of the persons involved (Hennessey & Amabile, 1988). Further work must be done to identify and understand these differences. Future research should also inquire into the issue of the person–environment fit in relation to other contingencies within the ecological framework. In terms of situational outlook, it may also be beneficial to examine other levels of analysis like groups, teams, and organizations.

Table 8.6 Sample narrative responses to open-ended questions of the SOQ[a] (in terms of VIEW[b] problem-solving style)

VIEW style	What helps your creativity at work?	What hinders your creativity at work?	Suggestions for ameliorating the conditions
Developer	My managers' and peers' encouragement to focus more on creativity and innovation	Lack of clear goals and information. The level of uncertainty	More management support and planning to build deliberate time for creative thinking
Explorer	The high amount of leeway to make my own decisions and set my course in my work	Long delays in senior management decision-making and a stringent planning process	I am not sure about the action, but the result I want is to motivate people to become more curious and committed to do things they have not done before
Internal	My personal work ethic and right of self-determination—the opportunity to work outside the office to avoid interruptions by e-mail, phone calls, visitors, and meetings	Time and energy I have to spend negotiating, reporting, relating with different people and different times—often on the same issue	We need time and space to let our body–mind–soul connect—with the right level of energy—to projects and work in order for ideas to rise from within
External	Constant dialog—working and discussing possibilities with many diverse teams	Not having all the people I work with at the same location. It's easier to communicate and ask questions when you can interact with them in person	More space and social events to encourage everyone to interact more with each other
People-oriented	High levels of trust, openness, and support—people with open minds	Poor collaboration—no shared agendas and ideas not treated with respect	Stay truthful to my notions of how to behave towards each other as human beings—keep my optimistic energy to solve creative tasks with a free flow of ideas
Task-oriented	Clear understanding of all the requirements to provide a solution	Not enough time to do great things—medium effort for everything	Gather all the initiative sponsors in one room, list all the projects and initiatives, and decide once and for all what we can realistically accomplish

[a] *Situational outlook questionnaire.*
[b] *View: an assessment of problem solving style.*

There are many limitations to exploratory research of this kind. For example, the sample size of the present study was limited and did not reflect a strict normal distribution for either measure. The findings must be considered tentative until larger samples are used and the results replicated. Nevertheless, both the demonstrated lack of any meaningful individual differences in perceptions of climate based on varying problem-solving styles and the nuances identified in the narrative data indicate the benefits of taking a multimethod approach to creativity research. Researchers studying ecological creativity should keep these advantages clearly in mind.

References

Albert, R. S. (1983). (Ed.), *Genius and eminence: The social psychology of creativity and exceptional achievement*. New York: Pergamon.
Amabile, T. M. (1983). *The social psychology of creativity*. New York: Springer.
Barron, F. (1969). *Creative person and creative process*. New York: Holt, Rinehart & Winston.
Boden, M. A. (1992). *The creative mind: Myths and mechanisms*. New York: Basic Books.
Boden, M. A. (1994). What is creativity? In M. A. Boden (Ed.), *Dimensions of creativity* (pp. 75–118). Cambridge, MA: MIT Press.
Brown, S. P. & Leigh, T. W. (1996). A new look at psychological climate and its relationship to job involvement, effort, and performance. *Journal of Applied Psychology*, *81*, 358–368.
Caplan, R. D. (1983). Person-environment fit: Past, present, and future. In C. L. Cooper (Ed.), *Stress research: Issues for the eighties* (pp. 35–77). New York: Wiley.
Caplan, R. D. (1987). Person-environment fit theory and organizations: Commensurate dimensions, time perspectives, and mechanisms. *Journal of Vocational Behavior*, *31*, 248–267.
Choi, J. N. (2004). Person-environment fit and creative behavior: Differential impacts of supplies–values and demands–abilities versions of fit. *Human Relations*, *57*, 531–552.
Clapp, R. G. & Kirton, M. J. (1994). The relationship between cognitive style and psychological climate: Some comments on the article by Isaksen & Kaufmann. *Studia Psychologica*, *36*, 129–134.
Cox, C. & Terman, L. M. (1926). *The early mental traits of three hundred geniuses: Genetic studies of genius* (Vol. 2). Stanford, CA: Stanford University Press.
Csikszentmihalyi, M. (1996). The creative personality. *Psychology Today*, *29*, 36–40.
Dacey, J. S. (1989). *Fundamentals of creative thinking*. Lexington, MA: D. C. Heath.
Davis, G. (1998). *Creativity is forever* (4th ed.). Dubuque, IA: Kendall/Hunt.
Denison, D. R. (1996). What is the difference between organizational culture and organizational climate? A native's point of view on a decade of paradigm wars. *Academy of Management Review*, *21*, 619–654.
Edwards, J. R., Cable, D. M., Williamson, I. O., Lambert, L. S., & Shipp, A. J. (2006). The phenomenology of fit: Linking the person and the environment to the subjective experience of person-environment fit. *Journal of Applied Psychology*, *91*, 802–827.
Epstein, S. & O'Brien, E. J. (1985). The person-situation debate in historical and current perspective. *Psychological Bulletin*, *98*, 513–537.
Florida, R. (2002). *The rise of the creative class: And how it's transforming work, leisure, community and everyday life*. New York: Basic Books.
Galton, F. (1869). *Hereditary genius: An inquiry into its laws and consequences*. New York: Appleton.
Gardner, H. (1993). *Creating minds*. New York: Basic Books.
Goertzel, M. G., Goertzel, V., & Goertzel, T. G. (1978). *Three hundred eminent personalities: A psychosocial analysis of the famous*. San Francisco, CA: Jossey-Bass.
Guilford, J. P. (1967). *The nature of human intelligence*. New York: McGraw-Hill.

Guilford, J. P. (1987). Creativity research: Past, present, and future. In S. G. Isaksen (Ed.), *Frontiers in creativity research: Beyond the basics* (pp. 33–65). Buffalo, NY: Bearly.

Hair, J. F., Anderson, R. E., & Tatham, R. L. (1987). *Multivariate data analysis with readings* (2nd ed.). New York: MacMillan.

Harrington, D. M. (1990). The ecology of creativity: A psychological perspective. In M. A. Runco & R. S. Albert (Eds.), *Theories of creativity* (pp. 143–169). Beverly Hills, CA: Sage.

Hennessey, B. A. & Amabile, T. M. (1988). The conditions of creativity. In R. J. Sternberg (Ed.), *The nature of creativity: Contemporary psychological perspectives* (pp. 11–38). New York: Cambridge University Press.

Hofstede, G. (2001). *Culture's consequences: Comparing values, behaviors, institutions, and organizations across nations*. Thousand Oaks, CA: Sage.

Holland, J. L. (1966). *The psychology of vocational choice: A theory of personality types and model environments*. Waltham, MA: Blaisdell.

Isaksen, S. G. (2004). The progress and potential of the creativity level–style distinction: Implications for research and practice. In W. Haukedal & B. Kuvas (Eds.), *Creativity and problem-solving in the context of business management* (pp. 40–71). Bergen, Norway: Fagbokforlaget.

Isaksen, S. G. (2007a). The climate for transformation: Lessons for leaders. *Creativity and Innovation Management, 16*, 3–15.

Isaksen, S. G. (2007b). The Situational Outlook Questionnaire: Assessing context for change. *Psychological Reports, 100*, 455–466.

Isaksen, S. G. & Ekvall, G. (2007). *Assessing the context for change: A technical manual for the Situational Outlook Questionnaire—Enhancing performance of organizations, leaders and teams for over 50 years* (2nd ed.). Orchard Park, NY: The Creative Problem Solving Group.

Isaksen, S. G. & Geuens, D. (2007). An exploratory study of the relationships between an assessment of problem solving style and creative problem solving. *The Korean Journal of Thinking and Problem Solving, 17*, 5–27.

Isaksen, S. G. & Kaufmann, G. (1990). Adaptors and innovators: Different perceptions of the psychological climate for creativity. *Studia Psychologica, 32*, 129–141.

Isaksen, S. G. & Lauer, K. J. (1999). Relationship between cognitive style and individual psychological climate: Reflections on a previous study. *Studia Psychologica, 41*, 177–191.

Isaksen, S. G. & Tidd, J. (2006). *Meeting the innovation challenge: Leadership for transformation and growth*. Chichester, England: Wiley.

Isaksen, S. G. & Treffinger, D. J. (2004). Celebrating 50 years of reflective practice: Versions of creative problem solving. *Journal of Creative Behavior, 38*, 75–101.

Isaksen, S. G., Puccio, G. J., & Treffinger, D. J. (1993). An ecological approach to creativity research: Profiling for creative problem solving. *Journal of Creative Behavior, 27*, 149–170.

Isaksen, S. G., Lauer, K. J., Ekvall, G., & Britz, A. (2001). Perceptions of the best and worst climates for creativity: Preliminary validation evidence for the Situational Outlook Questionnaire. *Creativity Research Journal, 13*, 171–184.

James, L. R. & Jones, A. P. (1974). Organizational climate: A review of theory and research. *Psychological Bulletin, 81*, 1096–1112.

James, L. R. & Sells, S. B. (1981). Psychological climate: Theoretical perspectives and empirical research. In D. Magnusson (Ed.), *Toward a psychology of situations: An interactional perspective* (pp. 275–295). Hillsdale, NJ: Erlbaum.

Kaufmann, G. (2004). Two kinds of creativity—But which ones? *Creativity and Innovation Management, 13*, 154–165.

Kirton, M. J. (1976). Adaptors and innovators: A description and measure. *Journal of Applied Psychology, 61*, 622–629.

Kirton, M. J. (1994). A theory of cognitive style. In M. J. Kirton (Ed.), *Adaptors and innovators: Styles of creativity and problem-solving (Updated paperback edition)* (pp. 1–33). London: Routledge.

Kirton, M. J. & McCarthy, R. M. (1988). Cognitive climate and organizations. *Journal of Occupational Psychology, 61*, 175–184.

Lewin, K. (1936). *Principles of topological psychology*. New York: McGraw-Hill.

Lewin, K. (1951). *Field theory in social science*. New York: Harper.
MacKinnon, D. W. (1978). *In search of human effectiveness: Identifying and developing creativity*. Buffalo, NY: Creative Education Foundation.
Mooney, R. L. (1963). A conceptual model for integrating four approaches to the identification of creative talent. In C. W. Taylor & F. Barron (Eds.), *Scientific creativity: Its recognition and development* (pp. 331–340). New York: Wiley.
Murray, H. A. (1938). *Explorations in personality*. New York: Oxford University Press.
Nyström, H. & Edvardsson, B., (1980). *Technological and marketing strategies for product development: A study of 20 Swedish food-processing companies*. Uppsala, Sweden: Innovation Research Group – Institute for Economics and Statistics.
Parnes, S. J. (1987). The creative studies project. In S. G. Isaksen (Ed.), *Frontiers in creativity research: Beyond the basics* (pp. 156–188). Buffalo, NY: Bearly.
Pervin, L. A. (1987). Person-environment congruence in the light of the person-situation controversy. *Journal of Vocational Behavior, 31*, 222–230.
Puccio, G. J. & Murdock, M. C. (1999). Creativity assessment: An introduction & overview. In G. J. Puccio & M. C. Murdock (Eds.), *Creativity assessment: Readings and resources* (pp. 5–24). Buffalo, NY: Creative Education Foundation.
Puccio, G. J., Joniak, A. J., & Talbot, R. J. (1995). Person-environment fit: Examining the use of commensurate scales. *Psychological Reports, 76*, 931–938.
Raven, J. (1984). *Competence in modern society: Its identification, development and release*. London: H. K. Lewis.
Rhodes, M. (1961). An analysis of creativity. *Phi Delta Kappan, 42*, 305–310.
Rodgers, C. L. & Kerstetter, R. E. (1974). *The ecosphere: Organisms, habitats, and disturbances*. New York: Harper & Row.
Runco, M. A. (1991). *Divergent thinking*. Norwood, NJ: Ablex.
Schein, E. H. (1992). *Organizational culture and leadership* (2nd ed.). San Francisco, CA: Jossey-Bass.
Schneider, B. (1987a). E = f(P,B): The road to a radical approach to person-environment fit. *Journal of Vocational Behavior, 31*, 353–361.
Schneider, B. (1987b). The people make the place. *Personnel Psychology, 40*, 437–453.
Schneider, B. & Reichers, A. E. (1983). On the etiology of climates. *Personnel Psychology, 36*, 19–39.
Selby, E. C., Treffinger, D. J., Isaksen, S. G., & Lauer, K. J. (2004). Defining and assessing problem-solving style: Design and development of a new tool. *The Journal of Creative Behavior, 38*, 221–243.
Selby, E. C., Treffinger, D. J., & Isaksen, S. G. (2007). *VIEW: An assessment of problem solving style—Technical Manual* (2nd ed.). Sarasota, FL: Center for Creative Learning.
Simonton, D. K. (1987). Genius: The lessons of historiometry. In S. G. Isaksen (Ed.), *Frontiers in creativity research: Beyond the basics* (pp. 66–87). Buffalo, NY: Bearly.
Stein, M. I. (1968). Creativity. In E. F. Boragatta & W. W. Lambert (Eds.), *Handbook of personality theory and research* (pp. 900–942). Chicago, IL: Rand McNally.
Stein, M. I. (1983). Creativity in Genesis. *Journal of Creative Behavior, 17*, 1–8.
Sternberg, R. J. & Vroom, V. (2002). The person versus the situation in leadership. *The Leadership Quarterly, 13*, 301–323.
Torrance, E. P. (1987). Teaching for creativity. In S. G. Isaksen (Ed.), *Frontiers in creativity research: Beyond the basics* (pp. 189–215). Buffalo, NY: Bearly.
Trompenaars, F. & Hampden-Turner, C. (2004). *Managing people across cultures*. West Sussex, England: Capstone.
Woodman, R. W. & Shoenfeldt, L. F. (1999). An interactionist model of creative behavior. In G. J. Puccio & M. C. Murdock (Eds.), *Creativity assessment: Readings and resources* (pp. 467–477). Buffalo, NY: Creative Education Foundation.

Chapter 9
Creativity in Cross-Cultural Innovation Teams: Diversity and Its Implications for Leadership

Ricarda Bouncken

Firms are increasingly in need of their employees' creativity for delivering novel products to the global marketplace (West, 2002; Westwood & Low, 2003). Creative employees experiment with new ideas and concepts and thus contribute to the firms' success. Prior research has demonstrated that the diversity of team participants' knowledge, behavior, and values can promote creativity (Craig & Kelly, 1999; Kurtzberg, 2005; Milliken & Martins, 1996). Creativity through diversity can be amplified by team members' different national cultures as well. Such national cultural diversity also increases through improved market knowledge about the match between expectations and products delivered to customers of global markets. Thus, culturally diverse innovation teams are an option for coping with the challenge of globalizing markets. Despite these advantages of heterogeneity, however, firms are confronted with the negative consequences experienced in international teams whose members come from differing educational backgrounds. The diversity of national background can cause problems in interpersonal understanding and the work atmosphere—precipitating excessive disagreement, for instance—and can thereby adversely affect team moral and efficiency (Jehn et al., 1999), possibly dampening creativity.

In short, it is important to understand how cross-cultural diversity influences creativity and the innovation process. Yet research has sorely neglected these aspects. Comparative research on cross-cultural teams has focused on discovering intercultural differences within teams (Kirkman & Shapiro, 2005; Sagie & Aycan, 2003; Westwood & Low, 2003). One also finds studies on how one cultural dimension (e.g., collectivism) relates to competition between groups (Triandis et al., 1988) or to self-efficacy for teamwork (Eby & Dobbins, 1997). However, it has not been analyzed how a given constellation of cultural profiles affects creativity, innovativeness, and effectiveness.

R. Bouncken (✉)
Ernst-Moritz-Arndt Universität, Lehrstuhl für ABWL und Organisation,
Friedrich-Loeffler-Str. 70, 17487 Greifswald, Germany
e-mail: bouncken@uni-greifswald.de

P. Meusburger et al. (eds.), *Milieus of Creativity*, Knowledge and Space 2,

To shine light into the black box, I aim in this chapter to analyze culturally diverse teams for their functioning and its effects on creativity. Because the concept of teamwork quality has been proven to foster creativity and innovation (Hoegl & Gemünden, 2001), possible influences of cross-culturality on teamwork quality are warrant to study. In view of the limited research on the cross-cultural effects on creativity, I take an exploratory approach to cross-cultural teamwork, aiming to find patterns that will guide subsequent exploration of influences that culturally diverse teams have on teamwork quality and, hence, on innovation. Propositions on this basis are derived at the end of this chapter. A special focus is on the discussion and formulation of implications for the leadership that has been found most critical to the success of such diverse teams.

Culturally Diverse Innovation Teams

Innovation projects in a global context are rarely completed by a single individual but rather by a team that has to be creative (Janssen et al., 2004). Cross-functional teams, which deliver heterogeneous knowledge (see Fig. 9.1), improve innovation performance through increased creativity and teamwork quality (Cooper & Kleinschmidt, 1995; Gupta & Wilemon, 1996; Hise et al., 1990). However, if teamwork due to cultural misunderstanding produces a climate of mistrust, threat, and anxiety, it damages the innovation process (Janssen et al., 2004). In this vein Kurtzberg (2005) concludes that cognitive diversity negatively affects satisfaction among team members. Therefore, teamwork does not always guarantee creativity and successful innovation. Its strong links to creativity and innovation performance are nonetheless important to clarify.

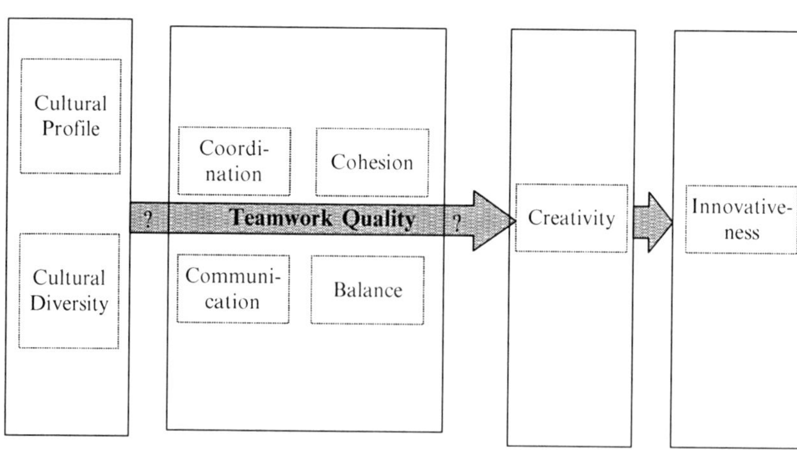

Fig. 9.1 Assumed chain of effects

Teamwork Quality in the Innovation Process

Hoegl and Gemünden (2001) present evidence that teamwork quality is related to the success of innovation projects. The two researchers define six components of teamwork quality: communication, coordination, balance of contributions by team members, mutual support, effort, and cohesion. Communication covers the frequency (time spent communicating), formalization (the degree of spontaneity with which members talk to each other), structure (communication with each other or via a mediator), and openness of the information exchange (nonwithholding of important information). Coordination describes whether the combination and status of individual tasks are synchronized and harmonized. By "balance," Hoegl and Gemünden (2001) mean an equal number of contributions by each team member. Mutual support requires intensive collaboration and cooperation between team members. Whereas mutual support can foster innovation, competitive behavior can lead to mistrust and frustration. Effort describes how important the team's work is considered to be in relation to other obligations of team members. Setting different priorities on the team's task can lead to conflict among team members, whereas a high level of effort shared by all team members will contribute to teamwork quality. Finally, cohesion refers to the strength of the team members' desire to remain on the team. Cohesion has three defined facets: (a) interpersonal attraction of team members, (b) commitment to the team's task, and (c) group pride and team spirit. Just as those three factors will promote teamwork quality, their absence will lead to a lack of togetherness and belonging. So far, the relationship between teamwork quality and the team members' dissimilar cultural backgrounds has not been researched.

Prominent Approaches to Describing National Cultures

National culture can be defined as the pattern of deep-level shared values by an interacting group and assumptions that influence societal effectiveness (Maznevski et al., 2002). So far, different approaches to understanding and describing national cultures have been developed. The most prominent ones are provided by Hofstede (1980, 1983), Hall and Reed Hall (1990), and Hampden-Turner and Trompenaars (2000). Their ideas can be investigated in terms of teamwork quality and creativity of innovation teams.

The concept by Hofstede (1980, 1983) consists of five dimensions:

- *Power Distance* (high/low) refers to how a society deals with differences in hierarchical status. Individuals with a high preference for power distance favor autocratic leadership and devalue a consultative leadership style.
- *Uncertainty Avoidance* (high/low) is based on an individual's motivation to avoid uncertainties and changes.
- *Individualism/collectivism* is a dimension in which values of leisure time, independence from one's company, and one's own activity are important.

- *Masculinity/femininity* describes how highly a society values income, recognition, and advancement. Whereas those aspects are key values in masculine societies, cooperation and safety are of higher concern in feminine societies.

Hall and Reed Hall's (1990) idea of difference between national cultures is based on numerous interviews with managers from France, the United States, and Germany. One important category for Hall and Reed Hall is context, that is, the degree to which information is directly or indirectly verbalized. If a person talks in an implicit manner, others have to draw the information out of the context of spoken words. These people are regarded as high-context persons. By contrast, low context means that the information is transferred directly. A second category for defining difference between national cultures—Hall and Reed Hall's (1990) concept of space—differentiates two kinds of individuals: those who prefer a strong spatial distance to others, and those who do not. Individuals who prefer distance to others will feel offended and disturbed if someone enters their personal space. Time (monochronic/polychronic) refers to the way people accomplish their tasks. Monochronic individuals prefer to finish one task before they start a new one. Polychronic individuals like to do many tasks simultaneously.

Hampden-Turner and Trompenaars (2000) distinguish between six dimensions of national cultures:

- *Universalism/particularism* refers to the way people in a society deal with rules. In some societies rules apply uniformly to everybody (universalism). In other societies rules affect people differently depending on their status, friendships, and loyalty (particularism).
- *Individualism/collectivism* has to do with the importance of the ego or the inner-group. It is a dimension that both Hampden-Turner and Trompenaars (2000) and Hofstede (1980, 1983) use.
- *Affectivity/neutrality* describes the degree to which a society accepts the expression of emotions. Individuals in neutral societies tend to control emotions, whereas individuals in affective societies are described as more impulsive.
- *Diffusiveness/specificiality* refers to the extent to which private life and work life are separated or intertwined. Diffuse individuals tend to connect work life and private life.
- *Achieved status/ascribed status* differentiates two kinds of society: one in which the status of an individual derives from his or her performance and effort in the past (achieved status) and one in which that status derives from the person's heritage (ascribed status).
- *The relationship with time* includes several dimensions. On the one hand, it separates sequential from synchronized time use. This dimension has a strong similarity to Hall and Reed Hall's (1990) time dimension (monochronic/polychronic use of time). In addition, Hampden-Turner and Trompenaars (2000) propose to differentiate societies in terms of whether they are oriented to their future, their past, or their present. The relationship between humans and nature differentiates people who believe they have control over nature from people who tend to feel that they are controlled by nature.

Methodology

This study on national cultural differences was designed to identify those national cultural dimensions that differentiate best between team members. The research, carried out by 32 people, was inspired by the thinking of Hofstede (1980, 1983), Hall and Reed Hall (1990), and Hampden-Turner and Trompenaars (2000). It is known that a qualitative design can be successfully used for teamwork in order to identify differences between cultures (Kirkman et al., 2006). For the present study of the often fuzzy and still largely unexplored nature of teamwork, creativity, and its antecedents embedded in national culture, the research in this chapter is therefore qualitative and exploratory in nature.

We selected firms that form global innovation teams with members from different national backgrounds. All of the teams that were studied had to be working on complex projects lasting at least two years. All projects had to be focused on disruptive innovation that required a large amount of creativity, and they all had to cover the entire innovation process, from idea generation to launch. Many of the 35 firms that were contacted did not have projects meeting these criteria. They either did not form culturally diverse teams or were not using such teams to pursue disruptive innovation. Ultimately, we interviewed six teams, each located in a different firm. The firms operated in various industries (food, plant engineering, pharmaceutical, safety, telecommunication, chemistry). All teams had five to seven members in the core-teams. All of the interviewees were working in projects that still had 12 months or more to run.

An interview template was developed from the literature on national cultures, teamwork, and creativity. One-on-one semistandardized interviews were then conducted with the culturally diverse team members from different countries (Brazil, Argentina, China, the Netherlands, Germany, the United States, and Sudan). In every team at least three core team members were interviewed by the same interviewer. Interviews that were carried out at the company's facilities lasted between 45 and 70 min. In total, 19 persons were interviewed within one month. The personal interviews were recorded, and two persons other than the interviewers later transcribed them. Both transcribers had to analyze every interview, producing two versions of each transcription. To master the large quantity of data, the transcriptions were categorized (Mayring, 2003). A short version of every interview was developed. Because two researchers were involved in this step two versions of the categorizations were produced. Both versions were compared for differences.

Drawing on the work by Hofstede (1980, 1983, Hall and Reed Hall (1990), and Hampden-Turner and Trompenaars (2000) on national cultural differences, the research team developed a deductive system of categories accommodating all the categories of national cultures. Statements that could not be classified into the named dimensions were categorized inductively in a second step. For that purpose other cultural concepts (e.g., Kluckhohn & Strodtbeck, 1961) were used. Categories of teamwork were informed by the framework provided by Hoegel and Gemünden (2001). Given the lack of models dealing with this topic, other antecedences and consequences of cross-cultural teamwork were inductively categorized. The two members of the research team also categorized whether the aspect of teamwork quality was valued as positive or rather as negative, and we counted the number of topics raised that related to national cultures, teamwork quality, and creativity. The two transcribers delivered and categorized the data. Both categorizations were then compared.

To evaluate whether there is a possible influence of cultural dimensions on teamwork quality and creativity, we analyzed how often cultural categories were articulated "near" statements concerning teamwork quality. That is, we counted statements in which cultural aspects and aspects of teamwork quality were both referred to within the same paragraph of the transcribed interview. The resulting categories were analyzed for their absolute topic frequency (ATF, i.e., the total number of times a topic is addressed by the interviewees) and their person frequency (PF, i.e., how many of the research subjects addressed a given topic). This type of analysis helps evaluate the topics raised during the interviews and gives a quantitative measure of how important the category is (Schilling, 2006).

Results of the Interview Analysis

Several cultural differences were perceived within the teams. Table 9.1 shows how often a topic was raised by one and by all interviewees. Also, interviewees perceived some cultural differences that did not correspond to the above-named cultural dimensions. Many of those statements were about how relaxed or work-oriented a certain culture was seen to be. In keeping with Kluckhohn and Strodtbeck (1961), those statements (ATF = 11) were categorized as a "doing/being orientation."

Aside from cultural differences, positive and negative consequences of cross-cultural teamwork were addressed. A positive consequence noted by interviewees was the wide range of knowledge in a cross-cultural team (ATF = 3, PF = 2). Most statements (ATF = 7, PF = 2) concerned personal benefits and insights, which lay

Table 9.1 Absolute Topic Frequencies (ATF)[a] and Person Frequencies (PF)[b] for cultural differences

Cultural dimension	ATF	PF
Power distance	17	8
Individualism	10	6
Collectivism	9	5
Affectivity/neutrality	8	5
Uncertainty avoidance	7	4
Polychronism/monochronism	4	2
Space	2	1
Universalism	1	1
Specificity	1	1

[a]The total number of times a topic was addressed by research subjects in semistructured one-on-one interviews conducted.

[b]The total number of research subjects who addressed a topic in semistructured one-on-one interviews.

Table 9.2 Absolute Topic Frequencies[a] and Person Frequencies[b] for teamwork quality

Component of teamwork quality	Number of positive statements		Number of negative statements	
	ATF	PF	ATF	PF
Communication	4	3	14	6
Cohesion	5	4	3	3
Effort	2	1	3	2
Balance	3	2	2	2
Mutual support	1	1	1	1
Coordination	4	4	5	4

[a]The total number of times a topic was addressed by research subjects in semistructured one-on-one interviews conducted.
[b]The total number of research subjects who addressed a topic in semistructured one-on-one interviews.

in enhanced composure and openness. Interviewees also perceived that cultural diversity increases their willingness to reflect on their own actions. A negative consequence mentioned by the interviewees in conjunction with cross-cultural teamwork was that certain cultural dimensions were causes of conflicts or dissatisfaction. There were five statements to the effect that differences in power distance caused difficulties (PF = 3). Differences in space, time, individualism, activity orientation, goal orientation (orientation to nature), and different cognitive styles were also found capable of eliciting difficulties. The qualitative analysis of teamwork quality in the team revealed that most problems occurred with communication (see Table 9.2).

Lastly, we analyzed aspects of differences in national cultural values and aspects of teamwork quality, again using ATF and PF. Overall, five statements (PF = 3) documented a relationship between activity orientation and effort. Team members can value differences in activity orientation negatively. Individuals, who were "being-oriented" were perceived as lazy by their "doing-oriented" colleagues. Conversely, strongly activity-oriented individuals were perceived as overly motivated.

In all statements concerning the category of context (ATF = 4, PF = 3), interviewees expressed preference for a direct use of language. There is also a statement that refers to both power distance and balance of team-member contributions. One statement was categorized as specificity and as cohesion. An interviewee with a diffuse cultural profile described how spontaneous activities involving all the team members after work hours positively influenced the team-building process. Coordination and relationship to nature (goal orientation) are linked in one of the statements.

Discussion and Conclusion

This study has addressed cross-cultural effects on teamwork quality, creativity, and the innovation process. Overall, it indicates that cultural values have unequal effects on teamwork and the innovation process. The main effects on teamwork quality and creativity are caused by power distance, context, and stimulation stemming from diversity.

The interviews and observations indicate that power distance has strong effects on teamwork quality and creativity. The recorded ATFs and PFs show that difference in power distance was the cultural dimension mentioned most often by the interviewees and was therefore strongest within cross-cultural teams. Power distance has additional effects on teamwork quality, especially communication, involvement, and, hence, the team's creativity and innovativeness. Differentials and a high preference for power distance were found to reduce the work quality and creativity of teams. One interviewee perceived differences in how strongly team members engage in a discussion. He attributed differences in behavior to a different hierarchical orientation. People with a strong sense of hierarchy do not participate as much in discussions as people with a low sense of hierarchy do. As pointed out by prior research suggesting that the sociocultural context influences leadership process and effectiveness (Elenkov & Manev, 2005), high preference for power distance is especially harmful when it occurs in team leaders. In any case, leadership behavior is a means of smoothing the harmful components of cultural differences. For example, different behaviors during discussions can be channeled by a sensitive team leader. Leaders whose behavior is based on the assumption of power distance will seldom respond in that way. Moreover, creativity is hampered when team members do not dare oppose the opinion of their supervisor. For achieving creativity, a project leader needs to integrate quiet or shy individuals in discussions and thereby create a balance of team-member contributions. I assume that leadership influences creativity in several ways: through (a) the internalization of values and ideas, (b) the motivation of subordinates, (c) the encouragement of diverse opinions, (d) the provision of protected work environments, (e) the expectation of compliance and, (f) establishment of a permanent frame of reference for the discontinuous and discrete generation of novel ideas that often permeate the whole team.

This assumption leads me to two propositions:

Proposition 1: Strong diversity in terms of power distance in teams has a negative effect on teamwork quality and creativity.

Proposition 2: Team leaders with a high preference for power distance decrease teamwork quality and creativity.

Power distance is a cultural value. Nevertheless, training can reduce the negative behavioral outcomes of power distance. Especially harmful to the teamwork quality and innovativeness is a team leader with a high preference for power distance. Training in leadership behavior therefore emerges as an important task.

The second main finding is related to communication, especially context. Because communication is one of the largest difficulties for cross-cultural teams, another important task of teams and their project leaders is managing information and assuring that team members have access to the information they need. In the interviews, a positive effect of low context was mentioned even by persons who use high-context language. (To interpret these statements correctly, recall that they were made by people who attended a team meeting in which the interviewer observed great difficulties with communication.) Though one Latin American participant who belongs to a high-context language culture cautioned that other team members with the same national background could easily be offended through direct language, all interviewees saw a positive relationship between direct language and communication in groups. Clear and direct communication is found to be important for the success of a project. Another interviewee noticed that goal definition requiring intense communication and discussion is much easier in Europe than in South America. He, too, considered goal clarity an important precondition of team performance. On the other hand, one participant explained that team members with a high-context background feel offended by a low-context language.

These considerations lead to another proposition:

Proposition 3: Communication improves as low-context communication increases, enhancing creativity and innovativeness.

If the number of statements pertaining to the advantages of intercultural work is compared to the number of statements relating to its difficulties and disadvantages, one must conclude that members of culturally diverse teams are less aware of the strengths of their groups than of their weaknesses.

When it comes to creativity, the effect of cultural diversity is not distinct. In sum, the findings of this study are related more to the effects that cultural diversity has on teamwork quality than to that diversity's effects on creativity. One of the greatest advantages of working in a cross-cultural team lies in personal self-reflection. Participants report enjoying the diverse perspectives and playing with dissimilar ideas and concepts in the culturally diverse teams. One person mentioned that cultural heterogeneity fosters creativity. It is also known that different cognitive styles, which affect creativity positively (Milliken & Martins, 1996; Murray, 1989), can be influenced by cultural norms. Many interviewees noticed that a cross-cultural team has a broader knowledge base than a culturally homogeneous team. Hence, I come to my final proposition.

Proposition 4: The *creativity of culturally diverse teams emerges through inspirational differences and improved self-reflection.*

To deepen the exploration of this aspect of working in culturally diverse teams, future studies need to find out whether there are systematic differences between cognitive styles from one country to the next. In that research, culture could be regarded as a mediating variable, and the influence on team creativity could be tested.

The propositions developed in this chapter should be tested empirically in future research. The results of this study can be generalized only on the basis of quantitative support. As with all studies, this one has limitations. First, the sample size of the interviewees is not large. (Qualitative data analysis does not usually require as large a sample as quantitative analysis does.) Still, König and Vollmer's (1997) recommended sample of 20–30 interviews was nearly achieved, and the group of interviewees was very heterogeneous. Each interview brought out many different aspects, and the point of saturation, which usually signals when to stop interviewing participants, is far from being reached. Furthermore, this study does not show how behaviors and cultural dimensions of cross-cultural teams change during the innovation process. It only hints at changes in the attitudes of persons. I am therefore seeking to expand the knowledge about how cultural profiles of cross-cultural teams change during the innovation process. Last but not least, implications that culture influences the concept of teamwork quality are not very strong—a crucial aspect of this study.

One avenue for future research could be the leadership of cross-cultural teams. To date, most leadership studies have been conducted in Western cultures, so it is not clear whether concepts such as transactional as opposed to transformational leadership or participatory as opposed to autocratic leadership apply to all cultures. Every leader exhibits behavior that can be characterized as transformational, transactional, and non transactional (Avolio & Bass, 1995). Most research on these topics focuses on how the different styles mediate or moderate leader effectiveness. But there are also individual and contextual antecedents of leadership behavior that need to be studied (Bass, 1997). For example, transformational leadership requires leaders to build trusting, warm relationships with subordinates through honest engagements and an agreeable personality. Keller (1992, 2006), stressing the positive outcomes that transformational leadership has on innovativeness, found that agreeableness is related to ratings of leader sensitivity. Not every leader from every country may have the personality antecedents to implement a specific leadership style. But a firm can either select leaders culturally highly sensitive and open to cross-cultural teams or, if it relies heavily on the specific technical competencies of leaders with low cultural sensitivity, it can invest in cross-cultural training programs. Both alternatives demand techniques and tools for choosing the right leaders for cross-cultural teams and for effectively training leaders in cross-cultural sensitivity. Although intercultural HR consultants offer relevant services in this area, the benefits and drawbacks of their instruments have not been empirically evaluated. Given the ever-growing pressures on leaders in the globalized marketplace, further study of precisely these issues would itself be a highly recommendable service.

References

Avolio, B. J. & Bass, B. M. (1995). Individual consideration viewed at multiple levels of analysis—A multilevel framework for examining the diffusion of transformational leadership. *Leadership Quarterly, 6*, 199–218.

Bass, B. M. (1997). Does the transactional-transformational leadership paradigm transcend organizational and national boundaries? *American Psychologist, 52*, 130–139.

Cooper, R. G. & Kleinschmidt, E. J. (1995). Benchmarking the firm's critical success factors in new product development. *Journal of Product Innovation Management, 12*, 374–391.

Craig, T. Y. & Kelly, J. R. (1999). Group cohesiveness and creative performance. *Group Dynamics: Theory, Research, and Practice, 3*, 243–256.

Eby, L. T. & Dobbins, G. H. (1997). Collectivistic orientation in teams: An individual and group-level analysis. *Journal of Organizational Behavior, 18*, 275–296.

Elenkov, D. S. & Manev, I. M. (2005). Top management leadership and influence on innovation: The role of sociocultural context. *Journal of Management, 31*, 381–402.

Gupta, A. K. & Wilemon, D. (1996). Changing patterns in industrial R&D management. *Journal of Product Innovation Management, 13*, 497–511.

Hall, E. T. & Reed Hall, M. (1990). *Understanding cultural differences - Germans, French and Americans*. Yarmouth, ME: Intercultural Press.

Hampden-Turner, C. & Trompenaars, F. (2000). *Building cross-cultural competence*. Chichester, England: Wiley.

Hise, R. T., O'Neal, L., Parasuraman, A., & McNeal, J. U. (1990). Marketing/R&D interaction in new product development: Implications for new product success. *Journal of Product Innovation Management, 7*, 142–155.

Hoegl, M. & Gemünden, H. (2001). Teamwork quality and success of innovative projects: A theoretical concept and empirical evidence. *Organization Science, 12*, 435–449.

Hofstede, G. (1980). *Culture's consequences: International differences in work-related values*. Beverly Hills, CA: Sage.

Hofstede, G. (1983). National cultures in four dimensions. A research-based theory of cultural differences among nations. *International Studies of Management and Organization, 13*, 46–74.

Janssen, O., van de Vliert, E., & West, M. (2004). The bright and dark sides of individual and group innovation: A Special Issue introduction. *Journal of Organizational Behavior, 25*, 129–145.

Jehn, K. A., Northcraft, G., B., & Neale, M., A. (1999). Why differences make a difference: A field study of diversity, conflict and performance in workgroups. *Administrative Science Quarterly, 44*, 741–763.

Keller, R. T. (1992). Transformational leadership and the performance of research-and-development project groups. *Journal of Management, 18*, 489–501.

Keller, R. T. (2006). Transformational leadership, initiating structure, and substitutes for leadership: A longitudinal study of research and development project team performance. *Journal of Applied Psychology, 91*, 202–210.

Kirkman, B. L. & Shapiro, C. (2005). The impact of cultural value diversity on multicultural team performance. *International Management, 18*, 33–67.

Kirkman, B. L., Lowe, K. B., & Gibson, C. B. (2006). A quarter century of Culture's Consequences. A review of empirical research incorporating Hofstede's cultural values framework. *Journal of International Business Studies, 37*, 285–320.

Kluckhohn, F. R. & Strodtbeck, F. L. (1961). *Variations in value orientations*. Evanston, IL: Row, Peterson.

König, E. & Vollmer, G. (1997). *Systemische Organisationsberatung. Grundlagen und Methoden* [Systematic organizational consulting: Principles and methods]. Weinheim, Germany: Deutscher Studien Verlag.

Kurtzberg, T. R. (2005). Feeling creative, being creative: An empirical study of diversity and creativity in teams. *Creativity Research Journal, 17*, 51–65.

Mayring, P. (2003). *Qualitative Inhaltanalyse—Grundlagen und Techniken* [Qualitative content analysis—Principles and techniques] (8th ed.). Weinheim, Germany: Beltz Verlag.

Maznevski, M. L., DiStefano, J. J., Gomez, C. B., Noorderhaven, N. G., & Wu, P.-C. (2002). Cultural dimensions at the individual level of analysis: The cultural orientations framework. *International Journal of Cross Cultural Management, 2*, 275–295.

Milliken, F. J. & Martins, L. L. (1996). Searching for common threads: Understanding the multiple effects of diversity in organizational groups. *Academy of Management Review, 21*, 402–433.

Murray, A. I. (1989). Top management heterogeneity and firm performance. *Strategic Management Journal, 10*, 125–141.

Sagie, A. & Aycan, Z. (2003). A cross-cultural analysis of participative decision-making in organizations. *Human Relations, 56*, 453–473.

Schilling, J. (2006). On the pragmatics of qualitative assessment: Designing the process for content analysis. *European Journal of Psychological Assessment, 22*(1), 28–37.

Triandis, H. C., Bontempo, R., Villareal, M. J., Asai, M., & Lucca, N. (1988). Individualism and collectivism: Cross-cultural perspectives on self–ingroup relationships. *Journal of Personality and Social Psychology, 54*, 323–338.

West, M. A. (2002). Sparkling fountains or stagnant ponds: An integrative model of creativity and innovation implementation in work groups. *Applied Psychology: An International Review, 51*(3), 355–424.

Westwood, R. & Low, D. R. (2003). The multicultural muse—Culture, creativity and innovation. *International Journal of Cross Cultural Management, 3*, 235–259.

Chapter 10
Space(s) of Innovation: Regional Knowledge Economies

Martina Fromhold-Eisebith

In terms of subnational entities such as urban agglomerations, provinces, and counties, regions are major arenas in which knowledge-producing and knowledge-exploiting activities are rooted, with marketable innovative products or processes emerging from the location-specific interplay of economic, social, cultural, technological, and political factors (Malecki, 1997, 2000; Meusburger, 2000, 2008). Presumably, there is something fundamentally "spatial" and "place-bound" that determines the abilities of corporate actors to innovate. There are also certain constellations of factors that shape spaces of innovation, understood as outstanding agglomerations of firms and other organizations that successfully engage in the generation and economic exploitation of knowledge-intensive activities. Because creativity is important in inducing innovativeness, the notions of spaces of innovation and milieus of creativity are, logically, strongly linked, as I show in this chapter.

Like regional economists, economic geographers have been exploring relevant dynamics and determinants of the knowledge-based economy for several decades, filling entire libraries with their writings (for recent synopses, see Cooke et al., 2007; Simmie, 2005). In the following pages I draw on a range of these works to present currently prominent chains of argumentation and perspectives. I discuss key terminological, conceptual, and empirical foundations, trying to open the black box of why and how innovative spaces and regional knowledge economies evolve in response to the dynamics of an environment that enhances creativity.

Evidence showing the significance of a geographical view on issues of economic creativity is easy to come by. Maps marking the distribution of knowledge-intensive activities across regions clearly convey the spatially distinct nature of these phenomena. One input indicator—the intensity of research and development (R&D) in the regions of the European Union (EU)—reveals an eminent set of knowledge-oriented places, notably in Germany, France, Scandinavia, and the United Kingdom (see Fig. 10.1). Indeed, just a dozen regions account for over one

M. Fromhold-Eisebith (✉)
RWTH Aachen, Geographisches Institut
Templergraben 55, D-52056 Aachen
e-mail: m.fromhold-eisebith@geo.rwth-aachen.de

P. Meusburger et al. (eds.), *Milieus of Creativity*, Knowledge and Space 2,

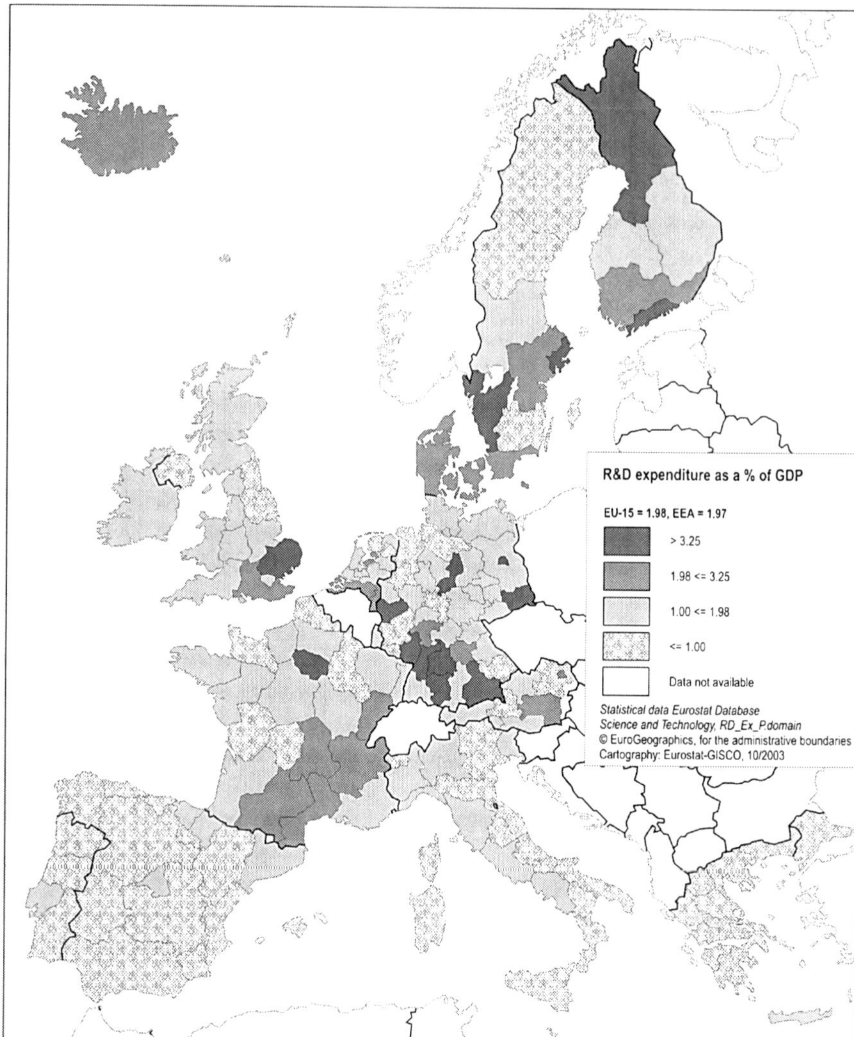

EU-15 = 1.98 and EEA = 1.97
refer to the EU-15 and the EEA averages, i.e. in 2001 and for all sectors, R&D expenditure in the EU and EEA amounted to 1.98% of their GDP respectively.

Exceptions to the reference year 2001
EU-15 and IS: 2002; FR, IT, LU, and NL: 2000; DE, EL and UK: 1999; AT: 1988.

Exceptions to data at the NUTS 2 Level
IE and UK: classified at NUTS 1 Level.

Fig. 10.1 R&D expenditure as a percentage of GDP, all sectors, by European Economic Area region, 2001 (NUTS 2 level) (From the European Commission and Eurostat, 2004, adapted with permission)

third of all R&D in the EU (European Commission & Eurostat, 2004). Similarly, an output indicator—patent applications—points to particularly creative localities, some of them in southern Germany (Greif, 2004). Outside the EU, too, the world has well-known, outstandingly dynamic technology hotspots such as Silicon Valley and Route 128 in the United States and Bangalore in India. Obviously, knowledge-based economic capabilities are very unevenly distributed across space and tend to agglomerate mainly in urban areas. As Simmie (2005) states, "there is a distinctive geography of innovation" (p. 790). I therefore closely examine place-specific assets in order to find out why.

The issue, however, is more complicated than it may first appear. Disparities of innovativeness exist on various interdependent scales of activity that can be described as a cascade of clustered creativity. Economic creativity differs between:

continents or economic blocks,
countries within a continent,
regions within a country,
cities within a region,
quarters within a city,
firms within a quarter,
teams within a firm, and
individuals within a team.

It is not easy to discern which level has the most influential constellation of identifying factors that aggregate to, and are reflected in, uneven spatial patterns of knowledge production and application. Disparities of innovativeness may thus emerge from the concatenation of different scale-specific sets of interdependent factors, influenced by economic and business issues; technological dynamics and infrastructure; and political, social, cultural, and psychological aspects. Moreover, constellations probably differ from one time or project phase to the next, for producing innovations is a rather discontinuous endeavor (Grabher, 2002). As suggested by the cascade depicted above, such complexities call for an approach that is both systemic and multidisciplinary if one is to explain convincingly why landscapes of innovativeness look as heterogeneous as they do.

These two features are what recommend economic geography in this context. Creativity, knowledge production, and innovation-oriented application are marked by a few clearly context-dependent qualities (see Cooke et al., 2007; Cooke & Schwarz, 2007; Meusburger, 2008). They include:

- The situatedness of innovation and knowledge creation, for processes take place and are bound to a given location or sets of interacting locations
- The relevance of local conditions and situations; and
- The role of spatial (physical) proximity of actors (or of economic, organizational, social, cultural, or epistemological elements)

Research on knowledge production and innovativeness can therefore substantially profit from geography's focus on regionally specific conditions and systemic sets of interdependent factors that shape economic creativity.

Embedding this chapter's topic into the overarching theme of the book, I aim in the next section to clarify logical relationships and spell out two notions—spaces of innovation and milieus of creativity. It is followed by a statement of basic assumptions about the nature of innovativeness and knowledge creation that matter with spatiality. I thereafter emphasize the facets that are the most relevant for considering innovativeness in a regional context and explore their conceptual dimension, outlining approaches that characterize ideal-type relationships of innovativeness and regional development. Attention then briefly turns to the difficult methodological task of empirically capturing the role that regional contexts have in processes of innovativeness and knowledge. The chapter closes with conclusions and open questions.

Spaces of Innovation and Creative Milieus: Mutual Dependencies

The question of how innovative spaces relate to milieus of creativity requires reflection on the fundamental logics of both notions. An additional complication is the fact that the term *creative milieu* can be associated with both a given conceptual framework (elaborated below) and a more common, rather broad definition as a socially and/or spatially constructed relational setting that is particularly conducive to creativity. This section draws more on the latter, less specific interpretation than on the former variant.

What sets creativity apart from innovativeness? The answer varies from one scientific discipline to the next. The definition that economists tend to offer, especially when it comes to creativity, differs from that by, say, psychologists (who look mainly at the individual level, though they do acknowledge the influence of social and cultural frameworks; see Amabile, 1996; Csikszentmihalyi, 1997). In this chapter's context of regional economic development, the two stances have common ground in that they resort mostly, albeit not necessarily, to interactive processes that produce previously nonexistent outcomes. They both express an attribute that may be ascribed to individuals, groups, or organizations. Creativity, however, subsumes a wider category than innovativeness; it includes any new combination of assets and ideas and the results of that combination (Buttimer, 1983). Innovativeness, by contrast, encompasses only certain manifestations of creativity: the generation of new processes and new commercially tradable products, of modes of organization, and of markets (as commonly stated by economists).

Accordingly, all innovations require some kind of creativity, whereas creativity may lead to a range of outcomes larger than just innovations. For example, creativity is necessary when one needs to solve an emergent problem of regional development, such as the sudden withdrawal of a major industrial investor. The response requires a collaborative search by various actors to find a feasible solution by creating sets of ideas and concepts for new developmental perspectives. The resulting suggestions, however, cannot actually be called innovations. Yet innovativeness may form part of the solution, for some ideas that develop in the creative process probably encourage the formation of new companies and the development of new products in existing ones. In fact, most economic geographers and economists usually mistake innovativeness for creativity, often blurring the difference between the two (see, for instance, Cooke & Schwarz, 2007; Malecki, 2000).

An even more discriminating take on creativity and innovativeness arises from the dissimilar functions that diversity and dissent have. As stressed in psychological research (Csikszentmihalyi, 1997), creativity (whether of individuals or social groups) essentially issues from dissenting ways of thinking and acting and from actors who deviate from common trajectories. By contrast, the objective of being economically innovative requires a range of actors to move collectively toward common goals and to share values and aspects of identity (Fromhold-Eisebith, 2004). These commonalities are understandably necessary to make ends meet, that is, to proceed from complex negotiations between actors to clear-cut innovations, but they also limit creative freedom (a contradiction debated in Jekel & Fromhold-Eisebith, 2003).

The notions of innovative spaces and milieus of creativity should not be regarded as synonymous but rather as different from and related to each other in certain ways. For example, just as creativity constitutes a category broader than innovativeness, spaces are units wider than milieus. The word *space* may refer to a supranational, national, regional, or local unit (which all of consequence for innovativeness). It can also denote any combination of those scales simultaneously (Massey, 2005; Meusburger, 2008), depending on the phases of an innovation process. A *milieu* is a unit below the national level because its size is fundamentally determined by social relationships whose density and intensity depend on the potential for face-to-face contact, on accessibility, and, hence, on proximity (Fromhold-Eisebith, 1995). Accordingly, scale, social underpinnings, and the importance of proximity distinguish creative milieus from innovative spaces, although both terms relate to functional, not normative units. *Innovative spaces* comprise *creative milieus* (but extend beyond them spatially). *Innovative milieus* (spaces in which innovations are actually produced by locally collaborating actors) form part of creative milieus (with various kinds of new outcomes emerging from the interaction of local actors). Whether there is something like a creative space, however, depends on the interpretation of space. A creative space should be thought of as a local-regional unit, as a level at which the social interaction necessary for creativity can be fulfilled. Creative spaces at macrolevels are hardly imaginable.

Basic Assumptions About Innovativeness and Knowledge Creation

Spatial differences in innovativeness can be partly due to basic modalities of the innovation process. The successful development and marketing of new products or processes is all but impossible without a number of activities and organizations, ideally ones linking inventors, applied researchers, commercial producers or service providers, and consumers (see Fig. 10.2). Given the continuously interactive, rather

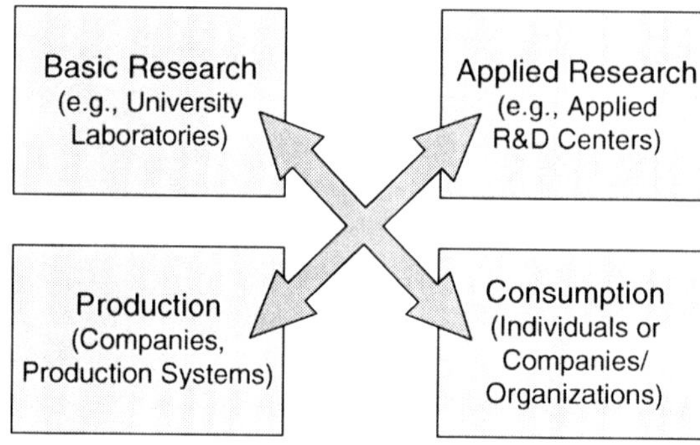

Fig. 10.2 Interactive innovation model—interdependent activities and actors

than simply linear nature of the innovation-related connection between these actor groups (Malecki, 1997), actors profit from proximity. One may therefore assume that the basic reliance of innovation on various forward and backward feedback loops between different contributors is best served when relevant organizations are colocated, with urban agglomerations being particularly fertile soil for constant innovativeness (Lambooy, 2002; Malecki, 2000; Simmie, 2003). Frequently, urban agglomerations also figure as functional and political centers and as centers of power, statuses that make them especially attractive to carriers of knowledge (Meusburger, 1998, 2000, 2008).

These observations are further corroborated by the realization that innovation processes rarely, if ever, occur in isolation. Most innovations are incremental rather than radical; they are considerable improvements on preexisting items instead of completely newly invented ones. They are part of a wave, a veritable swarm of interconnected innovations that together constitute a technological paradigm or trajectory (Cimoli & Dosi, 1995). In all probability, then, places of previous innovations also generate derived improvements, adding to spatially demarcated patterns of economic creativity (Fuchs & Shapira, 2005).

However, the assumption that communication needs make the innovation process depend on colocation, agglomeration, and proximity is assailable (Morgan, 2004). Proximity has various meanings and can exist even between partners who operate far away from each other. The word can convey not only physical but also social, organizational, or cultural closeness that may bridge wide geographic distances and animate innovation-related interaction. Direct collaboration during the innovation process may not be frequent enough to require colocation. Researchers, for instance, can easily travel to colleagues whenever deemed necessary. Innovativeness, too, is connected to certain places (Brown & Duguid, 2002; Malecki, 2000; Morgan, 2004; Simmie, 2005).

One basic assumption relates to types of knowledge that may be entailed when companies innovate. Different kinds of knowledge appear to be bound to a locality to different degrees. A distinction is commonly made between codified and noncodified (tacit) knowledge (although such simple dichotomies do not suffice to capture ambiguous, dynamic, and complex realities; see Schamp & Lo, 2003; Meusburger, 2008). Codified knowledge is manifested in texts and graphs and is easily transferable over great distances, whereas noncodified knowledge is regarded as being firmly embedded in people's minds and linked to experience, requiring face-to-face communication if it is to be shared and creatively recombined (Howells, 2002). Correspondingly, innovativeness draws on collective knowledge (such as that present in working groups) rather than on individual knowledge, which often calls for at least temporary colocation (Grabher, 2004). Recent discourse on knowledge foundations of innovativeness has introduced another category that initially appears to link relevant processes even more closely to places of copresence. Amin and Cohendet (2004), for example, distinguish between people's acquired competencies (possessed knowledge) and so-called practised knowledge or knowing (everyday routines of communities linked to cooperative practices of people). Paradoxically, however, this debate increasingly stresses the spatial volatility and mobility of knowledge and denies its place-bound nature. It posits spaces of knowledge as:

> organized spaces of varying length, shape, and duration, in which knowing, depending on circumstances, can involve all manner of spatial mobilizations, including placements of task teams in neutral spaces, face-to-face encounters, global networks held together by travel and virtual communications, flows of ideas and information through the supply chain, and transcorporate thought experiments and symbolic rituals. (p. 12)

Although Amin and Cohendet's viewpoint undermines simple assumptions about the geography of innovation, it dialectically establishes a new perspective on spatial determination that is more consistent with the spatiotemporal nature of knowledge production and application than earlier conceptions of spaces of knowledge are.

Embedding Innovativeness in a Regional Context

Beyond the terminological debate, several arguments support the assumption that innovativeness, in at least some phases of the innovation process, is partly fostered by regional features (Malecki, 1997, 2000; Morgan, 2004; Schamp & Lo, 2003; Simmie, 2005). In this respect, too, fundamental relationships exist between creativity and knowledge on the one hand and place-specific assets and advantages of proximity on the other hand (see Meusburger, 2008). The following six aspects are mostly emphasized by regional scientists and economic geographers.

First, economic theory holds that spatial concentrations of innovative activity are favored by conventional agglomeration economies and externalities and by related institutional frameworks that prevail in urban regions (Gordon & McCann, 2005; Lambooy, 2002). The colocation of various innovation-oriented organizations helps cut costs and gradually makes the entire setting evolve toward a suitable framework for specialized economic activities. Important factors are the existence and improvement of higher education and R&D infrastructure, the specialization of labor markets, and the adjustment of the regional industry mix to the needs of emerging lead activities. This seedbed encourages both the innovativeness of existing firms (first phase of the product life cycle) and the creation of new enterprises.

Second, processes governing the transfer of technology from academia and R&D laboratories to nearby industries are crucial and presumably at least somewhat sensitive to spatial proximity. Forms of cooperation such as joint R&D projects, consulting services, and the provision of human capital are seen as key inputs into regional innovativeness (Feldman & Desrochers, 2003; Malecki, 2000). Entrepreneurship spinning off from universities appears to be strongly oriented to the locality of the academic incubator, substantiating the widely known idea of a triple helix consisting of productively interacting academia, industry, and policy.

Third, the very colocation of firms, the sheer fact of "being there," is assumed to foster knowledge spillover that triggers innovative activity other than direct collaboration. By observing competitors nearby, by receiving fresh stimulus from labor market fluctuations, or by picking up interesting gossip (local buzz; Storper & Venables, 2004), companies may become inclined to venture into new directions, as others have before them (Audretsch & Feldman, 1996).

Fourth, residential proximity allows executives or staff of different firms to build up social relationships and trust based on casual face-to-face contact. This kind of communication amplifies private and informal ties, which amalgamate with and solidify professional collaboration (Fromhold-Eisebith, 2004; Malmberg & Maskell, 2002). Socially embedded interaction and learning can thus easily proliferate in the local community of firms. Their spread creates collectives marked by shared interests, values, and conventions—in sum, by "untraded interdependencies" (Storper, 1995), which substantially help vertically collaborating industrial partners develop flexible specialization and gain networking advantages. Consequently, particular "cultures of regional knowledge" closely linked to certain places emerge (Malecki, 2000).

Fifth, the self-reinforcing and cumulative nature of economic creativity keeps it rooted in certain places, continuity that is consistent with the intrinsically evolutionary character of innovativeness depicted above. The interconnectivity of innovation-related actions can be interpreted in different ways. One can say, for instance, that consecutive waves of new products or processes build upon each other and form technological trajectories over time (Cimoli & Dosi, 1995). In terms of time–space, sectorial and regional polarization concentrates factors, assets, and competence in certain centers and withdraw them from the periphery (Perroux, 1955). Yet another stance is that R&D investment cycles perpetuate the dominance of outstanding innovation centers, with the capital for continued R&D investment coming from those who earn from innovation (Simmie, 2005).

Sixth, policies that promote innovation eventually exert location-specific influence that shapes spatial patterns of knowledge production and application (Sternberg, 1998). The regional impact of policy can be quite explicit, as when local new R&D infrastructure is built or when regionally oriented promotion schemes are established. There are also implicit spatial effects of policy, as when it essentially pursues wider national goals of supporting certain kinds of innovative activities (e.g., in the defense sector or other high-tech fields) but proves to favor particular regions because they have sectorial concentrations in precisely those spheres.

Although these arguments seem persuasive enough to buttress geographical explanations for the emergence of limited sets of innovative spaces and regional knowledge economies, critical questions and caveats must be mentioned. Most of the stated assumptions are far from being unequivocally confirmed by empirical evidence. Answers to the following sets of open questions are especially uncertain.

First, to what extent does innovation really require interorganizational collaboration rather than the exploitation of a company's internal competencies? How effectively do unintended spillovers of local knowledge actually facilitate innovativeness? And is it legitimate to make such general assumptions given the heterogeneous nature of innovation behaviors exhibited by different types of firms (e.g., small firms as compared to large multinational players)? Is innovation-oriented interaction between firms or between firms and academia really nurtured by spatial proximity? If not, what other factors or kinds of proximity—organizational, social, cultural, or epistemological—potentially foster collaboration? Apparently, the term sometimes relates not to a clearly defined space in which interaction actually takes place but rather to a space of contact potential—contact that may or may not be used in the innovation process. Often the requirement of proximity does not refer to actors' mutual relationships but rather to the shared endeavor to be near a common source of reputation, such as a renowned research laboratory specializing in a field of technology (Meusburger, 2008).

Given the temporary nature of creative activities as projects, another acute question arises: For which phases or tasks in the innovation process is the relevance of proximity greatest? The debate over this issue in economic geography (Amin & Cohendet, 2004; Malecki, 1997, 2000) may also profit from insights offered by organization sociology (Mintzberg, 1979) and educational or social geography (Meusburger, 1998, 2000, 2001, 2008). The degree to which innovating companies need to draw on nearby knowledge partners appears to depend mainly on four factors.

- *The scope of autonomy and leadership qualities of the firm.* A global market leader and prime mover of innovativeness usually requires less external and, in particular, less local interaction with knowledge providers than a follower company.
- *The stability and dynamics of the operative environment.* In sectors with a rather stable, reliable framework for competition, firms appear to need less interaction with partners than in sectors operating under highly volatile, competitive conditions.
- *The age of an organization, a product, or a relationship.* In the phase of a firm's creation or at the start of a new product life cycle, companies tend to require more options for spontaneous face-to-face contact with knowledgeable partners nearby than in more mature stages of production and operation. The proximity of innovating actors is important mainly in the early stages of business relationships in order to overcome uncertainty and to build trust, whereas long-standing links based on trust may function across longer distances as well.
- *The orientation, planning, or purposes of operational routines.* As confirmed by research on the contact requirements of various office functions, local collaboration and information exchange seem to be of more consequence in orientation and planning than in the purposes of operation (see Goddard & Morris, 1976; Meusburger, 2008).

These suppositions are plausible, but remain debatable. With globalization and the growing influence of globe-spanning knowledge flows between branch locations of multinational companies, does local collaboration really matter more than national and international interaction, such as that of connecting sector-specific pockets of innovation all over the world? Considering the evolutionary nature of relevant processes, it is crucial to improve the understanding of what the starting points or initial events are and what the chief institutions, actors, or firms are that trigger self-reinforcing innovation-driven economic development at a location. Why do some formerly successful localities gradually lose innovative strength, whereas selected new technology regions emerge as hubs of innovation by venturing to open locational windows of opportunity (Fuchs & Shapira, 2005)? There is also the practical question of why regional knowledge-based, innovative interaction between firms or between academia and industry is often so difficult to promote effectively.

Economic geographers are still trying to shed additional light on these issues. There is a trend toward taking immaterial and soft factors into account. Researchers now even ask whether and how discourses on and perceptions of the knowledge economy influence the innovation behavior of actors and related spatial patterns. The attention to these aspects reveals traits of a cultural turn in economic geography (Coe et al., 2007), a shift that opens avenues for inspiring collaboration, especially with the social or cognitive sciences.

Conceptualizing Innovation-Oriented Regional Development

The preceding arguments concerning the spatiality of innovation processes have informed various concepts that serve as guidelines for researching and cultivating regional knowledge economies. These concepts, many of them having ideas in common, are logically consistent sets of assumptions about pivotal facets of innovation-oriented local development. They may therefore be subsumed under the overarching label of "territorial innovation models" that have logical and evolutionary links connecting various notions (Moulaert & Sekia, 2003). Only a few essential characteristics of the most popular concepts can be outlined in the following paragraphs of this section (which also contain important references for further reading). Although the focus and perspective of these approaches vary somewhat, all of them underscore the value of actor relationships for the regional embedding of innovativeness.

Clusters

The concept of clusters (and the related one of industrial districts) draws on basic ideas introduced by Alfred Marshall in the 1920s. It relates primarily to agglomeration and networking advantages expected from local concentrations of firms and other organizational arrangements (higher education, R&D, and public agencies) that belong or cater to the needs of the same sectorial group and the same kind of value chain (Porter, 2000). It is assumed that one can create specific assets on the spot by linking complementary and competing actors who operate in the same technoeconomic field and by jointly marketing that particular strength to the outside world. Major assets are collective innovativeness, the sharing of inspiring information, and other positive outcomes, such as joint initiatives in staff training, supply and service purchasing, and systemic marketing. The cluster notion has elicited more criticism among researchers and more enthusiasm among regional policy-makers than almost any other development model has (Asheim et al., 2006; Martin & Sunley, 2003). Its emphasis on shared interests, homogeneous goals, and joint action runs counter to the quests for diversity and dissent associated with the notion of creativity (Meusburger, 2008).

Innovative, or Creative, Milieus

The concept of innovative, or creative, milieus was established by a group of European scientists who investigated different types of regions on the basis of a shared conceptual and methodological framework (Aydalot, 1986; Ratti et al., 1997). The purpose of the concept is to spell out favorable conditions for corporate innovativeness. It attributes great influence to key actors' local interactions based on preexisting informal personal relationships. Innovation-enhancing creativity is said to derive partly from new combinations of previously separate regional knowledge assets embodied by the various participating actors (who represent firms, R&D laboratories, and other organizations) and partly from the social and professional interaction of those actors. It is the informal-cum-formal nature of relationships that provides a seedbed particularly conducive to socially embedded learning and collective innovativeness based on mutual trust, common developmental objectives, and shared values (Camagni, 1991; Fromhold-Eisebith, 1995, 2001, 2004). The concept offers a convincing set of suppositions, but it is hard to use as an analytical tool or policy guideline. Its merits lie in highlighting the significance of social and soft assets as foundations of regional knowledge economies.

Regional Innovation Systems

The notion of regional innovation systems encompasses all the actors and processes that affect industrial innovativeness at a locality. It is assumed that regions are distinguished by place-specific constellations of organizations, institutions, regulations, and interactions that shape the production and application of knowledge (Cooke, 1992; Cooke et al., 2004). Consequently, basic conditions, major dynamics, and supportive frameworks of industrial innovativeness are said to exist mainly on a regional scale (as opposed to a national system of innovation; Lundvall, 1992). By incorporating features of political regulation, R&D infrastructure, and patterns of collaborative relationships between business actors at a locality, the concept of a regional innovation system is an attempt to integrate all leading determinants of corporate innovativeness. It may be used analytically to characterize regional settings or normatively to suggest actor and policy constellations that appear to be the most conducive to regional innovativeness. Such assessment of innovation policies, however, requires consideration of national and international system scales as well (Fromhold-Eisebith, 2007).

Learning Regions

As a response to a range of basic ideas for a fruitful combination of regional interaction and knowledge-driven industrial development, the concept of the learning region centers on sets of interdependent evolutionary dynamics (Hassink, 2001; Morgan, 1997). More visionary than analytical, the idea is to establish localized capabilities for collectively organizing the continuous combination and recombination of internal and external knowledge. Regions are expected to act like organizations that manage knowledge acquisition and innovation-oriented uses in well-conceived ways. It is suggested that progress results from constant learning processes that connect firms, academia, and industry. The concept also expediently combines inflows of know-how with the local circulation and exploitation of newly acquired knowledge.

Knowledge Regions

The knowledge region is one of the most recent conceptual inventions in the field and stems mostly from regional policy makers and academics. It, too, serves as a guideline for regional promotion rather than as a descriptive or analytical tool. Knowledge regions are ostensibly spatial entities that are marked by a conscious regard for the creation and use of knowledge in all aspects of regional development (Buschmann, 2005). Unlike earlier concepts, that of the knowledge region includes increases in awareness and activity within the population at large, so it extends beyond the sphere of firms and other organizations. Because knowledge-oriented dynamics are broadly based, regions are expected to attract international attention as well.

Shadow Catching: Trying to Capture the Essence of Innovative Spaces Empirically

Despite the lively debate on determinants of innovation-oriented regional development and a range of publications that map and collect statistics on spatial differentiations of innovativeness (Malecki, 2000; Pro Inno Europe & Innometrics,

2007), relevant processes, interdependencies, and causal relationships are difficult to discern through empirical research. Case studies and narratives abound, but little thorough empirical work representing an attempt to confirm general assumptions on a relatively wide basis. One reason is that the idiosyncratic nature of regional knowledge-driven developments obscures overarching principles (Simmie, 2005). Each region is known for specific constellations of actors, activities, and agency—as communicated by the notion of regional innovation systems. Because of the complex patterns of self-reinforcing systemic processes and the unpredictable social dynamics involved, even similar initial conditions of regions that have been compared may eventually have quite different outcomes. Moreover, structural settings that truly allow for scientifically sound comparison are rare. Investigated regions therefore inevitably fall into different categories. Even when this variegation is taken into account, valid generalizations are highly elusive, as shown by comparative research on innovative milieus (Ratti et al., 1997) and regional innovation systems (Cooke et al., 2000).

In this section, I illustrate the complications of providing evidence for key determinants of innovative spaces when it comes to regional knowledge-intensive collaboration, which is assumed to be a principal anchor of the processes involved. Empirical research by economic geographers studying this topic have so far mainly addressed two things: the importance of regional as opposed to extraregional relationships for corporate innovativeness, and the role of regional institutional constellations in affecting the innovation behavior of firms. Achievements and shortcomings will be briefly discussed for two recent large-scale empirical projects, the European Regional Innovation Survey (ERIS), conducted by a German team (Koschatzky & Sternberg, 2000; Revilla Diez, 2000, 2001, 2002), and the international project entitled Regional Innovation Systems in Europe (REGIS) (Cooke et al., 2000; Tödtling & Kaufmann, 1999).

Both ERIS and REGIS aim to broaden insight by including fairly large numbers of regions and informants. Through postal questionnaire surveys, ERIS has encompassed 8,600 organizations, including 4,200 manufacturing firms, in 11 regions of the European Union. REGIS has gathered information on 833 firms in another set of 11 EU regions. The written surveys have been complemented by personal qualitative interviews. A standard objective has been to gain understanding of the role and spatial reach of collaboration in the context of innovation, particularly with respect to links to customers, suppliers, academia, R&D organizations, and other actors. In order to find out whether regional interaction really matters, researchers have compared the relative shares of regional, national, and international collaboration, providing a predominantly quantitative picture of relational orientations that respondents have. The spatial nature of relational systems has also been interpreted by means of interregional comparison. In addition, performance indicators of regionally collaborating firms have been compared with those of noncollaborating ones in order to detect significant and meaningful discrepancies.

But what have these efforts actually revealed? Most regional shares of collaboration have not been remarkably large relative to national and international links. The overall results have been quite mixed with regard to dominating scales of interaction. But low relative shares of local collaboration do not necessarily suggest that regional embedding is irrelevant; local partners in small numbers, too, may provide vital support to a firm. What the studies show is that small firms in particular are biased toward local relationships and that the firms which are generally more "cooperative" (though not necessarily in a regional sense) appear to have experienced more innovation-related success than the less integrated ones (Cooke et al., 2000; Revilla Diez, 2002).

These findings may not be fully convincing in light of theoretical and conceptual reasoning on regional knowledge economies. Whatever the case, both projects have a few notable shortcomings relating to basic obstacles faced by any regional researcher who tries to approach the issue empirically. One problem is associated with the task of suitably demarcating the right "region" to be studied. What type of region fits to the space of innovation being sought: one defined by administrative (official) or rather by functional (interaction-related) boundaries? Are compared regions truly comparable in size, institutional and economic structure, administrative competencies, and other features?

A second drawback has to do with the question of causality—in this context, the correlation between collaborative behavior and the innovative performance of firms as mentioned above. Are companies innovative because they collaborate with others (as theory suggests), or do they collaborate because they are innovative and generally more dynamic, like extroverts (Malecki & Poehling, 1999)? This question cannot be satisfactorily answered until logical links between interaction patterns and the dynamics of firms are explicitly explored (ideally by qualitative personal interviews).

Third, shares of local innovation-oriented relationships have not been adequately interpreted against the backdrop of institutional options for collaboration in specific industries and regions. When low shares of collaboration are recorded in a region, one reason may be that there is not much local embedding (as argued in some studies). Alternatively, it could simply be that the region lacks potential partners because it does not offer sufficient scope for local networking. In either case, conceptual interpretations and policy implications may have to change.

Lastly, social dimensions of spaces of innovation, which are important glue that makes knowledge (in the form of knowledgeable people) stick to a place, have been somewhat neglected by large-scale empirical projects (Malecki, 2000; Meusburger, 2008). Future research must vastly increase the attention devoted to the social underpinnings of observed patterns of innovative interaction and knowledge exchange.

Conclusions and Open Questions

The argumentation of economic geographers and regional economists who write about spaces of innovation and regional knowledge economies shows that the spatiality of such phenomena has been well perceived and taken as a challenge to scientific reasoning and application-oriented research. But despite convincing assumptions about the role of regions and proximity for innovativeness and the knowledge economy, there is still too little known about the actual mechanisms of creating and using interactive knowledge in and across regions. People know they are living in a space of interrelated and internally networked innovative places, in an "archipelago of situated knowledges" (Thrift, 1999, p. 303). But they are still searching for robust insights into relevant geographies of processes, innovation phases, and time, including the spatial nature and dynamics of innovation-oriented "project ecologies" (Grabher, 2002).

We researchers have immense difficulty empirically capturing general and overarching features that characterize spatial patterns of knowledge-based processes and geographical sides of actor behavior. We must probably keep relying on individual case studies about regions, organizations, and institutions because the path to theory-oriented generalizations is essentially blocked by the complex idiosyncrasies of the locational dynamics governing economic, social, cultural, technological, and political factors that bear on regional innovativeness.

Other disciplines can assist well with the geographer's tasks, affording welcome inspiration and methodological resources, especially for the following open questions:

- In what respects does individual or team creativity aggregate to regional innovativeness? How can insights from psychology or the cognitive sciences inform geographical views on regional innovation? To make progress in these areas, it would be useful to learn more about the contexts that encourage individual creativity and about the ideal size and other social-group factors of creative, innovative teams.
- How do cultural characteristics and ethical norms contribute to the regional creation and economic application of knowledge? Approaches and findings of sociology, philosophy, or anthropology could help explain such influences.
- With which methodological approaches can researchers empirically capture the spatiality, relational aspects, and regional embedding of innovativeness more adequately than has been the case thus far? In all disciplines, whether quantitatively or qualitatively oriented, this question pushes the range of applicable methods to its very limits.
- What means of promotion can proliferate innovativeness beyond large urban agglomerations toward more peripheral regions? This issue deserves additional joint exploration by economic geographers, political economists, political scientists, and others scholars.

Economic geography, although offering some meaningful insights into regional milieus of economic creativity, inevitably has a few weaknesses in its research perspectives and procedures. They reveal that we researchers need to learn from other views or approaches and to continue widening our horizons of reasoning. May this chapter open avenues for increased interdisciplinary collaboration in this realm.

Acknowledgements The author cordially thanks Peter Meusburger, the participants of the September 2006 workshop on "Knowledge and Space," and an anonymous referee for their instructive and perceptive comments on an earlier version of this chapter.

References

Amabile, T. M. (1996). *Creativity in context: Update to The social psychology of creativity*. Boulder, CO: Westview.

Amin, A. & Cohendet, P. (2004). *Architectures of knowledge: Firms, capabilities and communities*. Oxford, England: Oxford University Press.

Asheim, B., Cooke, P., & Martin, R. (2006). *Clusters and regional development: Critical reflections and explorations*. London: Routledge.

Audretsch, D. B. & Feldman, M. P. (1996). R&D spillovers and the geography of innovation and production. *The American Economic Review, 86*, 630–640.

Aydalot, P. (1986). *Milieux innovateurs en Europe* [Milieus of innovators in Europe]. Paris: GREMI.

Brown, J. S. & Duguid, P. (2002). Local knowledge: Innovation in the networked age. *Management Learning, 33*, 427–437.

Buschmann, B. (2005). *Bedeutung von Wissen für Regionen* [The importance of knowledge for regions]. Position paper in online conference documentation for Perspektiven für Land & Leute: Regionale Kooperationen für Existenzgründungen, Berlin, February 1, 2005. Retrieved July 28, 2008, from http://www.bmfsfj.de/Publikationen/perspektiven/3-Arbeitsgruppen/Arbeitsgruppe-regionales-prozess-und-wissensmanagement/positionspapier-dr-birgit-buschmann.html

Buttimer, A. (Ed.) (1983). *Creativity and context* (Lund Studies in Geography, Series B, No. 50). Lund, Sweden: The Royal University of Lund.

Camagni, R. (1991). *Innovation networks: Spatial perspectives*. London: Bellhaven.

Cimoli, M. & Dosi, G. (1995). Technological paradigms, patterns of learning and development: An introductory roadmap. *Journal of Evolutionary Economics, 5*, 243–268.

Coe, N., Kelly, P., & Yeung, H. W. (2007). *Economic geography. A contemporary introduction*. London: Blackwell.

Cooke, P. (1992). Regional innovation systems: Competitive regulation in the new Europe. *Geoforum, 23*, 365–382.

Cooke, P. & Schwarz, D. (Eds) (2007). *Creative regions—technology, culture and knowledge entrepreneurship*. London: Routledge.

Cooke, P., Boekholt, P., & Tödtling, F. (2000). *The governance of innovation in Europe—Regional perspectives on global competitiveness*. London: Pinter.

Cooke, P., Heidenreich, M., & Braczyk, H.-J. (2004). *Regional innovations systems: The role of governance in a globalized world*. London: Routledge.

Cooke, P., De Laurentis, C., Tödtling, F., & Trippl, M. (2007). *Regional knowledge economies: Markets, clusters and innovation*. London: Routledge.

Csikszentmihalyi, M. (1997). *Creativity: Flow and the psychology of discovery and invention*. New York: HarperCollins.

European Commission & Eurostat (2004). *Statistics on science and technology in Europe: Data 1991–2002* (p. 26). Luxembourg: Office for Official Publications of the European Communities.

Feldman, M. P. & Desrochers, P. (2003). Research universities and local economic development. *Industry and Innovation, 10*, 5–24.

Fromhold-Eisebith, M. (1995). Das "kreative Milieu" als Motor regionalwirtschaftlicher Entwicklung. Forschungstrends und Erfassungsmöglichkeiten [The "creative milieu" as an engine of regional economic development: Research trends and survey possibilities]. *Geographische Zeitschrift, 83*, 30–47.

Fromhold-Eisebith, M. (2001). Kreatives Milieu [Creative milieu]. In E. Brunotte, H. Gebhardt, M. Meurer, P. Meusburger, & J. Nipper (Eds.), *Lexikon der Geographie* (Vol. 2, pp. 269–272). Heidelberg, Germany: Spektrum Akademischer Verlag.

Fromhold-Eisebith, M. (2004). Innovative milieu and social capital—Complementary or redundant concepts of collaboration-based regional development? *European Planning Studies, 12*, 747–765.

Fromhold-Eisebith, M. (2007). Bridging scales in innovation policies: How to link regional, national and international innovation systems. *European Planning Studies*, *15*, 217–233.
Fuchs, G. & Shapira, P. (2005). *Rethinking regional innovation and change: Path dependency or regional breakthrough?* New York: Springer.
Goddard, J. B. & Morris, D. (1976). The communications factor in office decentralization. *Progress in Planning*, *6*, 1–80.
Gordon, I. R. & McCann, P. (2005). Innovation, agglomeration, and regional development. *Journal of Economic Geography*, *5*, 523–543.
Grabher, G. (2002). Cool projects, boring institutions: Temporary collaboration in social context. *Regional Studies*, *36*, 205–214.
Grabher, G. (2004). Learning in projects, remembering in networks? Communality, sociality and connectivity in project ecologies. *European Urban and Regional Studies*, *11*, 103–123.
Greif, S. (2004). Wissen als Ressource: Patentaktivitäten [Knowledge as a resource: Patent activities]. In Leibniz-Institut für Länderkunde (Ed.), *Nationalatlas Bundesrepublik Deutschland—Unternehmen und Märkte* (pp. 82–83). Munich, Germany: Elsevier Spektrum.
Hassink, R. (2001). The learning region: A fuzzy concept or a sound theoretical basis for modern regional innovation policies? *Zeitschrift für Wirtschaftsgeographie*, *45*, 219–230.
Howells, J. (2002). Tacit knowledge, innovation and economic geography. *Urban Studies*, *39*, 871–884.
Jekel, T. & Fromhold-Eisebith, M. (2003). Identität und regionalwirtschaftliche Innovativität. Diskussion eines hypothetischen Zusammenhangs [Identity and regional economic innovativeness: Discussion of a hypothetical link]. *Geographische Zeitschrift*, *91*, 115–129.
Koschatzky, K. & Sternberg, R. (2000). R&D cooperation in innovation systems—some lessons from the European Regional Innovation Survey (ERIS). *European Planning Studies*, *8*, 487–501.
Lambooy, J. G. (2002). Knowledge and urban economic development: An evolutionary perspective. *Urban Studies*, *39*, 1019–1035.
Lundvall, B. Å. (1992). *National systems of innovation*. London: Pinter.
Malecki, E. J. (1997). *Technology and economic development: The dynamics of local, regional and national competitiveness*. Harlow, England: Longman.
Malecki, E. J. (2000). Knowledge and regional competitiveness. *Erdkunde*, *54*, 334–351.
Malecki, E. J. & Poehling, R. M. (1999). Extroverts and introverts: Small manufacturers and their information sources. *Entrepreneurship & Regional Development*, *11*, 247–268.
Malmberg, A. & Maskell, P. (2002). The elusive concept of localization economies: Towards a knowledge-based theory of spatial clustering. *Environment and Planning A*, *34*, 429–449.
Martin, R. & Sunley, P, 2003, Deconstructing clusters: Chaotic concept or Policy panacea? *Journal of Economic Geography*, *3*, 5–35.
Massey, D. (2005). *For space*. London: Sage.
Meusburger, P. (1998). *Bildungsgeographie. Wissen und Ausbildung in der räumlichen Dimension* [Geography of education: Knowledge and education in the spatial dimension]. Heidelberg, Germany: Spektrum Akademischer Verlag.
Meusburger, P. (2000). The spatial concentration of knowledge: Some theoretical considerations. *Erdkunde*, *54*, 352–364.
Meusburger, P. (2001). Geography of knowledge, education, and skills. In N. J. Smelser & P. B. Baltes (Eds.), *International encyclopedia of the social & behavioral sciences* (Vol. 12, pp. 8120–8126). Amsterdam: Elsevier.
Meusburger, P. (2008). The nexus of knowledge and space. In P. Meusburger (Series Eds.) & P. Meusburger, M. Welker, & E. Wunder (Vol. Eds.), *Knowledge and space: Vol. 1. Clashes of knowledge* (pp. 35–90). Dordrecht, The Netherlands: Springer.
Mintzberg, H. (1979). *The structuring of organizations: A synthesis of the research*. Englewood Cliffs, NJ: Prentice Hall.
Morgan, K. (1997). The learning region: Institutions, innovation and regional renewal. *Regional Studies*, *31*, 491–503.
Morgan, K. (2004). The exaggerated death of geography: Learning, proximity and territorial innovation systems. *Journal of Economic Geography*, *4*, 3–21.

Moulaert, F. & Sekia, F. (2003). Territorial innovation models: A critical survey. *Regional Studies, 37*, 289–303.

Perroux, F. (1955). Note sur la notion de pôle de croissance [Note on the notion of a growth pole]. *Économie Appliquée, 7*, 307–320.

Porter, M. E. (2000). Locations, clusters, and company strategy. In G. L. Clark, M. S. Gertler, & M. P. Feldman (Eds.), *The Oxford handbook of economic geography* (pp. 253–274). Oxford, England: Oxford University Press.

Pro Inno Europe & Innometrics (2007). *European innovation scoreboard 2006: Comparative analysis of innovation performance*. Brussels: Commission of the European Union (Available at http://www.proinno-europe.eu/doc/EIS2006_final.pdf, last accessed November 26, 2007).

Ratti, R., Bramanti, A., & Gordon, R. (1997). *The dynamics of innovative regions: The GREMI-approach*. Aldershot, England: Ashgate.

Revilla Diez, J. (2000). The importance of public research institutions in innovative networks—empirical results from the metropolitan innovation systems Barcelona. *European Planning Studies, 8*, 451–463.

Revilla Diez, J. (2001). Innovative links between industry and research institutes–how important are they for firm start-ups. In K. Koschatzky, M. Kulicke, & A. Zenker (Eds.), *Innovation networks—Concepts and challenges in the European perspective* (pp. 93–108). Heidelberg, Germany: Physica.

Revilla Diez, J. (2002). *Betrieblicher Innovationserfolg und räumliche Nähe—Zur Bedeutung innovativer Kooperationsverflechtungen in metropolitanen Verdichtungsregionen, die Beispiele Barcelona, Stockholm und Wien* [Industrial innovation success and spatial proximity—on the significance of innovative cooperation networks in regions of urban agglomerations as exemplified by Barcelona, Stockholm, and Vienna]. Münster, Germany: LIT-Verlag.

Schamp, E. W. & Lo, V. (2003). Knowledge, learning, and regional development: An introduction. In V. Lo & E. W. Schamp (Eds.), *Knowledge, learning, and regional development* (pp. 1–12) (Wirtschaftsgeographie 24). Münster, Germany: LIT.

Simmie, J. (2003). Innovation and urban regions as national and international nodes for the transfer and sharing of knowledge. *Regional Studies, 37*, 607–620.

Simmie, J. (2005). Innovation and space: A critical review of the literature. *Regional Studies, 39*, 789–804.

Sternberg, R. (1998). *Technologiepolitik und High-Tech-Regionen—ein internationaler Vergleich* [Technology policy and high-tech regions—an international comparison]. Münster, Germany: LIT-Verlag.

Storper, M. (1995). The resurgence of regional economies, ten years later: The region as a nexus of untraded interdependencies. *European Urban and Regional Studies, 2*, 191–221.

Storper, M. & Venables, A. J. (2004). Buzz: Face-to-face contact and the urban economy. *Journal of Economic Geography, 4*, 351–370.

Thrift, N. J. (1999). Steps to an ecology of place. In D. Massey, J. Allen, & P. Sarre (Eds.), *Human geography today* (pp. 295–322). Cambridge, England: Polity.

Tödtling, F. & Kaufmann, A. (1999). Innovation systems in regions in Europe—a comparative perspective. *European Planning Studies, 7*, 699–717.

Chapter 11
The Unconscious City: How Expectancies About Creative Milieus Influence Creative Performance

Jens Förster

Paris: City of love. New York: The city that never sleeps. Hamburg: The Reeperbahn. Jerusalem: The Holy City. These pairings are only few examples of specific contents people associate with cities or other places. True or not, correspondences of this kind are represented in memory and may pop into mind when people are asked about their knowledge of certain cities. For example, if you are asked to tell a friend what you think about New York, you may recall the many galleries in Chelsea, the roaring nightlife and the clumsy, Woody-Allen-like neurotic genius who avoids your glances.

Over the last decades, researchers in social psychology have made enormous progress in understanding how these sorts of stereotypes are represented in memory (e.g., Collins & Loftus, 1975; Higgins, 1996; Higgins et al., 1977; Huber et al., 2001; Wyer, 2004; Wyer & Radvansky, 1999). More relevant for this chapter, social psychology shows that representations of this nature influence people's feelings, thinking, and behavior (for reviews, see Dijksterhuis & Bargh, 2001; Förster & Liberman, 2007). Maybe the most provocative insight from this research is that representations like these can influence the behavior of people even without them knowing or desiring it (Moskowitz et al., 2004). Such outcomes are called "priming effects." I first summarize classic research focusing on human judgments and behavior and suggest that even creative thinking can be affected by unconscious activation of stereotypes. I then outline research that social psychologists have conducted on creative thinking and continue by arguing that some cities are associated with creativity. I suggest that such thinking works like a self-fulfilling prophecy, that is, that the creativity of people increases when they are reminded of a creative place. I then recount an experiment in which undergraduate participants were exposed to the names of particular cities so briefly that conscious recognition of

J. Förster (✉)
Jacobs University Bremen and Universiteit van Amsterdam
Campus Ring 1, PO BOX 750561, D-28725 Bremen, Germany
e-mail: j.a.forster@uva.nl

P. Meusburger et al. (eds.), *Milieus of Creativity*, Knowledge and Space 2,

the names was impossible (subliminal presentation). I also report the results of a posttest showing whether this exposure influenced the participants' performance on a creativity task, the prediction being that the creativity of the participants would be automatically increased when they were subconsciously reminded of cities that they associated with a creative milieu.

Classic Research on the Effects of Priming on Human Judgments and Behavior

Higgins et al. (1977) invited participants to an experiment that consisted of two apparently unrelated studies. The first study was a color-naming task for part of which participants had to memorize words, purportedly to increase the difficulty of the task. Some participants were given positive words (self-confident, independent, persistent, adventurous), whereas other participants were given negative words (conceited, aloof, stubborn, reckless). The second study was an impression-formation task in which participants read a description of a person named Donald, who performed a series of ambiguous behaviors that could be regarded as adventurous or reckless (e.g., Donald thought about crossing the Atlantic in a sailboat). Participants then wrote a free description of the target person and marked on a rating scale how much they liked him. Results indicated that Donald was rated as more adventurous and less reckless by the positive priming group than by the negative priming group.

A typical priming experiment has two phases: (a) priming in which participants are exposed to some information, and (b) an ostensibly unrelated perception, memory, or judgment task in which participants' responses to a target stimulus are examined. Psychologically, in the first phase of a priming experiment a stimulus presentation (e.g., reckless or adventurous) leads to activation of associated memory structures. In the second phase this activated memory structure influences the evaluation of a specific target or task performance.

From the perspective of the participant, there is no relation between the two phases (the procedure described above is often referred to as the unrelated-task paradigm). Any effect that the first stage has on the second stage is not noticed by the participant, a fact that rules out conversational effects (e.g., the person thinks that the first stage's information is useful and thus uses it) and motivational effects (e.g., the person wants to help the experimenter find the expected results). Thus, one can conclude that the activated information unconsciously influences the second phase.

It is argued that the first stage enhances the accessibility of the primed construct and that constructs with higher accessibility are more likely to be used than those with lower accessibility (Higgins, 1996). In the study by Higgins et al. (1977) the behavior "thinking about crossing the Atlantic in a sailboat" could be perceived as both adventurous and reckless. A higher accessibility of one of these constructs leads to the perception of the target in terms of that construct. In other words, accessible knowledge may help disambiguate complex stimulus sets.

In a typical priming task people are not aware that they were affected by the first task. If directly asked how they formed their judgments, they would most likely reply that they based their judgments on the stimulus itself. Participants in the "Donald" experiment, for instance, said that the target person was, indeed, involved in reckless or adventurous activities (i.e., the judgment was "about" the target's behavior and not about the priming event). In logical terms, the influence of the prime is not justifiable; nobody would maintain that the target should be perceived differently because of a previous task of color-naming. Yet these effects occur reliably and have been replicated and used in many experiments by social psychologists (for reviews see Förster & Liberman, 2007; Higgins, 1996).

A different, now classic study by Devine (1989) showed that subliminally priming associations with the category of African Americans (e.g., Negroes, Blacks, lazy) influenced hostility ratings of an ambiguously aggressive target. Because the stereotype against African Americans in the United States is typically related to aggressive concepts, Devine argued that subconscious activation of the social category of Blacks would render such concepts accessible, and would thereby influence judgments about other, unrelated targets. More specifically, participants in the studies were exposed to word sets that were either highly related to the categories or not highly related to them. The stimuli were presented at the computer subliminally, that is, for a duration so short that participants were not even aware of the fact that words had appeared. In the second phase the participants received a description of a person who behaved in an ambiguously aggressive way. In other words, the person could have been described as assertive or aggressive. Participants who had been exposed to words related to the African-American stereotype rated the person to be more aggressive than did participants who had not been exposed to the stereotype. Notably, Devine never displayed words that were directly related to aggression, so it can be assumed that the mere activation of the category Blacks also activated associated information that had an additional influence. It was assumed that social categories are stored in memory in the shape of associative or semantic networks (Collins & Loftus, 1975) in which activation spreads from the category node (e.g., Black) to more specific information (e.g., aggressive).

More recently, Bargh and his research group have reported data showing that even behavior can be unconsciously affected by priming procedures. For example, Bargh, Chen, and Burrows (1996) subliminally primed participants by showing them either African-American faces or Caucasian faces. The former condition was intended to prime aggression, for many U.S. Americans have been shown to associate African Americans with aggression. After priming, participants were filmed as they completed a boring computer task during which the computer unexpectedly crashed. Participants' behavior was more aggressive after exposure to African-American faces than after exposure to Caucasian faces. Researchers have found such priming effects with other behaviors as well, including slowed walking by participants after they had been primed by information associated with the elderly (Bargh et al., 1996), enhanced helping behavior after semantic priming of helping-related concepts (Walther et al., 2001), and conformity after priming of conformity-related information (Epley & Gilovitch, 1999).

Does priming also influence people's creative thinking? In order to make such an assumption, one first needs to accept that creativity can change within the social context. Is there any reason to believe that creativity is not only a fixed personality trait (i.e., differs between people) but also something that can change depending on the situation? I now summarize research on creativity in context.

The Social Psychology of Creativity

Human creativity has traditionally been considered a personality trait (see e.g., Eysenck, 1993; Simonton, 1991). As Guilford (1950) put it: "In its narrow sense, creativity refers to the abilities that are most characteristic of creative people" (p. 444). However, social psychologists have demonstrated that creative cognition is significantly responsive to situational and social contextual variation, suggesting that the notion of creativity as a personality trait is too narrow. In her pioneering work, Amabile and her colleagues (for a review see Amabile, 1996) has shown that creativity is undermined by the provision of extrinsic rewards and the expectation of social evaluation. Thus, a person is not always living up to his or her creative potential; creative thinking can be enhanced by situationally induced instructions.

A different extensive research program examining the role of moods on creative thought showed that individuals demonstrate more creativity under conditions of situationally induced positive mood than under conditions of a neutral mood (e.g., Isen et al., 1987; Murray et al., 1990; for reviews see also Clore et al., 1994; Hirt et al., 1996; Isen, 2000; Wyer et al., 1999). In a similar vein, Seibt and Förster (2004) showed that negative expectations about one's own group in relation to a certain task undermines creativity, whereas positive expectations enhance it. Another extensive research program has shown that the creative thinking of people is diminished when they dwell on security while performing a creativity task but that creative thinking is enhanced when they focus on ideals (Förster et al., 2006; Friedman & Förster, 2000, 2001, 2002, 2005, 2008; for reviews see Förster & Friedman, 2003). In situations of threat or lack of security, people usually adopt an analytic and detail-oriented processing style that impedes creative thinking, whereas they start exploring their environments and use a more global and flexible processing style when security issues are replaced by ideals, hopes, and aspirations (Friedman & Förster, 2008).

All the studies reported above used manipulations that were set up in a specific situation and demonstrate that a person's creative performance can be influenced by transient states. The studies thus question the notion of traits or "talents" as the only factor influencing creative and analytic thinking style (see also Simonton, 2000). Research documents influences that priming procedures have on creative thinking as well. For example, Förster, Friedman, Butterbach, and Sassenberg (2005) argued that cues of deviancy facilitate creative thinking in a relatively automatic fashion. Given that creativity varies situationally, it seems only reasonable to infer that certain conceptual primes activate behavioral plans or procedures that serve to facilitate creative cognition.

Indeed, there are several reasons why creative thinking should be enhanced when concepts of deviancy are accessible. First, creative products are, by definition, more unusual than others and thus deviate from standard or customary solutions (for a review see Amabile, 1996; Sternberg, 1999). To give some prominent examples, Bruner (1962) sees the creative product as anything that produces "effective surprise" in the observer (p. 5) and Barron (1955) stresses that an original response "should have a certain state of uncommonness in the particular group being studied" (p. 479). Second, the producers of creative outcomes, such as artists, product developers, and scientists, are usually *perceived* as different and unusual in many respects even though this stereotype is sometimes not true (for example, Franz Kafka and Thomas Mann had quite ordinary lives). However, as Sternberg and Lubart (1995) put it, creative people usually "defy the crowd" and produce products that are good but not exactly the kind of thing people would expect. Again, although this notion might be a myth and not a necessary condition for being creative, the stereotype (right or wrong) of creative people and creativity might therefore be associated with deviancy. Third, people might themselves have experienced that creative solutions they have produced were socially nonconforming and sometimes at risk of nonacceptance by some members of society (Sternberg & Lubart, 1991, 1995).

Last but not least, the *processes* that lead to creative thinking might be experienced as somewhat unusual compared to other situations. McGraw (1978) and Taylor (1960) argue that creative tasks are heuristic as opposed to algorithmic. According to Hilgard and Bower's definition (1975), algorithmic tasks are those for which the path to solution is straightforward and clear. In other words, they are tasks for which an algorithm already exists and just has to be retrieved from memory (see Förster & Denzler, 2006). For heuristic tasks, however, no algorithm yet exists; they do not have a clear, readily available, and identifiable solution or solution path (see Amabile, 1996). This creative process might itself therefore be less frequently used, less conventional, more unusual or deviant, and newer than the algorithmic approach. For all these reasons the authors already referred to in this section have assumed the existence of an associative link between deviancy and creativity. In situations that cue the mental representation of deviancy, the supposition is that cognitive procedures usually leading to creative solutions are likely to be activated, a response that facilitates creative performance.

In a series of experiments, the existence of a semantic deviancy-creativity link and its consequences for creative performance was examined. For example, in one experiment, participants were asked to think about members of social groups that are usually perceived to be equally creative but that differ in perceived conformity. Specifically, one group of participants was asked to think about a punk for a short period, whereas the other group was asked to think about an engineer. It was predicted that creative insight would be promoted after thinking about the more deviant figure, the punk. In this experiment, an analytical reasoning (i.e., algorithmic) task was also administered. One may speculate that the stereotype of an engineer is associated with analytical thinking, a link that might facilitate tasks of this variety. Consistently, the results showed that the participants primed with punk outperformed participants primed with engineer in the creativity task but that the reverse was true for the analytic task. Even thought priming in this experiment was not subliminal; participants consciously thought about the punk, and the experimental procedure ruled out conscious adoption of such algorithmic or heuristic thinking styles by the participants. The two phases were introduced to the participants as two different studies that had been coupled for economic reasons. When asked afterwards, participants showed no suspicion that the tasks were related. Thus, one may conclude that social categories such as "punk" or "engineer" can remind participants of certain ways of thinking that are automatically triggered upon perception of such a group.

To minimize the possibility of conscious effects, the authors used an even more subtle manipulation in a different experiment. While completing a creativity task, some of the participants were incidentally exposed to an abstract painting that symbolically represented the concept of deviancy, whereas the others were exposed to a painting representing the concept of conformity. More specifically, the professionally framed painting in the nondeviancy cue condition was 100 × 70cm and had a light green background with four rows of three darker green Xs (20 × 15cm) symmetrically arranged in the foreground. The poster resembled common works of abstract art. In the deviancy cue condition, the X at the far right side in the third row appeared in yellow instead of dark green, symbolically conveying the concept of deviancy. It was predicted and found that incidental exposure to the "deviant" painting would facilitate creative generation. Notably, participants were not aware of this influence, so the experiment demonstrated that creativity can in fact change within the social context by mere exposure to "reminders" of creative thinking, which in this case were cues of deviancy that are part of the creative thinking process. Inspired by the literature on geography and creative milieus, I wondered whether associations with creative cities could also enhance creative thought.

City and Creativity

Some places are indisputably associated with creativity. Montmartre, Trastevere, Berlin's Prenzlauerberg (now instead of Kreuzberg), the Schwabing district in Munich, or New York's Chelsea (now instead of SoHo), for example, might come to mind or ring a bell when people think about creative places. Artists have in fact been attracted by such "creative" places. Some of these persons have moved there, and the tourism industry has fostered such expectancies to attract visits by average people. Similarly, scientists and product developers may be attracted by "creative" places, institutions, or colleges and universities (e.g., Ivy League schools and production centers such as Silicon Valley). The reasons for such accumulation of creative power may be manifold, including the desire to build networks; to experience input from other creative people; to be accepted within a diverse, nonconformist community; to create research teams; and to feel special, to name a few. However, the aforementioned research suggests that such environments may facilitate creative ways of thinking by virtue of conscious and unconscious reminders. In other words, the mere thought that one lives in Chelsea may trigger creative processing styles, and this effect may additionally bolster creative processing. I even go a step further: Because cities have to be represented in memory to have an effect, the mere thought of a creative place is likely to trigger creative strategies to solve problems, so one does not even need to be in that city in order to increase creativity.

To test this hypothesis, I devised an experiment in which participants were exposed to the names of different cities (Amsterdam, London, New York, and Cologne). I found in a pretest with 40 undergraduate students that these four cities were all associated with creativity by some people in the population. For the study reported below, I also asked the participants whether they associated these cities with creativity. I predicted that only those people who have such an association stored in memory would show enhanced creativity, whereas people for whom such stereotypes did not exist would not exhibit influence of the prime.

This reasoning was based on previous research showing (for male participants) relations between alcohol cues and expectancies of sexual arousal—relations that, in turn, influence arousal-related judgments (Friedman et al., 2005). More specifically, it was found that unconscious exposure to alcohol-related cues, relative to neutral cues, increased the tendency to judge women as sexually attractive but had no effect on judgments of their intelligence. This finding however, was true only for those who held a belief in the aphrodisiac qualities of alcohol. Those who did not expect alcohol to increase their sexual desire were unaffected by the alcohol-related primes. Thus, I suggest similar boundary conditions for the following study, which uses city names as primes. Unconscious reminders will trigger creative thinking only in those participants who have stored a link between creativity and a certain city in their long-term memories.

Study

Participants and Design

We recruited 54 students majoring at the University of Bremen in different disciplines except psychology (27 women; 27 men; average age: 23.40 years; whereby gender had no effects) for an experimental battery on psychological tasks that lasted 2 h. The following study was part of this battery. The experiment had a two-factorial design: 4 City (Amsterdam vs. London vs. New York vs. Cologne) × 2 Expectancy (primed city is creative vs. primed city is not creative). All conditions were tested between participants, and participants were randomly assigned to the priming conditions. Participants were paid €14 for their participation.

Procedure

Participants were asked first whether or not they found the cities of Amsterdam, London, New York, and Cologne creative on a scale from 1 (*not creative at all*) to 9 (*very creative*). This measure was used to decide whether participants had certain associations with or expectations of particular cities. My team and I then used a subliminal priming task (see Chartrand & Bargh, 1996; Mussweiler & Förster, 2000) in which participants, depending on their assigned group, were primed with the word "Amsterdam," "London," "New York," or "Köln" (Cologne) for 70ms on the pretext of an attention task. The names were flashed on the screen at unpredictable places and times, and the participants had to press designated keys to indicate as quickly and accurately as possible whether the flash had appeared on the right or the left side of the screen. Previous research in our laboratories had shown that 70ms was long enough to ensure that the stimulus words and masks always appeared to the participants but short enough to rule out conscious recognition. The priming stimuli were therefore unlikely to evoke controlled processes. We closely followed recommendations by Chartrand and Bargh (1996), including all suggested precautions for preventing conscious awareness of these stimuli, including very brief exposure to the primes, immediate masking, and placement of stimulus content in the parafoveal processing area (for details see Mussweiler & Förster, 2000). All participants completed 48 experimental trials, which took approximately 2min.

After priming, participants were asked to do another, ostensibly unrelated task. They were shown a cartoon picture of a dog sleeping on a sofa, asked to find the most creative caption for the picture, and told that their solution should not be something that was completely meaningless. Participants had 5min to do the task, but none of the participants needed that much time.

Two experts who did not know the identities of the participants were given the solutions and asked to evaluate them for creativity on a scale from 1 (*not creative at all*) to 5 (*very creative*). The averaged ratings served as the main dependent variable of creativity. On the basis of the expectancy measures, it was decided whether participants believed that the city they were primed with was more creative than the other cities that had served as primes. More specifically, the mean rating of all the cities with which the participants had not been primed was subtracted from the mean of the primed city. A resulting difference higher than zero indicated that participants thought the city they were primed with was more creative than the other three cities. A 4 (City) × 2 (Expectancy) ANOVA was used to test the hypothesis statistically.

Debriefing and interviews took place at the end of the study. Participants did not notice any connection between the tasks, nor did they notice that the priming task contained words. Most of them reported having seen "flashes," some stated that "letter strings" may have appeared, but none of them was able to state the city names that actually had been presented to them.

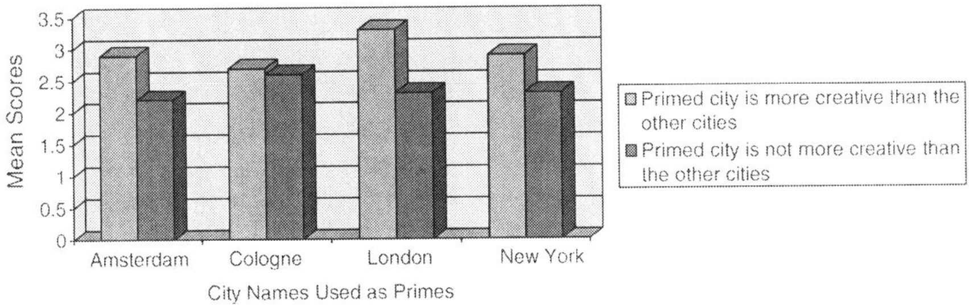

Fig. 11.1 Mean creativity scores (expert judgments) as a function of city name and expectancy: Means on a scale from 1 (*not creative at all*) to 5 (*very creative*)

Results

The mean creativity ratings are shown in Fig. 11.1. Analysis showed no main effects for city and no interaction effect ($F < 1$). However, the predicted main effect of expectancy was highly significant ($F\,(1,46) = 7.47$, $p < .01$). Having been primed with a city, participants who generally thought of that city as being relatively creative generated a qualitatively more creative solution to the cartoon of the dog sleeping on the couch (average creativity score: $M = 3.0$ across City conditions) than did participants primed with a city that they generally thought was not more creative than the other cities ($M = 2.3$).

Discussion

The results show that beliefs about the creativity of a given city led to an unconscious activation of processes that bolstered creative thinking, thereby enhancing creative generation relative to situations in which no such belief existed. Before far-reaching conclusions can be formulated, though, the experiment by my team and me obviously needs to be replicated with different participant samples and different creativity tasks. Ordinarily, experimental research by social psychologists examining the feelings, thinking, and behavior of *average* people centers on rather mundane kinds of creativity, such as finding unusual uses for a brick or creating titles for objects and cartoons. Because that research does not take into account more extreme expressions of creativity, it is a problematic basis on which to make generalizations to creative geniuses.

However, our experiment does provide initial evidence that cities can be mentally associated with creativity and that this association can increase creative generation processes. Interestingly, none of the cities we used as primes was specifically attributable to increased creativity. For some participants London was the most creative city, whereas for others it was Cologne, Amsterdam, or New York. This finding may reflect our participants' individual differences with respect to associations with these cities. Thus, replications in other labs should take into account that their population may associate creativity with rather different cities. We can conclude, though, that merely being reminded of a creative place enhanced participants' creative thought. The preceding experiment adds to research in social psychology by showing how environmental cues can change creative performance. If creativity were only a talent or personality trait, one would not be able to explain why situational manipulations as used in our study led to higher creativity for some participants than for others. We can also conclude that city names and other reminders of creativity can unconsciously trigger processes of creative thinking if a link between those reminders and creativity exists in peoples' long-term memory.

The following paragraphs address some of the issues that may arise from the results we have reported. I first discuss whether the notion of unconscious processes interferes with the notion of free will. I also explore the question of which factors can intensify or change the relation between cities and creativity and then examine some of the implications that these considerations have for real life.

Controlling Behavior by Priming: Myth or Reality?

The idea that priming may change a person's behavior and even such important processes as creative thinking is profound and unsettling. The possibility of altering behavior by using subliminal priming is even frightening. But even though the research community no longer doubts that such effects can occur in real life (Hassin et al., 2005; Wyer, 2008), there are important limitations on them.

The first is the factor of awareness. As used in the study presented above, subliminal primes preclude conscious correction because people are not aware of the prime. But people who do know that such influences are occurring can counteract them. To the best of my knowledge, there is no research on how people control their automatically elicited thinking styles, but investigations on priming related to evaluations show that correction for this influence may well occur. When participants in experiments were made aware that they were primed with the African-American stereotype, for instance, they showed no influence of the primes and had thus corrected for it (Devine, 1989).

The second important limitation on the effects of subliminal priming is the motivation and the time to correct it. A person who lives in a "creative city" and does not want to think creatively because he or she needs to concentrate on analytic tasks instead can consciously focus on the details rather than permit distraction by excessively global, heuristic processes that interfere with goal pursuit. The question is whether he or she wishes to correct for undesired known influences. Such a desire is unlikely in the case of creative thinking because in modern societies creative thinking is a highly desirable skill most people want to improve. In our study, however, we did not include cities that are associated with *lack* of creativity. Those cities might impede creative thinking, so a person who becomes aware of such a detrimental influence may start to counteract it. Correction processes that need cognitive resources also need time to overcome the automatic influence. Thus even though one may have the intention to correct for an unconscious influence, one may not have the time to correct. More research is needed to test these hypotheses derived from priming research (see Förster & Liberman, 2007).

As implied by the first two limitations on the effects of subliminal priming, the third one is the direction of the influence (see Strack, 1992; Strack & Hannover, 1996; for a recent review see Förster & Liberman, 2007). Knowing the direction of influences is important because people may have incorrect lay theories about what increases or decreases their creative thinking processes. For example, research on whether positive mood enhances creativity contradicts the typical stereotype that creative processes need to be based on depression, suffering, and the like. Thus, if a person thinks positive mood decreases creativity, a correction process may fail because of this incorrect theory.

Aside from limitations on effects, another aspect of subliminal priming is the question of whether manipulations by, say, the media can enhance or decrease thinking. Although there are no studies showing such influences on creativity, research in the domain of consumer research suggests that manipulations are possible. In a series of experiments, Strahan, Spencer, and Zanna (2002) demonstrated that priming people with thirst-related cues such as pictures of soft drinks made them consume more beverages. In these studies, the primes (cans of soft drinks) were presented subliminally in a film, showing that exposure to drink-related stimuli intensified experienced thirst. However, the study also showed an important limitation of such automatic effects, namely, the fact that they were found only for people who came to the lab thirsty. They experienced increased thirst after exposure to the flashed images of soda cans. Thus, a preexisting motive to pursue the goal (in this case, to drink) seems to be necessary for priming to have an effect on behavior.

Other studies, too, have found that priming affects behavior only if it is consistent with an already existing motivation. For example, alternative focal goals may block the effects of priming on behavior. Macrae and Johnston (1998) found that participants who had been primed with helpfulness but who were in a hurry to get to the next experimental session did *not* stop to help a confederate pick up his pens (scattered on the floor). By contrast, participants who had no conflicting goal did give help after helpfulness priming. Concurrent activation of incompatible goals might also explain the fact that people do not get up in the movie theatre to buy soft drinks when they are subliminally primed with these refreshments during the film: Even though some of these people may be thirsty, as demonstrated in Strahan et al. (2002), they have the focal goal of watching the movie and therefore do not get up to buy a soft drink (see Bargh et al., 1996).

Moreover, the same primed concepts can have different behavioral implications, depending on the situation in which the behavior is relevant. New York, for instance, may prime creativity when a person is watching a family blockbuster movie, but it may prime fear when a person is reminded of 9/11—the aerial terrorist attack that destroyed the city's twin towers, killing thousands of people on September 11, 2001 (see Wyer, 2004).

Further research is needed to examine similar effects in the domain of creativity priming. It seems reasonable to suggest that these kinds of influence can be found and produced, though it is beyond my expertise to judge how strong they are. This conjecture may be seen as bad news for the idea of humans as rational thinkers who decide according to their free will what they think and do. On the other hand, the rational thinker may acknowledge the fact of being able to shape the environment and the associations of a society actively. If one no longer wants New York to be associated with 9/11, one may well replace the association by calling attention to other features of this city.

Factors Enhancing the Impact of Cities on Creativity

From the research summarized in this chapter, one may conclude that a city's name is not the only factor with an impact on creativity, that many natural aspects of certain cities stimulate creative thinking as well. It is important to note that experimental studies like the one described above do not reflect reality but rather test a theory, in this case whether mere reminders of creative cities are able to increase creative performance. The study indicates that images, media reports, narratives, or any other reminders of creative cities may have a similar influence. In real life, however, a variety of factors may independently influence how and what people think when they are in or reminded of certain places.

As research on mood and creativity shows (see Isen, 2000), in cities that enhance mood it is likely that creativity is also enhanced. Thus, all the factors involving art festivals, music, ballet, funny parties, mood enhancing colors, architecture, and landscaping, to name only a few, may improve creativity. This connection does not mean that it is necessary to promote all places, occasions, and endeavors to create a "fun" society. At memorials, of course, one wants exactly the opposite of fun and comic humor. In this regard, city planners seeking to shape a creative milieu may wish to keep in mind the insights from research in social psychology.

Similarly, research shows that deviancy (see Förster et al., 2005) may increase creativity. Exposure to deviancy may, of course, occur in cities that allow for and even attract diverse individuals and groups. For example, neighborhoods with mixed ethnic backgrounds may increase the likelihood of encountering uncommon and dissimilar opinions, behaviors, and people, and such exposure can stimulate creative thinking. Research on multicultural work teams shows their superiority to relatively uniform work teams when it comes to creative production (see Crosby et al., 2003). Likewise, diverse architecture and cultural events from diverse backgrounds may create a challenging environment. I hasten to add, though, that exposure to exotic events needs to occur in a nonthreatening way. People are usually afraid of uncommon events (Berlyne, 1974), so diversity needs to be experienced as pleasant rather than frightening, as when people are educated for diversity or when novelty is introduced in a friendly and nonoffensive way.

Emphasizing ideals rather than security should promote creative behavior (Friedman & Förster, 2008; Higgins, 1997) because that sort of focus opens processing up to new, but risky, events and triggers holistic processing that bolsters creative thought (Förster & Higgins, 2005). It might be instructive to examine whether police guards, fences, multiple warning signs, and other constant reminders of security (or lack thereof) decrease creativity in certain places. Historical events such as 9/11 or sudden economic slumps endangering employment and wealth may also immediately change the creative milieu by inducing fear and stressing security and shelter.

Lastly, encouraging environments should promote creativity, as Seibt and Förster's (2004) study shows. People such as caretakers, leaders of organizations and institutions, and politicians can mentally shape the creative atmosphere of a city or other place. If they want a creatively rather than analytically thinking society, they should focus on the positive aspects of performances and should encourage activities, sometimes even risky ones, rather than try to prevent them.

Although these implications are truly speculative, they may inspire further research. All the predictions I have made are based on well-established theories in social psychology that have been confirmed in the laboratory. It is now time to test the implications of these findings in natural and more complex environments.

Acknowledgements The writing of this chapter, and part of the research discussed therein, was supported by grants FO 244/6-3 and FO 392/8-2 from the Deutsche Forschungsgemeinschaft. Sarah Horn and Konstantin Mihov are thanked for editing the manuscript. Laura Dannenberg, Markus Denzler, Amina Özelsel, and Janina Marguc are thanked for invaluable discussions.

References

Amabile, T. M. (1996). *Creativity in context: Update to The social psychology of creativity*. Boulder, CO: Westview.

Bargh, J. A., Chen, M., & Burrows, L. (1996). Automaticity of social behavior: Direct effects of trait construct and stereotype activation on action. *Journal of Personality and Social Psychology, 71*, 230–244.

Barron, F. (1955). The disposition towards originality. *Journal of Abnormal and Social Psychology, 51*, 478–485.

Berlyne, D. E. (1974). *Studies in new experimental aesthetics*. New York: Wiley.

Bruner, J. (1962). Introduction: The new educational technology. *American Behavioral Scientist, 6*, 5–7.

Chartrand, T. L. & Bargh, J. A. (1996). Automatic activation of social information processing goals: Nonconscious priming reproduces effects of explicit conscious instructions. *Journal of Personality and Social Psychology, 71*, 464–478.

Clore, G. L., Schwarz, N., & Conway, M. (1994). Affective causes and consequences of social information processing. In R. S. Wyer & T. K. Srull (Eds.), *Handbook of social cognition, 1*, 323–418. Hillsdale, NJ: Erlbaum.

Collins, A. M. & Loftus, E. F. (1975). A spreading activation theory of semantic processing. *Psychological Review, 82*, 407–428.

Crosby, F. J., Iyer, A., Clayton, S., & Downing, R. A. (2003). Affirmative action: Psychological data and the policy debates. *American Psychologist, 58*, 93–115.

Devine, P. G. (1989). Stereotypes and prejudice: Their automatic and controlled components. *Journal of Personality and Social Psychology, 56*, 5–18.
Dijksterhuis, A. & Bargh, J. A. (2001). The perception-behavior expressway: Automatic effects of social perception on social behavior. In M. Zanna (Ed.), *Advances in experimental social psychology* (Vol. 33, pp. 1–40). San Diego, CA: Academic.
Epley, N. & Guilovich, T. (1999). Just going along: Nonconscious priming and conformity to social pressure. *Journal of Experimental Social Psychology, 35*, 578–589.
Eysenck, H. (1993). Creativity and personality: An attempt to bridge divergent traditions. *Psychological Inquiry, 4*, 238–246.
Förster, J. & Denzler, M. (2006). Selbst-Regulation. In W. Bierhoff & D. Frey (Eds.), *Handbuch der Psychologie: Vol. 3. Sozialpsychologie* (pp. 128–132). Berlin: Hogrefe.
Förster, J. & Friedman, R. (2003). Kontextabhängige Kreativität [Context-dependent creativity]. *Zeitschrift für Psychologie, 211*, 149–160.
Förster, J. & Higgins, E. T. (2005). How global versus local perception fits regulatory focus. *Psychological Science, 16*, 631–636.
Förster, J. & Liberman, N. (2007). Knowledge activation. In E. T. Higgins & A. W. Kruglanski (Eds.), *Social psychology: Handbook of basic principles* (2nd ed., pp. 201 – 231). New York: Guilford.
Förster, J., Friedman, R., Butterbach, E. M., & Sassenberg. K. (2005). Automatic effects of deviancy cues on creative cognition. *European Journal of Social Psychology, 35*, 345–360.
Förster, J., Friedman, R., Özelsel, A., & Denzler, M. (2006). Enactment of approach and avoidance behavior influences the scope of perceptual and conceptual attention. *Journal of Experimental Social Psychology, 42*, 133–146.
Friedman, R. S. & Förster, J. (2000). The effects of approach and avoidance motor actions on the elements of creative insight. *Journal of Personality and Social Psychology, 79*, 477–492.
Friedman, R. S. & Förster, J. (2001). The effects of promotion and prevention cues on creativity. *Journal of Personality and Social Psychology, 81*, 1001–1013.
Friedman, R. S. & Förster, J. (2002). The influence of approach and avoidance motor actions on creative cognition. *Journal of Experimental Social Psychology, 38*, 41–55.
Friedman, R. S. & Förster, J. (2005). Effects of motivational cues on perceptual asymmetry: Implications for creativity and analytical problem solving. *Journal of Personality and Social Psychology, 88*, 263–275.
Friedman, R. S. & Förster, J. (2008). Activation and measurement of motivational states. In A. Elliott (Ed.), *Handbook of approach and avoidance motivation* (pp. 233–246.). Mahwah, NJ: Erlbaum.
Friedman, R. S., McCarthy, D. M., Förster, J., & Denzler, M. (2005). Automatic effects of alcohol cues on sexual attraction. *Addiction, 100*, 672–681.
Guilford, J. P. (1950). Creativity. *American Psychologist, 5*, 444–454.
Hassin, R. R., Uleman, J. S., & Bargh, J. A. (2005). *The new unconscious*. New York: Oxford University Press.
Higgins, E. T. (1996). Knowledge activation: Accessibility, applicability, and salience. In E. T. Higgins & A. W. Kruglanski (Eds.), *Social psychology: Handbook of basic principles* (pp. 133–168). New York: Guilford.
Higgins, E. T. (1997). Beyond pleasure and pain. *American Psychologist, 52*, 1280–1300.
Higgins, E. T., Rholes, W. S., & Jones, C. R. (1977). Category accessibility and impression formation. *Journal of Experimental Social Psychology, 13*, 141–154.
Hilgard, E. R. & Bower, G. H. (1975). *Theories of learning*. Englewood Cliffs, NJ: Prentice-Hall.
Hirt, E. R., McDonald, H. E., & Melton, J. R. (1996). Processing goals and the affect-performance link: Mood as main effect or mood as input? In L. L. Martin & A. Tesser (Eds.), *Striving and feeling: Interactions among goals, affect, and self-regulation* (pp. 303–328). Mahwah, NJ: Erlbaum.
Huber, D. E., Shiffrin, R. M., Lyle, K. B., & Ruys, K. I. (2001). Perception and preference in short-term work priming. *Psychological Review, 108*, 149–182.
Isen, A. M. (2000). Positive affect and decision making. In M. Lewis & J. Haviland-Jones (Eds.), *Handbook of emotions* (2nd ed., pp. 417–435). New York: Guilford.

Isen, A. M., Daubman, K. A., & Nowicki, G. P. (1987). Positive affect facilitates creative problem solving. *Journal of Personality and Social Psychology, 52*, 1122–1131.
Macrae, C. N. & Johnston, L. (1998). Help, I need somebody: Automatic action and inaction. *Social Cognition, 16*, 400–417.
McGraw, K. O. (1978). The detrimental effects of reward on performance: A literature review and a prediction model. In M. R. Lepper & D. Greene (Eds.), *The hidden costs of rewards* (pp. 33–60). Hillsdale, NJ: Erlbaum.
Moskowitz, G. B., Li, P., & Kirk, E. R. (2004). The implicit volition model: On the preconscious regulation of temporarily adopted goals. In M. P. Zanna (Ed.), *Advances in experimental social psychology* (Vol. 36, pp. 317–404). New York: Academic.
Murray, N., Sujan, H., Hirt, E. R., & Sujan, M. (1990). The influence of mood categorization: A cognitive flexibility interpretation. *Journal of Personality and Social Psychology, 59*, 411–425.
Mussweiler, T. & Förster, J. (2000). The sex → aggresion link: A perception-behavior dissociation. *Journal of Personality and Social Psychology, 79*, 507–520.
Seibt, B. & Förster, J. (2004). Risky and careful processing under stereotype threat: How regulatory focus can enhance and deteriorate performance when self stereotypes are active. *Journal of Personality and Social Psychology, 87*, 38–56.
Simonton, D. K. (1991). Emergence and realization of genius: The lives and works of 120 classical composers. *Journal of Personality and Social Psychology, 61*, 829–840.
Simonton, D. K. (2000). Creativity: Cognitive, personal, developmental, and social aspects. *American Psychologist, 55*, 151–158.
Sternberg, R. J. (1999). *Handbook of creativity*. New York: Cambridge University Press.
Sternberg, R. J. & Lubart, T. I. (1991). An investment theory of creativity and its development. *Human Development, 34*, 1–31.
Sternberg, R. J. & Lubart, T. I. (1995). An investment approach to creativity: Theory and data. In S. M. Smith, T. B. Ward, & R. A. Finke (Eds.), *The creative cognition approach* (pp. 271–302). Cambridge, MA: MIT Press.
Strack, F. (1992). The different routes to social judgments: Experiential versus informational strategies. In L. L. Martin & A. Tesser (Eds.), *The construction of social judgments* (pp. 249–275). Hillsdale, NJ: Erlbaum.
Strack, F. & Hannover, B. (1996). Awareness of influence as a precondition for implementing correctional goals. In P. M. Gollwitzer & J. A. Bargh (Eds.), *The psychology of action: Linking cognition and motivation to behavior* (pp. 579–596). New York: Guilford.
Strahan, E. J., Spencer, S. J., & Zanna, M. P. (2002). Subliminal priming and persuasion: Striking while the iron is hot. *Journal of Experimental Social Psychology, 38*, 556–568.
Taylor, D. (1960). Thinking and creativity. *Annals of the New York Academy of Sciences, 91*, 108–112.
Walther, E., Müller, D., & Schott, O. (2001). Automatisches soziales Verhalten. Wie wirkt sich die Aktivierung der Konzepte Altruismus und Egoismus auf Hilfeleistung aus? *Zeitschrift für Experimentalle Psychologie, 48*, 248–257.
Wyer, R. S., Jr. (2004). *Social comprehension and judgment: The role of situation models, narratives, and implicit theories*. Mahwah, NJ: Erlbaum.
Wyer, R. S., Jr. (2008). The role of knowledge accessibility in cognition and behavior: Implications for consumer information processing. In C. Haugvedt, F. R. Kardes, & P. M. Herr (Eds.), *Handbook of consumer research* (pp. 31–76). Mahwah, NJ: Erlbaum.
Wyer, R. S., Jr. & Radvansky, G. A. (1999). The comprehension and validation of social information. *Psychological Review, 106*, 89–118.
Wyer, R. S., Jr., Clore, G. L., & Isbell, L. M. (1999). Affect and information processing. In M. P Zanna (Ed.), *Advances in experimental social psychology* (Vol. 31, pp. 1–77). San Diego, CA: Academic.

Chapter 12
Conceptual Spaces

Margaret A. Boden

Spaces exist in the mind as well as on land and sea. They cannot be seen, or traversed by ship or on foot, not even with seven-league boots. For they are abstract spaces, or styles of thinking. But although the geographer's methods of dealing with terrestrial space do not reliably apply to the conceptual realm, there are other methods with which to navigate, explore, and, occasionally, even transform them. In this sense, one has the advantage over the geographer: How many explorers of oceans and continents are able to change the nature of the territory they traverse?

These mental spaces cover all domains of thought (and thoughtful action), from chemistry to choreography, from pottery to poetry. They enable us human beings to come up with ideas that are new, surprising, and valuable; they enable us to be creative.

Creativity can happen as a result of three different types of mental process: combinatorial, exploratory, and transformational (Boden, 2004, pp. 40–146). As I show in this chapter, two of these forms of creativity are defined in terms of conceptual spaces. Because conceptual spaces vary across cultures, the exercise and appreciation of creativity often depend upon specific cultural knowledge. Someone from a different culture may not even be able to recognize the novelty involved, let alone understand and appreciate it. In other words, only someone who inhabits the relevant conceptual spaces, only someone whose mind contains those spaces, can generate these novelties and appreciate their significance.

By the same token, someone who seeks to understand what creativity is, and how it is even possible, needs to consider the mental geography of conceptual spaces. That person also needs to ask how they can be negotiated: What are the equivalents of seven-league boots for the creative thinker?

M.A. Boden (✉)
Sussex University, Department of Informatics, Falmer, Brighton BN1 9QN, United Kingdom
e-mail: m.a.boden@sussex.ac.uk

P. Meusburger et al. (eds.), *Milieus of Creativity*, Knowledge and Space 2,

Ambiguities Galore

To say that creativity involves the generation of ideas that are new, surprising, and valuable is to say something highly ambiguous. Each of the four key terms explored in this chapter (idea, new, surprising, and valuable) has more than one meaning. This lack of clarity explains why disagreements about creativity are often carried on at cross-purposes.

As implied by the preceding illustrations, I use the term *idea* as a convenient catch-all in this context. The new idea is sometimes an idea in the normal sense: a concept, or a scientific theory, for example. But it need not be. It may be a new method for producing artifacts, such as a novel way of casting bronze. Or it may be a general style of painting or sculpting. Or it may be a musical composition, a new dance-step, and so on. Indeed, it is customarily up to the individual to decide the "size," or scope, of the idea in a particular case. Consider Samuel Taylor Coleridge's poem "The Rime of the Ancient Mariner," for instance. Is one to count the *poem* as the idea, each individual image in it, each rhetorical ruse employed to communicate the poet's meaning, or some combination of these elements? The first option is certainly inadequate. Speaking lazily of "the poem" as the creative idea does not actually get very far. If the purpose is to understand the creative processes responsible for "The Rime of the Ancient Mariner," or any other poem, one needs to consider the many specific images, the various literary conceits, and the rhetorical structure of the poem as a whole (Boden, 2004, pp. 125–146).

Each of the three criteria of creativity is ambiguous, too. To begin with, novelty has two importantly different senses. On the one hand, an idea may be new to the person who had it. It is a first-time occurrence within his or her particular mental biography. It illustrates what may be called *P-creativity*, with *P* standing for "psychological." On the other hand, an idea may be new to the whole of human history as far as anyone knows. That is, it illustrates *H-creativity*.

The psychological point of view, from which one seeks to understand how creativity happens and how it is even possible, reveals that P-creativity is the more important concept, for every case of H-creativity is a special case of P-creativity. (If the occurrence of a given idea is the first in human history, then it must also be the first occurrence to the mind of its originator.) From the historical point of view, H-creativity is the focus of special interest. But because every H-creative idea is P-creative, too, one can always ask what type of P-creativity was involved. And that question pertains no matter what type of conceptual space is involved.

The second criterion of creativity, surprise, has three meanings. One may be surprised because something is statistically unusual or contrary to commonsense expectations—like an outsider winning the Derby. Surprise may also result because one had not realized that the new idea had been a possibility all along, like discovering a beautiful village tucked away in a hollow between two spurs of the motorway. (Its location had always been marked on the map, but one hadn't examined the whole map closely.) Third, one may be surprised by something previously thought to be impossible and still regarded as utterly counterintuitive.

Think, for example, of events that religious people categorize as miracles, or imagine the impact that the introduction of television or wireless radio has on nonphysicists.

The third criterion, that of being valuable, has even more meanings than the other two do. For various reasons, they cannot be wholly pinned down. What is valuable in music is not necessarily valuable in, or even applicable to, architecture. What is valuable in a baroque fugue may not be valuable in the blues. And what was thought valuable in the 1860s may have been scorned in the 1960s.

As that remark suggests, values can change (sometimes virtually overnight) as a result of shifts in fashion. Some shifts of that kind are deliberately engineered for commercial purposes, whereas others arise from unpredictable events (such as what an admired celebrity chooses to wear to a party). There may be some universal or near-universal values (symmetry and shininess) grounded in shared evolutionary history (Boden, 2006, Section 8.iv.c). But even these values can be deliberately transgressed and their opposites admired in their stead (as with the highly asymmetrical architecture of Frank Gehry or Daniel Libeskind).

To make matters worse, it is possible to know *that* one finds something valuable without knowing just what it is that one is valuing about that thing. What is it about a Bach fugue or a mathematical proof that instills people's admiration? You may answer "complexity, rigor, elegance...." But just what do these terms mean? And to just which aspects of the fugue or the proof do they apply?

In the following sections, I assume that *novel* means P-creative (which sometimes includes H-creativity) and that being *valuable* means being determined or contested, depending on the context and domain. Lastly, I show that the three sorts of surprise correspond to the three types of creativity—combinatorial, exploratory, and transformational.

The Three Types of Creativity

Saying that there are three types of creativity does not mean that every creative idea or artifact results from only one of them. It may involve all three. The brief preceding comments about "The Rime of the Ancient Mariner," suggest several different kinds of creativity were involved in the composition of those verses. The tripartite division of creativity is intended for analytical purposes. In real life, more than one of those processes may be involved in the generation of what is normally regarded as "one" creative product. Rather than asking whether a theory or artwork as such is creative, yes or no, it is thus ordinarily more sensible to ask whether this or that *aspect* of a new theory or artwork is creative, and *in just what way* it is creative.

Combinatorial creativity involves the generation of unfamiliar (and interesting) juxtapositions of familiar ideas. It tends to give rise to the first type of surprise mentioned above. Just as one does not expect an outsider to win the Derby, for such upsets do not normally happen, one does not expect ideas X and Y to be combined, for they seem to be mutually irrelevant. Everyday examples of combinatorial creativity include visual collage (in advertisements and MTV videos, for instance), much poetic imagery, all types of analogy (verbal, visual, or musical), and the unexpected juxtapositions of ideas found in political cartoons in newspapers.

Combinatorial creativity requires mental resources, of course: *ex nihilo nihil fit* (nothing comes of nothing). Indeed, these resources are richly complex. However, they are neither tightly structured nor confined to a particular domain or style of thought. On the contrary, they are highly general. They include all known concepts and conceptual linkages (e.g., superordination, similarity, and contradiction), commonsense knowledge of the world, individual experience, and knowledge of the sounds and spellings of words (e.g., synonyms, homonyms, and syllable counts).

If you doubt the generality of the mental resources that people invest in combinatorial creativity, consider these riddles:

> *Question*: What do you call a depressed train? *Answer*: A low-comotive.
>
> *Question*: What do you call a strange market? *Answer*: A bizarre bazaar.
>
> *Question*: What kind of murderer has fiber? *Answer*: A cereal killer.

Even a little thought suffices to show that one needs a wide range of knowledge about the world and language in order to understand these riddles or to come up with them in the first place. Much more thought is needed, however, to appreciate the detail of what is going on here. For example, just how do people manage to fill in familiar schemas (like *What do you call an X that is Y?* and *What kind of X is Y?*) to produce these riddles and similar ones? Clearly, combinatorial creativity requires concepts—legions of them. (Potentially, it involves the entire range of concepts and world knowledge in someone's head.) But it does not require conceptual spaces, which are both more limited and more tightly structured.

Exploratory and transformational creativity are significantly different. Both are grounded in some previously existing, culturally accepted, structured style of thinking. It may be a board game (chess or Go, perhaps), an area of theoretical chemistry (e.g., aromatic molecules), or a particular type of music or sculpture. In exploratory creativity, the existing stylistic rules or conventions are used to generate novel structures (ideas), whose possibility may or may not have been realized before the exploration took place. (You may or may not have had reasons to expect to find that village nestling between the Motorways.) It can also involve the search for and testing of the specific stylistic limits concerned. Just which types of structure can be generated within this space, and which cannot?

Transformational creativity is what leads to "impossibilist" surprise. The reason is that some defining dimension of the style or conceptual space is altered—so that one can generate structures that could not be generated before. Imagine altering the rules in chess according to which pawns can not jump over other pieces. If pawns were allowed to move in that manner, as knights have always been permitted to do, then it would be possible to play games of chess that were literally impossible before. The greater the alteration and the more fundamental the stylistic dimension concerned, the greater the shock of impossibilist surprise.

However, not every dimension will have been changed. (Otherwise, why call it a new form of *chess*?) There will be both structural continuities and structural discontinuities between the untransformed space and its seemingly impossible successor. If you retain some feature of the game that you enjoyed before the change, then you will find something to enjoy in the transformed version. You may, however, be so averse to jumping pawns that you decide to revert to old-style chess. In art and science, where some relevant cultural style is presupposed, there will be continuities and discontinuities, too. And the discontinuities may or may not be regarded as valuable.

After the transformation, the artist or scientist may add new rules, defining *and exploring* the new style more fully. One clear example is the composer Arnold Schoenberg (Rosen, 1976). He transformed the space of Western tonal music by dispensing with the fundamental home-key constraint. In other words, it was no longer necessary for every composition to favor one of a finite number of sets of seven notes (the major and minor scales). Atonality was born. In addition to abandoning this constraint, Schoenberg experimented by introducing new ones. At one point, for instance, he said that each composition should contain every note of the chromatic scale. Musical exploration could then ensue on this basis. But the radical transformation was the decision to eliminate the constraint of a home key. Similarly, the space of benzene derivatives continued to be explored in detail long after the transformational leap to benzene ("ring") chemistry had been taken.

In Praise of Exploration

In general, the most highly valued type of creativity is the transformational. (This preference is less pronounced in literature than in other arts and sciences because language offers scope for especially rich creative combinations, and the theme of human motivation offers huge exploratory potential.) However, novel transformations are relatively unusual. Novel exploration is much more common.

All artists and scientists spend most of their working time engaged in combinatorial creativity, exploratory creativity, or both. When pursuing exploratory creativity, they are experimenting with accepted conceptual space, asking what new structures lie within it, how it can be traversed, what its limits are, and whether it can be marginally tweaked to stretch its boundaries a little. Only very seldom are the limits made to vanish through radical transformation of one or more dimensions of the space. That rarity becomes abundantly clear during a visit to a retrospective exhibition of a painter's work—especially if the canvasses are displayed chronologically. One discerns a certain style being adopted and then explored, clarified, and tested. It may be superficially tweaked (e.g., adoption of a different palette). But it is highly uncommon to see a transformation taking place. The artists whose names are recorded in the history books are usually remembered above all for changing the accepted style.

Typically, stylistic change meets initial resistance, often taking quite some time to be accepted. After all, transformational creativity by definition involves the breaking and ignoring of culturally sanctioned rules. The history of science and of art abounds with examples of creative prophets not honored in their own countries during their lifetimes—if even then. The German-Hungarian physician

Ignaz Semmelweiss (1818–1865), who recommended that obstetricians wash their hands in disinfectant to prevent puerperal fever, and the artist Vincent van Gogh (1853–1890) are just two examples of the many visionaries whose new style of thinking was not celebrated until many years after they had died. But these occupational hazards are not apt to deter creative scientists and artists in their drive to explore. Van Gogh, despite his failure to impress most people with his art, focused largely on testing the potential of his new way of painting. Similarly, the chemist Friedrich von Kekule (1829–1896), having transformed the theoretical chemistry of benzene, spent much of his time thereafter studying its derivatives.

That overriding desire is entirely understandable. Someone coming across a new conceptual space, especially one with untapped and possibly huge potential, is likely to want to explore it. This observation is true whenever a teenager or adult adopts a new interest or falls prey to a new craze. (In this context "new" typically means P-new. The person eagerly explores a conceptual space that is new to him or her but that may actually be hundreds of years old.) But such intensity is also the case even when the conceptual space did not exist yesterday, prior to that someone's radical transformation of yesterday's space today.

How Computational Psychology Can Help

Computers are notoriously unforgiving. They do not allow you to wave your hands and say "You know what I mean!" They do not know; you really do have to tell them. (I am speaking informally. Whether computers can "really" understand anything at all is not relevant here; see Searle, 1980.) It follows that if one wants to test the power and coherence of a psychological theory by implementing it in a computer model, one must express it with exceptional clarity and great detail. Having done so, one can see whether or not the program does what it was expected to do when it is actually run. Power and coherence can be tested in this way, but not truth. To discover whether the human mind does things in the same sort of way as a particular computer does requires empirical evidence.

In other words, psychology can be helped by work in artificial intelligence. Artificial intelligence is the attempt to make computer systems of various kinds do things that human (and/or animal) minds can do. They include using natural language, recognizing visual images, making medical diagnoses, and avoiding obstacles when walking across a room. In addition, of course, they include creative thought.

With respect to psychological theories of creativity, the need for clarity and detail forces people to inquire into mental processes by asking questions that simply would not have occurred to them otherwise. When discussing combinatorial creativity above, for example, I presented three riddles and said that they could be generated or appreciated only by someone possessing a rich store of concepts and world knowledge. I also noted that understanding just how to do so is a highly complex matter but did not elaborate. I am prepared to bet a large sum that you, the reader, did not go into much detail, either, when considering the questions I asked at that point. And if you had, you would not have gotten very far, no matter how hard you tried. However, some preliminary, but already impressively complex, answers have been offered in computational terms. All three riddles were generated by a computer model that used a semantic network consisting of over 30,000 items and several kinds of links (Binsted & Ritchie, 1997). That work at least partly answers my question—How do people manage to fill in familiar riddle schemas (like *What do you call an X that is Y?* and *What kind of X is Y?*).

I say only "partly" for two reasons. First, as already pointed out, no computer model alone suffices to explain human thought or behavior. At best, it is a theory or a hypothesis suggesting how human thought and behavior *could* or *might* be generated. Additional evidence about human beings is required in order to discover whether they do in fact generate it in anything like that way. In many cases, there has been no such theory or hypothesis in the first place. Second, this riddle-generating computer model, like most "psychological" computer models, has limitations that human minds do not have. There are many riddles that it cannot cope with. In other words, the theory implemented in it is flawed, incomplete, or both. But there is no shame in that condition. The philosopher Karl Popper (1963) famously argued that *science in general* progresses not only by conjectures (hypotheses) but also by refutations. The specific nature of the refutations of a computer model may be clear enough to suggest how it might be improved, or at least to show just where the major inadequacies lie.

At least three computer models of creativity have been continuously refuted and improved over the last quarter of a century or more (Boden, 2006, Sections 13.iv.b–c). One is a line-drawing and coloring program whose author, a highly acclaimed artist since the 1960s, now described as a "world-class" colorist, whereas he himself is only a "first-class" colorist. The second model of creativity is a computer composer that can generate scores in the style of a wide range of famous composers and that can amaze even highly skeptical musicians (Cope, 2001, 2006; Hofstadter, 2001). The third computer model designs alphabetic fonts. It can already generate a new 26-letter font on the basis of five seed letters, and its programmers hope to enable it to use only one seed letter and even to generate its own seeds (Hofstadter & McGraw, 1995; McGraw, 1995).

It is no coincidence that all these three programs and projects are studies of exploratory, not combinatorial, creativity. In a sense, exploratory creativity is more manageable, more limited, than the combinatorial variety is. Whereas riddle-makers can call on their entire body of concepts and world knowledge, artists or scientists working within a given thinking style are constrained, but also enabled and guided, by the conceptual space they are exploring. Psychologists seeking to model their creativity must define the space in computational terms and must provide ways of exploring and perhaps even tweaking it.

That work is usually not easy. It requires expert knowledge of the relevant domain (space, style) on the part of the programmer or some collaborator. Often, discoveries made in the process of defining the space are news, perhaps very interesting news, to the scholars specializing in the area concerned. For example, one computer model managed to distill the essence of Frank Lloyd Wright's

Prairie Houses, which a leading architectural historian had declared to be "occult" (Koning & Eizenberg, 1981, especially p. 322).

For the record, transformational creativity can in principle be modeled, too. Specifically, it is possible through evolutionary computing. In that approach, "genetic algorithms" enable a program to change its own rules—that is, to modify its own conceptual space—by random alterations inspired by point mutations and crossovers in biology. In each "generation," one or two products are selected to be the "parents" of the next generation. These methods have been applied in various domains of science and art, including colored graphics (Sims, 1991; Todd & Latham, 1992). The major problem lies not in generating transformations but rather in generating *valuable* transformations and in defining an automatic rather than an interactive fitness function to effect the selections. Transformational creativity *always* breaks some preexisting rule, so is always subject to rejection. As one might say, "You can make a conceptual space transform, but you cannot always make it *create*."

Conclusion

The exploits of explorers such as Marco Polo, James Cook, and David Livingstone are rightly applauded. They showed courage, insight, and persistence in the face of failure. They pushed contemporary geographical knowledge to its limits—and beyond. Indeed, it is that "beyond" that has won them such acclaim.

The exploits of creative thinkers are rightly applauded, too. But it is too seldom realized that much of what they do is also exploration, albeit of conceptual spaces, not physical ones. If researchers can define those spaces and specify ways of navigating and even transforming them it will be possible not only to map the contents of the mind but also to understand how it is possible to generate novel, surprising, and valuable ideas. Concepts drawn from artificial intelligence can help do just that.

References

Binsted, K. & Ritchie, G. D. (1997). Computational rules for punning riddles. *Humor: International Journal of Humor Research, 10*, 25–76.

Boden, M. A. (2004). *The creative mind: Myths and mechanisms* (revised and expanded 2nd ed.). London: Routledge.

Boden, M. A. (2006). *Mind as machine: A history of cognitive science*. Oxford, England: Oxford University Press.

Cope, D. (2001). *Virtual music: Computer synthesis of musical style*. Cambridge, MA: MIT Press.

Cope, D. (2006). *Computer models of musical creativity*. Cambridge, MA: MIT Press.

Hofstadter, D. R. (2001). Staring Emmy straight in the eye—and doing my best not to flinch. In D. Cope (Ed.), *Virtual music: Computer synthesis of musical style* (pp. 33–82). Cambridge, MA: MIT Press.

Hofstadter, D. R. & McGraw, G. (1995). Letter spirit: Esthetic perception and creative play in the rich microcosm of the Roman alphabet. In D. R. Hofstadter & The Fluid Analogies Research Group (FARG), *Fluid concepts and creative analogies: Computer models of the fundamental mechanisms of thought* (pp. 407–466). New York: Basic Books.

Koning, H. & Eizenberg, J. (1981). The language of the prairie: Frank Lloyd Wright's prairie houses. *Environment and Planning B, 8*, 295–323.

McGraw, G. (1995). *Letter spirit (part one): Emergent high-level perception of letters using fluid concepts*. Unpublished doctoral thesis, Department of Cognitive Science, Indiana University, Bloomington, Indiana.

Popper, K. R. (1963). *Conjectures and refutations: The growth of scientific knowledge*. London: Routledge & Kegan Paul.

Rosen, C. (1976). *Schoenberg*. Glasgow: Collins.

Searle, J. R. (1980). Minds, brains, and programs. *Behavioral and Brain Sciences, 3*, 473–497.

Sims, K. (1991). Artificial evolution for computer graphics. *Computer Graphics, 25*, 319–328.

Todd, S. C. & Latham, W. (1992). *Evolutionary art and computers*. London: Academic.

Chapter 13
Looking at the Present Through the Future: Science-Fiction Urbanism and Contingent and Relational Creativity[1]

Rob Kitchin

Writing fiction is a creative act. It involves the production of a narrative that tells a fictional story. Much fiction is derivative of stories that have preceded it, and although much of it is clichéd, shallow, and uninspired, there is a steady stream of new works that continue to push boundaries with respect to style, substance, and foci. They are stories that are creative in ways that extend beyond the act of simply making something. Rather than being citational, imitative, and stereotypical (where the plot lines and characters are similar to much of the fiction that had preceded), they are genuine attempts to challenge conventional tropes and styles and to say something meaningful about the world (rather than simply entertain). They are works that are insightful, surprising, educational, interesting, exciting, and enlightening; they interpolate (fill in holes) and extrapolate (make fragments into a whole); and they might be intertextual, but in knowing, clever, witty, and meaningful ways. They make their readers look at the world afresh with new perspectives.

Such creative acts, I argue, do not arise out of nowhere, from some innate product of a novelist's biological make-up (and thus are measureable in some reductionist way through psychological testing). Instead, their creativity is a product of the writer's skills and talents coupled with their embeddedness in networks of people, things, and places. These networks profoundly shape the fiction of creative acts. Writers learn the various facets of how to write—literacy, grammar, punctuation, composition, observation, translation (the process of taking knowledge of the world and converting it into a narrative), imagination, and speculation—of how to engage critically with philosophy, ideology, aspects of the human condition, and so on. Whereas some individuals might possess great talent and skill, these supposed "gifts" are nurtured, shaped, and encouraged by diverse factors such as schooling, tutoring in literary theory and praxis, exposure to other writers' work, and encouragement and critical feedback from peers. And although some writers might claim

R. Kitchin (✉)
National Institute for Regional and Spatial Analysis, Hume Building, NUI, Maynooth, Co. Kildare, Ireland
e-mail: Rob.Kitchin@nuim.ie

[1] Parts of this chapter are based on previously published work (see Dodge & Kitchin, 2000; Kitchin & Kneale, 2001; Kneale & Kitchin, 2002).

P. Meusburger et al. (eds.), *Milieus of Creativity*, Knowledge and Space 2,

to have had no formal training in creative writing, their abilities to craft a story has nonetheless been nurtured in informal ways. Nobody sits down to write as a fully formed writer. And a story derives its inspiration, focus, and politics from its writer's life experiences and engagements with people and places.

Take the novel *Frankenstein*, written by Mary Shelley and published 1818. It is a profoundly creative and imaginative work that provided the genesis for the genre of science fiction (Malmgren, 1991). The story and Shelley's ability to write it did not come from nowhere; they were not the product of an innately talented mind. Rather, the book was a product of Shelley's schooling, her engagements with other fiction, and her relations and discussions with the set of literary figures who constituted her circle of friends (Lord Byron, Dr. John Polidori, and Percy Shelley). It sprang from her travels around Europe and her reading of the cultural landscape. (Frankenstein was written in Geneva, with the Alps and locales such as Chillon Castle providing inspiration.) It issued from her knowledge and understanding of the radical changes occurring around her: the age of Enlightenment, the fledging industrial revolution, the development of rational scientific practice, and a growing sense of how science could advance society and how the future could be extrapolated from the present. Indeed, Shelley herself acknowledges in the introduction to the 1831 edition that the idea for the novel stemmed from a challenge to write a ghost story after she and her friends had read *Fantasmagoriana, ou Recueil d'histoires d'apparitions de spectres, revenants, fantômes,* etc. (Eyries, 1812), a French translation of a German book of ghost stories. A subsequent set of conversations about the scientific work of Erasmus, Darwin and Luigi Galvani contributed the scientific underpinnings for the story. At later stages various drafts would have been read by friends, editors, and others, with edits then being applied to the text. *Frankenstein* was therefore the product of a complex engagement between Shelley and the world at a particular time and place.

Just as other actors and actants (objects and items that have agency, such as various technologies and tools) shape writers, fiction itself does work in the world. Not only does it entertain, it affords a discursive medium of ideas that act as sources of insight and inspiration. This discursive work is sometimes acknowledged explicitly by others, perhaps through statements or interviews ("I have been profoundly influenced by the writings of …," as in Mary Shelley's case). In other instances it is implicitly acknowledged through intertextuality or imitation. Fiction is therefore an actant in the creative processes of other actors (e.g., teachers, journalists, engineers, urban planners, artists, and other novelists), often in ways that were never intended by the novelist. Nearly 200 years after the publication of *Frankenstein*, the novel and its ideas are still actively at work in the world. The ideas within the story serve as sources of film adaptations, derivative stories (e.g., *The Bride of Frankenstein*), and inspiration for a slew of other horror stories and films, not to mention the biotechnology sector—including the opponents of gene modification (GM) and animal testing. It is no coincidence that this protest movement has labeled genetically modified crops "Frankenstein crops" to highlight their "unnatural" and "monstrous" qualities (see Bingham, 2002). This kind of science and the protests it incites are fresh stimuli and substantive issues for fiction writers. In other words, a recursive relationship can develop between fiction and fact.

This chapter is an exploration of the creative writing by several science-fiction writers. I illustrate ways in which their writing emerges from diverse engagements with the world, how their fiction does work in the world, and how a recursive relationship has in some cases developed between novelists and people who read and act upon their stories. The empirical material for my argument comes from a project that involved analysis of 34 novels and four collections of short stories with plots focusing on the development and use of cyberspatial, virtual reality, other information and communication technologies (ICTs), and issues such as telemediation, computer intelligence, surveillance and governance, person–machine relations (cyborgs), and the changing nature of work and urbanism (see Dodge & Kitchin, 2000; Kitchin & Kneale, 2001). All but two of the novels were by North American writers, all but two were by men, all were published between 1982 and 1998, and many belonged to the genre known as "cyberpunk." Of particular interest in this project were the manners in which the novelists dealt with notions of space and time, given the supposed ways in which ICTs "destroy distance." It also described the new geographies of the near future. In this chapter I confine my focus to asking how the authors of these stories imagine urban environments and life and the nature of future cities. I argue that these novelists' visions of the near future serve as a powerful cognitive lens on urbanism in the present, extrapolating from spatial processes that are already at work. This lens is no coincidence; fiction is a product of its place and time, and in some cases ideas are drawn from contemporary urban theory. As I show, this urban theory has, in turn, drawn inspiration from these novels, creating a recursive relationship between novelists and urban theorists.

Introducing Science Fiction and Cyberpunk

Since the time that science fiction emerged as a specific literary genre with the publication of *Frankenstein* (Malmgren, 1991), it has grown to become a very large and popular genre with many subgenres focused on particular realms or technologies. Focusing on the near or far future, but rarely set in the present day, science-fiction writers create imaginative worlds in which to explore new sciences and the meaning and nature of life. Suvin (1979) argues that they create a totalizing novum (novelty, innovation)—entire new worlds, either fully imagined ones or this one in the future—by employing extrapolation and speculation. Suvin argues that these tactics create a sense of estrangement for readers by making the familiar strange. Science fiction's appeal is that it opens readers to new ways of thinking and knowing, but in ways that are tempered by scientific rationale and explanation and by social and spatial metaphors that domesticate the implausibility of the narrative. These realms are not purely fantastical worlds, separated from what people understand as reality and what might seem rationally possible (such as with fantasy writing). Instead, they are worlds that seem plausible given where science seems to be heading. "S[cience] F[iction] rigorously and systematically 'naturalizes' or 'domesticates' its displacements and discontinuities" (Malmgren, 1991, p. 6).

By grounding its science and society in the realities of peoples' experiences, however tenuously, science fiction thus has something to say about the present and the human condition. It creates a cognitive space, an estrangement between real and fictional worlds, which the reader must negotiate to link this world with that of the author's (Malmgren, 1991). Science fiction thereby creates sites of contemplation and accommodation from which to examine the world and what it might become, a discursive field through which to critically think through the present and the possibilities of the future. This point is acknowledged by science-fiction writers themselves. As stated by Gibson (1989, p. 32), a writer referred to several times in this chapter: "What's most important to me is that it's about the present. It's not really about an imagined future. It's a way of trying to come to terms with the awe and terror inspired in me by the world in which we live". He has also declared that his original ideas for the San Francisco of *Virtual Light* was "some permutation of the city as it exists today, that might be remotely possible" (Gibson, 1993, p. 32).

In writing such salutary stories, as with Shelley's *Frankenstein*, science-fiction writers draw on their engagements with the world—their knowledge; experiences; and networks of people, things, and places. It can be no other way. To create feelings of estrangement and defamiliarization, the science and societies they generate have to be grounded in what is presently known and what scientists think is technically possible in the future. Science-fiction writers are therefore often acutely aware of scientific and technological developments. They read scientific journals or follow stories written by science journalists and ask themselves what this technology could mean. What might it make possible? For example, Gibson has noted that his initial inspiration for *Neuromancer* in the early 1980s came as he watched children play computer-arcade games and subsequently imagined these machines as being connected across distributed networks that could be consciously entered. At the time, such distributed technologies had started to be built beyond the purview of the military forming the nascent Internet, and there was much talk of artificial intelligence and telemediation. Gibson thereupon looked around at the changing nature of cities and processes of globalization and projected the whole lot 30 years into the future. Coupled with a writing style that seemed in tune with the processes and technologies being described, the result was a story that seemed strange and yet familiar, fantastical, but plausible. It opened up new ways to think about the emerging information society.

Gibson's novels were both popular and influential and formed part of a subgenre that developed throughout the 1980s, cyberpunk. Cyberpunk focused almost exclusively on the exploration of the effects that ICTs might have on social, economic, political, and spatial relations. But unlike other types of science fiction that explored similar themes, it was decidedly postmodern in its focus and style. Indeed, literary analysts have argued that cyberpunk was "postmodernist SF" (McCaffery, 1991, p. 1) for three reasons. First, cyberpunk was one of the first forms of literary genre to recognize, reflect, and explore the postmodern condition (e.g., the transformation into a postindustrial society, the creation of hyperreal places and simulacra, the merging of technology and nature). Second, cyberpunk was a decidedly posthumanist orientation exploring the interconnectedness and contingencies in the relationship between people and technology (rather than treating them as separate ontological domains). Third, the narrative style was itself decidedly postmodern in character. For example, it often broke with traditional conventions, being discontinual and stitching together different styles and motifs. Cyberpunk was thus seen to destabilize the basic modernist assumptions lying at the core of nineteenth- and twentieth-century science fiction: the dichotomy between self–other; self–society; nature–technology; nature–civilization; rational–irrational; order–chaos; life–death, real–imaginary, and the privileged central position occupied by humans (Hollinger, 1991).

Just as cyberpunk was a product of the world, it did work in the world. In addition to entertaining its readers, cyberpunk challenged them to examine the development of ICTs and the ways in which they had an impact on social and economic relations on a variety of scales. Indeed, the often dystopian stories of the genre prompted readers to think politically and ethically about the new worlds coming about through new technologies that had potentially far-reaching consequences for social formation and modes of governance. As a result, cultural critics such as Jameson (1991) argued that cyberpunk offered privileged insights into contemporary culture, furnishing cognitive maps of the postmodern condition. Many of these readers, in their own working lives, further drew on the ideas and ideologies contained within the novels. In cyberpunk's case this interaction included engineers, politicians, and academics. Cyberpunk was seen as delivering not only powerful social commentary on the present (by extrapolating into the future) but also blueprints for future technologies. This dual outcome was particularly strong in the case of work by Gibson. Some scholars even claim that recent developments in both computing and society can be regarded as an attempt to put his fictional visions into practice. For example, Tomas (1991) and Stone (1991) suggest that Gibson has significantly shaped the information society. As Stone (1991) has stated, *Neuromancer* "provided … the imaginal public sphere and reconfigured discursive community that established the grounding for the possibility of a new kind of [person–computer] interaction" (p. 95).

For example, Gibson's trilogy (*Neuromancer* [1984], *Count Zero* [1986], and *Mona Lisa Overdrive* [1988]) inspired John Walker to launch the Autodesk (leading virtual reality developers) "Cyberpunk Initiative" in 1988 (Chesher, 1994). In a white paper entitled "Through the Looking Glass: Beyond User Interfaces," he invoked Gibson to propose a project to produce a visual 3D cyberspace within 16 months. Similarly, Al Gore and other politicians have, somewhat ironically, drawn on Gibson to formulate their own visions and policies of the future and to justify investment in ICT. Gibson himself has noted how his often dystopian fiction is often misread by others who use it to justify other ends:

> I was delighted when scientists and corporate technicians started to read me, but I soon realized that all the critical pessimistic left-wing stuff just goes over their heads. The social and political naiveté of modern corporate boffins is frightening, they read me and just take bits, all the cute technology, and miss about fifteen levels of irony. (Gibson, 1989, as cited in Hayward, 1993)

Urban planners have similarly (and rather worryingly) drawn inspiration from cyberpunk's posturban landscapes, which, in turn, partially mirror those created by Ph. Dick (*Do Androids Dream of Sheep?* (1968/1996), which was made into film *Blade-Runner*): "In February, 1990, at a public lecture series on art in Los Angeles, three out of five leading urban planners agreed that they hoped someday Los Angeles would look like the film *Blade Runner*" (Klein, 1991, p. 147). Here, I want to examine the latter in more detail. But rather than examine the relationship between cyberpunk writers and urban planners, I want to focus on their understanding of cities and their engagement with urban theorists.

Writing Urban Futures

> The themes and processes which a symptomatic reading of cyberpunk reveal are a good deal more insightful than those offered by what now passes for the theoretical and empirical mainstream.... I think that one gets a clearer analytical understanding of contemporary urban processes from a reading of Gibson or Stephenson than one does from a reading of Sassen or Castells. (Burrows, 1997, p. 38, 45)

Cyberpunk authors were almost exclusively American, and by and large they were writing about American and Asian cities. They were doing so in the 1980s and early 1990s, a time of massive deindustrialization, growth in the service economy and ICT-based industries, and rapid globalization. The period was marked by the increasing importance of supranational structures (e.g., the United Nations and the European Union) and specific global cities tied together by distributed, transnational information and by financial, business, and entertainment networks (e.g., New York, Los Angeles, London, Paris, Toyko, Singapore, and Hong Kong). It was also a time of growth in surveillance technologies, new forms of governance, erosion of public space and ideals, and a redrawing of the political map (e.g., the collapse of the Soviet Union and the division of Yugoslavia). The fiction of cyberpunk writers reflected these trends to create a near future in which the world has been reconfigured by libertarian capitalism, globalization, and social Darwinism. Their writing captured a milieu in which the economy is dominated by a few large multinationals and a panoply of informal businesses, where countries have divided into weak nation-states and where fractured and fragmented cities are tied into a new global order. It was a place where the middle-class has been eliminated and the population neatly divided into haves and have-nots, with the wealthy living in private and defensible spaces, public space no longer existing, and the poor subsisting in ungoverned, anarchic, lawless spaces.

For example, in the fiction of Neal Stephenson and William Gibson, processes of decentralisation, fueled by a collapse of place-based politics and the destruction of the middle class,[2] produce sprawling, centerless, heterogeneous urban landscapes composed of small enclaves where "old cities were doomed, except possibly as theme parks" (Stephenson, 1995, p. 71). Urban space becomes a large, decentralized sprawl with pockets of highly centralized and dense city spaces: "Home was

[2] "There's only two kinds of people. People can afford hotels like that, they're one kind. We're the other. Used to be, like, a middle class, people in between. But not anymore" (Gibson, 1992, p. 123).

BAMA, the Sprawl, the Boston–Atlanta Metropolitan Axis" (Gibson, 1984, p. 57). In such environments, places away from the center have become financially unviable and form twenty-first-century ghost towns, "fallen-in edge-cities, the kind of place that went down when the Euro-money imploded" (Gibson, 1992, p. 245), and decaying rust-belt areas. For the poor, locked out of the gleaming, ordered, private, and regulated spaces of defense, spaces become "jury-rigged and jerry-built from scraps" (Gibson, 1988, p. 31), jumbled, heterogeneous spaces growing in upon themselves. In *Neuromancer*, Case resides in Chiba City, a seedy, low-rent, criminalized Toyko edge-city; in *Count Zero*, Bobby lives in The Projects, which are run-down, forgotten, and disenfranchised large-scale public housing, home to the underclass and gang culture. In *Virtual Light* the (Golden Gate) Bridge can no longer carry vehicles because it has been damaged by an earthquake. It is left to decay until the city's homeless take it over and begin to squat there:

> The integrity of the span was as rigorous as the modern program itself, yet around this had grown another reality, intent upon its own agenda. This had occurred piecemeal, to no set plan, employing every imaginable technique and material. The result was something amorphous, startlingly organic. At night, illuminated by recycled neon, by torchlight, it possessed a queer medieval energy. By day, seen from a distance, it reminded him of the ruin of England's Brighton Pier, as though viewed through some cracked kaleidoscope of vernacular style.
>
> Its steel bones, its stranded tendons, were lost within an accretion of dreams: tattoo parlours, gaming arcades, dimly lit stalls stacked with decaying magazines, sellers of fireworks, of cut bait, betting shops, sushi bars, unlicensed pawnbrokers, herbalists, barbers, bars. Dreams of commerce, their locations generally corresponding with the decks that had once carried vehicular traffic; while above them, rising to the very peaks of the stable towers, lifted the intricately suspended barrio, with its unnumbered population and its zones of more private fantasy. (Gibson, 1992, pp. 58–59)

In the case of the bridge, Gibson's visions were drawn from Kowloon Walled City (or Hak Nam), an anarchic space within pre-handover Hong Kong where up to 33,000 people were packed into little more than 6.5 acres, the shantytowns of developing countries, and the projects of developed countries. Such visions reappear in Williams (1996). He also envisaged Western cities fragmenting into spaces of the haves and have nots, depicted spaces where capital and state invested and regulated or where free-market survival exists, and imagined what would happen if slum areas were left to their own devices and if homelessness continued to grow unchecked:

> Orlando scrunched down in his seat so he could see the hammock city. He had long been fascinated by the multi-level shantytowns, sometimes called 'honeycombs' by their residents—or 'rats' nests' by the kind of people who lived in Crown Heights.... Long ago, he had discovered, during the first great housing crisis at the beginning of the century, squatters had begun to build shantytowns beneath the elevated freeways, freeform agglomerations of cardboard crates, aluminum siding, and plastic sheets. As the ground beneath the concrete chutes filled up with an ever-thickening tide of the dispossessed, later arrivals began to move upward into the vaulting itself, bolting cargo nets, canvas tarpaulins, and military surplus parachutes onto the pillars and undersides of the freeway. Rope walkways soon linked the makeshift dwellings, and ladders linked the shantytown below with the one growing above. (p. 510)

In contrast to the sprawling suburbs of the urban centers, the value of space forces development both upwards and underground, producing a vertical spectrum of stylized, mirrored, postmodern architecture—a riot of glass and steel. Besher (1994) thus describes Toyko:

> Sure enough, immense mounds dotted the landscape as far as the eye could see. Gobi guessed these were underground cities. The freeway suddenly dipped. To Gobi's surprise, they were now traveling through the guts of one of these mound cities. The elevated maglev freeway had suddenly become a transparent artery. They flew through a tube at a height about 30 stories above base level. All along both sides of the tube were rows of internal high rises. These high rises were spread-eagled over a series of parks and urban work-play centers.... He caught his breath. They had finally arrived in downtown Neo-Toyko, the circuit-board heart of the rim. Gobi saw wave after wave of towers. Some of them were 500 stories tall, soaring to a point almost above the earth's atmosphere. He saw the famous Aeropolis sky-rise, much larger than life but no different than the postcard image that was famous all over the world. Like a skeletal Mt. Fuji constructed of living tubes, it was a man-made volcano that pulsed and breathed in an awesome symmetry of life and death. Half-a-million people lived on its top floors and commuted from one vector to another. (p. 211, 213)

Portraying a futuristic Singapore, Sterling (1988) writes:

> It was like downtown Houston. But more like Houston than even Houston had ever had the nerve to become. It was an anthill, a brutal assault against any sane sense of scale. Nightmarishly vast spires whose bulging foundations covered whole city blocks. Their upper reaches were pocked like waffle irons with triangular bracing. Buttresses, glass-covered superhighways, soared half a mile above sea level. ... Story after story rose silent and dream-like, buildings so unspeakably huge that they lost all sense of weight; they hung above the earth like Euclidean thunderheads, their summits lost in sheets of steel-gray rain. (p. 215)

These buildings are more than mere glass and steel, however. They are virtualized through the incorporation of computer networks that render them "smart." They are "buildings with advanced infrastructure, buildings with the late twenty-first century embedded in their diamond bones and fiber-optic ligaments" (Sterling, 1996, p. 139).

These images are visions of the future that strongly resonate in academic observations of the time. Academics were writing about the new international division of labor; the globalization of trade and labor; deregulation, strategic takeovers and buyouts, backofficing, and teleworking; the growth in neoliberalization and the privatization of state functions; the development of entrepreneurial cities and new, postmodern cityscapes; the new global ordering and connectivity between cities; the fragmentation of cities along wealth lines; the growth in the new poor; the erosion of public space; deterritorialization; the rescaling of politics; and contested spaces from the local level to the global.

This resonance should come as no surprise. Science-fiction writers and academics were observing and writing about the same things. They were also reading each other's work. For example, Mike Davis, one of the foremost urban theorists when writing his classic text on the development of Los Angeles, City of Quartz (1990), drew inspiration and explanation from Gibson: "William Gibson ... has provided stunning examples of how realist, 'extrapolative' science fiction can operate as prefigurative social theory, as well as an anticipatory opposition politics to the cyber-fascism lurking over the horizon" (Davis, 1992, p. 3). Interestingly, Gibson drew inspiration for his dystopian visions of urban and political economy in a future San Francisco from Davis's analysis of the urban politics of Los Angeles and from journals such as The Architectural Review (Featherstone & Burrows, 1995). In the case of these two authors a recursive relationship between fact and fiction, theorist and novelist, developed. For both types of writer, creativity is contingent and recursive—fiction is produced in relation to fact and vice versa. Neither would have emerged as it did without the other.

Conclusion

In this chapter I have sought to challenge notions of creativity that suggest it is something innately held and wielded by individuals. I have forwarded a conception of creativity that is contingent-, relation-, and context-driven. As argued in these pages, creativity is a product of skill and talent that is embedded within complex networks of people, things, and places. These networks so profoundly shape creativity through training, inspiration, and critical feedback that creativity cannot be divorced from them. Nobody is a fully formed thinker, and nobody works in isolation. Instead, creative acts are a product of the milieu in which they were conceived and created. Along this line of reasoning, I have suggested that fiction is a product of the world (not just of a single author) and that it does work in the world (in diverse and unforeseen ways). To support my argument, I have examined a set of science-fiction writings, highlighting how the authors draw on a diverse range of sources of inspiration, how the texts are shaped by multiple conversations, and how such works then have influence in the world. This process is particularly well illustrated by the recursive relationship that has developed among some science fiction writers, academics, and others. It is only by examining the full milieu in which a text was created that one can begin to understand its production and its work in the world.

References

Besher, A. (1994). *RIM*. London: Orbit.

Bingham, N. (2002). In the belly of the monster: Frankenstein, food, factisches and fiction. In R. Kitchin & J. Kneale (Eds.), *Lost in space: Geographies of science fiction* (pp. 180–192). London: Continuum.

Burrows, R. (1997). Virtual culture, urban social polarisation and social science fiction. In B. Loader (Ed.), *The governance of cyberspace* (pp. 38–45). London: Routledge.

Chesher, M. (1994). Colonizing virtual reality: Construction of the discourse of virtual reality, 1984–1992. *Cultronix* 1(1). Retrieved 1999 from http:// eserver.org/cultronix/chesher/

Davis, M. (1990). *City of quartz*. London: Verso.

Davis, M. (1992). *Beyond blade runner: Urban control, the ecology of fear*. Westfield, NJ: Open Magazine Pamphlet Series, No. 23.
Dick, P. K. (1996). *Do androids dream of sheep?* New York: Ballantine Books (Original work published 1968).
Dodge, M. & Kitchin, R. (2000). *Mapping cyberspace*. London: Routledge.
Eyries, J-B. (Ed.) (1812). *Fantasmagoriana, ou Recueil d'histoires d'apparitions de spectres, revenants, fantômes*, etc. Published in English as *Tales of the Dead* (1813). London: White, Cochrane.
Featherstone, M. & Burrows, R. (1995). *Cyberspace/Cyberbodies/Cyberpunk: Cultures of technological embodiment*. London: Sage.
Gibson, W. (1984). *Neuromancer*. London: HarperCollins.
Gibson, W. (1986). *Count zero*. London: HarperCollins.
Gibson, W. (1988). *Mona Lisa overdrive*. London: HarperCollins.
Gibson, W. (1989). High tech high life: William Gibson and Timothy Leary in conversation. *Mondo 2000*, No. 7 (Fall), 58–64.
Gibson, W. (1992). *Virtual light*. London: Penguin.
Gibson, W. (1993). Cyberpunk rides again? From Los Angeles to Hong Kong in search of urban futures, interviewed by Bob Catterall. *Regenerating Cities*, *6*, 29–33.
Hayward, P. (1993). Situating cyberspace: The popularisation of virtual reality. In P. Hayward & T. Wollen (Eds.), *Future visions: New technologies of the screen* (pp. 180–204). London: British Film Institute.
Hollinger, V. (1991). Cybernetic deconstructions: Cyberpunk and postmodernism. In L. McCaffery (Ed.), *Storming the reality studio: A casebook of cyberpunk and postmodern fiction* (pp. 203–218). London: Duke University Press.
Jameson, F. (1991). *Postmodernism, or, the logic of late capitalism*. Durham, NC: Duke University Press.
Kitchin, R. & Kneale, J. (2001). Science fiction or future fact? Exploring the imaginative geographies of the new millennium. *Progress in Human Geography*, *25*, 17–33.
Klein, M. N. (1991). Building blade runner. *Social Text*, *28*, 147–152.
Kneale, J. & Kitchin, R. (2002). Lost in space. In R. Kitchin & J. Kneale (Eds.), *Lost in space: Geographies of science fiction* (pp. 1–16). London: Continuum.
Malmgren, C. D. (1991). *Worlds apart: Narratology in science fiction*. Bloomington, IN: Indiana University Press.
McCaffery, L. (1991). Introduction: In the desert of the real. In L. McCaffery (Ed.), *Storming the reality studio: A casebook of cyberpunk and postmodern fiction* (pp. 1–16). London: Duke University Press.
Stephenson, N. (1995) *The diamond age*. New York: Bantham Books.
Sterling, B. (1988). *Islands in the net*. New York: Ace Books.
Sterling, B. (1996). *Holy fire*. New York: Bantham Books.
Stone, A. R. (1991). Will the real body please stand up? Boundary stories about virtual cultures. In M. Benedikt (Ed.), *Cyberspace: First steps* (pp. 81–118) Cambridge, MA: MIT Press.
Suvin, D. (1979). *Metamophoses of science fiction: On the poetics and history of a literary genre*. New Haven, CT: Yale University Press.
Tomas, D. (1991). Old rituals for new space: Rites de passage and William Gibson's cultural model of cyberspace. In M. Benedikt (Ed.), *Cyberspace: First steps* (pp. 31–48). Cambridge, MA: MIT Press.
Walker, J. (1988). Through the looking glass: Beyond user interfaces. White Paper. Retrieved November 21, 2007, from http://www.fourmilab.ch/autofile/www/ chapter2_69.html
Williams, T. (1996). *Otherland*. London: Orbit.

Chapter 14
Teleology, Contingency, and Networks

Barney Warf

This chapter is concerned with some implications of how time, space, and social change have been powerfully retheorized under the impetus of poststructuralism. Social science has steadily jettisoned long-standing teleological conceptions of social structure and change that pervaded Marxist and Weberian accounts. Inspired by structuration theory and philosophical realism, disciplines such as geography and sociology have increasingly come to emphasize the contingent nature of social reality, that is, the manner in which it could be different. As part of this transformation, poststructural theorizations have focused on the rejection of simplistic dichotomies such as individual/society, culture/economy, nature/society, objective/subjective, global/local, and time/space, all of which thwart their effective integration.

I assert that the theorization of social and spatial life necessarily involves the rejection of an additional dichotomy, that between the real and the imaginary, the actual and the possible, the ontological and the epistemological. If what is defined as the "real" is not simply equated with the observed, the definition of "reality" broadens to include not only what is, but what might be, and the lines between the real and the possible become blurred in productive and imaginative ways. Poststructuralism elevates unmaterialized possibilities to the level of ontology. In other words, what is taken to be real is not simply what is observable or actual but forms one outcome secreted from a broader universe of possibilities. Social reality includes events that never happened in fact, but *could* have happened plausibly as defined by theory. Thus, the distinction between what did happen and what could happen is not obvious or unproblematic. History and geography are the understanding of not only why things happen, but why they do not.

The turn toward contingency has been accompanied by an associated retheorization of the role of time and space in the constitution and unfolding of social life: They are viewed as social constructions. In particular, older Cartesian and Newtonian absolutist notions of space have given way to a decided emphasis on relative space and, simultaneously, the reassertion of space into social thought via

B. Warf (✉)
Department of Geography, University of Kansas, Lawrence, KS 66045-7613,
Florida State University, Tallahassee, FL 32306, USA
e-mail: bwarf@ku.edu

P. Meusburger et al. (eds.), *Milieus of Creativity*, Knowledge and Space 2,

the "spatial turn." In the retreat from modernist metaexplanations, many theorists have focused on path-dependent analyses, which substitute contingent historical and spatial contexts for universal explanatory laws. Lines of thought such as the space of flows, commodity chains, and actor networks all circumvent either/or binaries and emphasize time and space as social constructions rather than simply presocial, given containers of human life.

Unrooting Teleology in the Social Sciences

Western scholarship has long reflected deeply embedded assumptions about time and social causation, with far-reaching analytical consequences. Since the Enlightenment, the idea of progress became so firmly entrenched in the social sciences that time effectively became equated with change, and change became synonymous with improvement (Nisbet, 1980). Causation and determination in history have frequently been attributed to teleological forces. Teleology may be defined as "a theory that events can only be accounted for as stages in the movement towards a pre-ordained end" (Johnston et al., 1994, p. 617) and traditionally assumed eschatological or Marxist forms, which portray history as having a purpose or final endpoint. Teleology has a long history in social analysis; indeed, until the modern age, fate was taken as a given in the course of human affairs (Giddens, 1991). Hegel saw history as the unfolding of the spirit of progress, moving from Asia to Europe and culminating in the nineteenth-century Prussian state. Marxism adopted an evolutionary determinism with its modes of production marching in succession throughout the ages. Weberian sociology—idealist and pessimistic compared to Marx—saw the "iron cage" (Weber, 1905/1958, p. 25) of rationalism inevitably descending over the West, squeezing religion into the domain of the irrational. Braudel (1992) and the structuralists in the Annales school advocated a "*long durée*" (p. 12) that shaped the course of historical events regardless of the daily lives of the people who inhabited it. The social ecology of the Chicago School, borrowing from Darwinian social biology, stressed suburbanization as the final product of invasion and succession in urban housing markets. Modernization theory, wedded to Parsonian functionalism, likewise embodied an ethnocentric teleology that culminated in the twentieth-century West as the final product of world history, a view stated most explicitly in Rostow's (1971) famous stages of development or the Fisher–Clark thesis of transition (see Tobin, 1985) from agriculture to manufacturing to services-based economies. Contemporary teleological interpretations continue to assume that the present is the endpoint of history, and, implicitly, the only endpoint: The present is as it is precisely because it could be no other way (e.g., Fukuyama, 1992).

In all of these cases, contingency is made subservient to a predetermined end, which, in hindsight appears inevitable. *Yet only in retrospect does one find inevitability*. By conferring inevitability, hindsight can be as destructive as it is useful.

Frequently, the conceptual logic deployed in teleological accounts is functionalist, as in the Marxist notion that capitalism needs a reserve army of labor. Teleological views elevate social structures above events, emphasize long-run over short-run processes, and fetishize social relations, giving them a power they otherwise could not, and do not, have. Determinism in history is the assumption that each set of causes has one and only one outcome (Carr, 1961); for an outcome to be different, its causes must therefore necessarily be different. Determinism is closely associated epistemologically with the search for laws of explanation, in which explanation consists of showing the unique to be an outcome of the general (Sayer, 1992). In the denial of the significance of human agency and its creative capacities, teleological interpretations draw a sharp line between the potential and the real, the possible and the impossible, the contingent and the necessary, what was and what might have been, between what occurred empirically and what *could* have occurred theoretically.

By obscuring the contingent nature of social life, such accounts present space and time as external to human action, as containers that "hold" society but are not produced by it. Although time and space appear as "natural" presocial categories, they are in fact social constructs (Harvey, 1990). But there is an important political and analytical repercussion to teleology: The real problem with teleological interpretations is not that they are simplistic or wrong, but that they are reactionary. By reifying history, teleological accounts deny the importance of human actors in the construction of space and time, reducing people to finders of a world already made (Duncan & Ley, 1982). To deny human agency is to deny contingency and vice versa; to dethrone teleology is to accept the bounded flow of agency through time and space as the motor of historical and spatial change. Debates about teleology, therefore, are important because they have ramifications in the present; they shape the ways in which people come to understand how societies are reproduced and transformed. As discourses, they do not simply mirror the world, they help constitute it.

More broadly, the dichotomy between the real and the contingent reflects the modernist tendency to divide reality into convenient dualisms (Cloke & Johnston, 2005), bifurcations that thwart their creative integration. Binary categorizations of the world do more than help one deal with its complexity, they come to have lives of their own, shaping human identities and behavior individually and collectively. Far from constituting politically neutral divisions, classifications such as A/not-A typically ascribe power and significance to one term at the expense of its opposite (e.g., male/female and white/black). Binary divisions have thus become increasingly viewed as systems of social domination. Once seen as imparting clarity and certainty to intellectual analysis, conceptual oppositions have been thoroughly demonized as static instruments of Othering, simplistic vehicles of misrepresentation and the obliteration of diversity. Maintaining differences among categories—race, for example—typically relies upon the annihilation of differences within them. As Cloke and Johnston (2005) put it, "the basic polarities of the binary are either too incoherent or too weak to bear the explanatory weight which is placed upon them" (p. 13). Any social science that seeks to escape these limitations, therefore, must move beyond simplistic dichotomies.

Constructing Alternatives to Teleologies

The transformation sketched out in this chapter necessitates two substantial analytical steps: (a) taking human consciousness—and thus the contingent nature of social reality—seriously, and (b) jettisoning the dualities that have plagued teleology.

Contingency is a notion that has suffered from its popularity. Jones and Hanham (1995) offer a plethora of definitions of contingency, all of which signal "the possibility of multiple outcomes derived from similar causal processes due to the complexity of social relations embedded in spatially differentiated contexts" (p. 187). Contingency can denote accident or chance, simple indeterminacy and unpredictability, or opposition to all-determining forces that dictate the course of human affairs. Contingency in whatever form is a vital antidote to teleological readings that hold history rolls along a single track of possibility. To ignore contingency is to assume that historical outcomes are also logical necessities. Rather, in a contingent sense, the present is precipitated out of a probability distribution of possible worlds. This view has important epistemological implications: To know a society and a geography is to know how it could be different, a key foundation for any critical social science.

Reconciling determinism and voluntarism has long been one of the most intractable problems of social science. Giddens's (1984) structuration theory has become the most popular resolution to this dilemma. This theme begins with the recognition that only human beings are sentient (i.e., only they have consciousness about themselves and their world) and draws upon the rich phenomenological tradition of perception, cognition, symbolic form, and language, all of which are fundamental to any understanding of the human subject (Buttimer, 1976; Smith, 1984). Because social science is concerned with sentient beings, it is fundamentally, qualitatively different from analyses of the nonhuman world, so metaphors imported from physics and biology (e.g., the gravity model) are automatically suspect when applied to the study of humans. Moving beyond the usual elementary definitions of culture as the sum total of learned behavior or a way of life (e.g., religion, language, mores, traditions, and roles), structuration theory allows for an understanding of culture as what humans take for granted: common sense, the matrix of ideologies that allow people to negotiate their way through their everyday worlds (Giddens, 1984, 1991). Culture defines what is normal and what is not, what is important and what is not, what is acceptable and what is not, within each social context. Culture is acquired through a lifelong process of socialization: individuals never live in a social vacuum but rather are socially produced from cradle to grave (Giddens, 1984). The socialization of the individual and the reproduction of society and place are two sides of the same coin. In other words, the macrostructures of social relations are interlaced with the microstructures of everyday life (Pred, 1984, 1990; Thrift, 1983). People reproduce the world, largely unintentionally, in their everyday lives, and, in turn, the world reproduces them through socialization. In this view, people are always conscious, acting subjects, capable of "doing otherwise." If someone is held responsible for an action, it is presumed that he or she had alternatives; responsibility is measured by the degree of risk one takes (Demandt, 1993). However, the human subject is a social product, for individuals live through a life-long process of socialization. In forming their biographies everyday, people unintentionally reproduce and transform their social worlds; the counterpart to the unacknowledged preconditions to action is the unintended consequences, which means that social relations continually escape the intentions of their creators. In this way, structuration theory sutured the macrostructures of social relations to the microstructures of everyday life (Pred, 1990; Thrift, 1983). Everyday thought and behavior hence do not simply mirror the world, they constitute it (Sayer, 1992). Such a view asserts that cultures are always intertwined with political relations and are continually contested. That is, dominant representations and explanations that reflect prevailing class, gender, and ethnic powers are often challenged by marginalized discourses from the social periphery. Thus, if political economy centers on production systems, the analysis of culture involves the corresponding analysis of reproduction.

Even in the natural sciences, in which the social construction of reality is generally not an issue, contingency has become an increasingly accepted part of how the world works. Whereas the Darwinian revolution led to the widespread view that nature inevitably evolves from simple to ever-more complex and hence "better" life forms, many contemporary biologists stress the role of environmental catastrophes and accidents that disrupt the process of natural selection. Gould (1989), for example, in his famous study of Canada's Burgess shale, sketched seven alternative possible evolutionary worlds and concluded that the eighth, the only one to include humans, arose only because of the unlike success of one phylum. In physics, Einstein argued that God does not play dice with the universe, but he was wrong: Heisenberg's uncertainty principle of quantum mechanics holds that there are multiple outcomes for any particular observation, introducing an element of unpredictability at the most fundamental subatomic level (Hawking, 1988; Prignogine, 1996). Contemporary chaos theory reveals much the same conclusion, positing that the "arrow of time" is not unidirectional, that causality can never be separated from context. Chaos, or complexity theory, applies stochastic forms of mathematics to demonstrate that the impacts of small, seemingly trivial events can become magnified over time, creating ripples and eddies with large, unpredictable consequences.

Nonteleological explanation centers on the distinction between necessary and contingent relations (Sayer, 1992). Necessary relations, identified through theoretical abstraction, concern the mechanisms that produce change, not events or observations *per se*. Contingent relations, in contrast, are specified in concrete empirical contexts. Whereas necessary conditions must exist in every social explanation, the causal properties they describe may or may not get activated empirically. Thus, causal properties are detached from empirical regularities. As Sayer (1992) notes, "Not surprisingly then, depending on conditions, the operation of the same mechanism can produce quite different results and, alternatively, different mechanisms may produce the same empirical result" (p. 108). Only if causal laws are equated with empirical regularities does necessity triumph over contingency, a move that gives rise to teleology. In the open, contingent systems of the contemporary, poststructuralist social sciences, causation concerns necessity, not universality; explanations do not concern events, but what produces them, and the same causal mechanism may produce many different events, or not produce them, as the case may be. In the language of realism, the relations between necessary and contingent events are inescapably geographically specific: "Whether an object's causal mechanisms are activated and with what effect depends on the presence of certain contingently related conditions; this in turn depends on the spatial form" (Sayer, 1992, p. 148). Thus, to understand how alternative trajectories through time and space are constructed is to identify, theoretically, the specific causal properties involved, their contingent manifestations in particular places and times, and the plausible alternatives that never materialized empirically but are nonetheless real.

But what, exactly, is meant by the real? If reality consists only of what is observed, as in the positivist and empiricist conceptions, one is deprived of any understanding of that which is not directly observable yet still very real—causal properties included. Yet empirical outcomes are only contingently related to their causes. If, however, the real is not simply equated with the observed, the definition of reality broadens to include not only what is, but what might have been, and the lines between the real and the might-have-been become blurred in productive and imaginative ways. Realism thus elevates unmaterialized possibilities to the level of ontology. That is, what is taken to be real is not simply what is observable or actual but forms one island surrounded by a sea of possibilities. Thus, reality includes events that never happened in fact but *could* have happened. The task of social theory is to identify those outcomes that are realistic and plausible—and thus real—but never actually materialized empirically. By focusing on historical alternatives that are plausible, one resolves the dilemma of choosing between a single deterministic past and an infinite number of pasts. At times the line between plausibility and implausibility is difficult to draw. Does "plausible" mean that which departs the least from known historical reality? After all, the number of possibilities increases exponentially with distance from the plausible. Perhaps only minor departures from what "really" happened can be considered realistic. One may have the most faith in counterfactual scenarios that most closely mimic the known past, yet they are also the least interesting. Unrealized possibilities can be theorized based on people's understanding of the past and of how social relations behave over time and space. To understand an event is to know the probability that it took place as well as the probability that it might not have taken place. Thus, the distinction between what did happen and what could have happened is not so clear-cut after all. Social science becomes as concerned with possibilities that seemed probable in the past as it is with possibilities in the future. History—and geography—are the understanding not only of why things happen but why they do not.

This line of thought owes much to a rich philosophical tradition concerned with "plausible worlds" (see Bennett, 1974; Kvart, 1986; Nozick, 1981; Nute, 1975). The philosopher, mathematician, and political advisor Gottfried Wilhelm Leibniz (1646–1716), who started this train of thought, argued that of all the possible worlds that might exist, God chose to create only the one people inhabit. Contemporary theorists such as Lewis (1973, 1986) have argued that the known world is but one among an infinite number, unique only in that we humans happen to inhabit it. He argues (1986): "There are so many other worlds, in fact, that absolutely *every* way that a world could possibly be is a way that some world *is*" (p. 2). Which worlds are real and which are ersatz is a matter of actualization; *all* worlds are actualized somewhere, that is, within different spatiotemporal coordinates.

As poststructuralist social science has gained in popularity, it has initiated a wide-ranging retheorization of time and space. For example, Soja (1989, 1993) argued that a truly postmodern social science overcomes the legacy of historicism by putting space on a par with time. Historicism, the doctrine holding that time, not space, is important in social explanation, triumphed in the nineteenth century, when, under the white-hot waves of time–space compression unleashed by the industrial revolution, time became intimately associated with dialectics and change whereas space was relegated to the fixed and unchanging. In the postmodern age, Berger (1974, p. 40) claims, "it is not time but space that hides consequences from us." In other words, geographies appear to freeze the contradictions of advanced capitalism by reifying and naturalizing them via teleology. In this light, alternative geographies expose the contingent, politically pregnant construction of space that lay hidden under the patina of commodity fetishism, the "geohistory of otherness." As an alternative to historicism, Soja's *Thirdspace* (1996) advocates "journeys to real-and-imagined places," a trip through what Borges (1970) calls the "garden of forking paths" (p. 27), a dizzying net of divergent, convergent, and parallel times in which all possibilities are examined synchronically.

When the human past and present are viewed as one possible path among many, there is no privileged path to the present but rather a large (though not infinite) number of possibilities. The selection of one history over its alternatives is not a purely intellectual choice, but one grounded in the politics of knowledge. Discourse analysis sheds light on why some interpretations gain legitimacy and others do not. The contingency of social life is thus partly a function of the discourses employed to understand it. If the past, and the present, is truly, essentially, chaotic and random, then the probability of every alternate history is, by definition, equal. If history and geography are defined as stories told about the world, one can always tell different stories with the same set of facts; whatever coherence the past has is a function of the discourses employed to make sense of it. Contingency is inherent not just in the social construction of the world, but in its interpretation as well. Perhaps precisely because politics is the struggle for power, the "art of the possible," it lends itself well to this sort of interpretation. Every nationalist or secessionist movement, successful or not, is fueled by dreams of an alternative political geography (Anderson, 1983).

On what epistemological grounds does one accept that which is real and that which is not? Why is it so essential to differentiate between these realms? Analyses of social life that take contingency seriously teach how this world came to be by illuminating how it did not become. To understand who we humans are, we must also understand who we are not. By demonstrating that history and geography could always "be otherwise," they reveal our world as anything but the culmination of inevitabilities, as a palimpsest of unintended consequences. Every set of actions is path-dependent, open-ended, and creatively negotiated over time and space by human beings through unacknowledged preconditions and unintended consequences to action. Anything less than this stance is reactionary reification; anything more is naïve idealism. If reality is taken to be simply that which is empirically observable through the historical record—a highly suspect notion—then the world as it appears to people could have been no other way. To appreciate the real as something more than that which is observed—to blur the artificial line between the empirical and the possible—is to recognize its deeply contingent nature. Reality includes the alternative trajectories that never actually occurred, but plausibly might have. In this light, social structures seem less structural, less cohesive, stable, or deterministic. Every trajectory can pursue a finite number of paths, none of which leads inevitably to the present. This conclusion compels clarification of one's assumptions and lines of causation, one's "ways of worldmaking" (Goodman, 1978). When it is acknowledged that the past and the present are mutable, plastic products of human effort, even if largely unintended ones, the future, too, is up for grabs.

Networks and the Contingent Construction of Space

As geographers, among others, grappled with these issues, it became apparent to many people that a new conception of space was urgently needed. Essentially, this shift entailed the jettisoning of absolute space and the focus on relative and relational notions. Harvey (2006) offers a concise summary of these differences:

> Absolute space is fixed and we record or plan events within its frame. This is the space of Newton and Descartes and it is usually represented as a pre-existing and immovable grid amenable to standardized measurement and open to calculation. Geometrically it is the space of Euclid and therefore the space of all manner of cadastral mapping and engineering practices. The relative notion of space is mainly associated with the name of Einstein and the non-Euclidean geometries that began to be constructed most systematically in the 19th century. The relational concept of space is most often associated with the name of Leibniz, who objected vociferously to the absolute view of space and time so central to Newton's theories. By extension, the relational view of space holds there is no such thing as space or time outside of the processes that define them. Processes do not occur *in* space but define their own spatial frame. (pp. 272)

Rather than the closed surfaces of modernism, which portray absolute space in static terms of fixed boundaries (e.g., the nation-state), many turned to networks as a model of relational space to shed light on how human beings are intertwined with one another geographically, that is, as "power-geometries" (e.g., Massey, 1993, p. 55). Similarly, Castells (1996, 1997) emphasizes the "network" society of contemporary capitalism, dominated by a "space of flows" rather than a "space of places" (p. 412), which has given birth to new political formations, forms of identity, and spatial associations. He notes, for example, that "people live in places, power rules through flows" (p. 415). This view has become important in conceptions of multinational corporations, global finance, and geographies of cyberspace. In an age of intense globalization, hypermobile capital, and time–space compression, this notion indicates the great degree to which people and localities have become intertwined with one another over vast distances, a distanciation (Giddens, 1984) that has become global in scope.

The latest product of this line of reasoning is actor-network theory (Law & Hassard, 1999; Murdoch, 1995). Inspired by the work of Latour (1993) and Serres (1997; see Serres & Latour, 1995), actor-network theory incorporates sociological understandings of structuration theory (Giddens, 1984) with a poststructuralist, French, social-constructivist philosophy of science. The essence of actor-network theory is the linkages among different ontological categories, a departure from the Enlightenment focus on pure essences that created dualities such as individual and society, people and nature, human and nonhuman, western and nonwestern, urban and rural, micro and macro, and local and global. Rather, it takes as its point of departure the linkages among these categories as actors draw upon and combine them in various forms of hybridity (Murdoch, 1997). As Latour (1993) notes, "How are we to gain access to networks, those beings whose topology is so odd and whose ontology is even more unusual, beings that possess both the capacity to produce both time and space?" (p. 77).

Networks involve the mobilization of rules, resources, and power, including information, that actors must draw upon in order to accomplish tasks. They function analogously to Giddens's notion of structure, which simultaneously enables and constrains action. Networks are created by a net of intended and unintended consequences that stretch across the spatiotemporal boundaries of the network. To maintain network functionality, actors must perform by being engaged with one another recursively, interpreting and translating one another's behavior. Actors and networks are thus twin, mutually presupposing aspects of one phenomenon, simultaneously enabling and constraining actions in time and space. Because actor-network theory strives to overcome the artificial boundaries between culture and nature (Latour, 1993), actors in this sociotechnically seamless "nature–culture" nexus need not be human, but may include inanimate objects such as books, papers, or computer systems (Bingham, 1996; Murdoch, 1997), which are necessary to the maintenance and operation of networks.

Networks obviously can and do change over time, sometimes dramatically, and actors may enter and exit networks, but if the network is to achieve the purposes for which it is designed, actors must be able to draw upon its resources and effect the necessary consequences. Thus, it is not simply actors in everyday life who constitute the primary focus in networks; it is their relative positionality and powers within integrated systems of power and information that matters most. For example, Thrift and Leyshon (1994) employed actor-network theory to examine the dynamics of global capital markets as they are structured by firms, nation-states, the media, and telecommunications, all of which are deployed simultaneously to produce, transmit, and consume knowledge about markets and other actors. Such a perspective has helped to humanize even the most abstract of economic processes by revealing them to be the products of agents enmeshed in webs of power and meaning, not disembodied processes that operate independently of the people who create them (Law, 1994). The strategy of embodiment goes a long way toward demythologizing teleological interpretations of globalization, which present it as natural and inevitable, and reveal global processes to be the contingent outcomes of decisions made by human actors tied up in networks that cross multiple spatial scales.

Such a view elides the conventional focus on spatial scale, for networks operate across many scales simultaneously, creating as Latour (1993) puts it, "an Ariadne's thread that allows us to pass with continuity from the local to the global, from the human to the nonhuman. It is the thread of networks of practices and instruments, of documents and translations" (p. 121). Likewise, Massey's (1999, 2005) well-received notion of power-geometries has called attention to the intertwined scales of the global, national, and local, refusing to see them as a simple hierarchy in which the global determines the local. The distinctions among these scales are as misleading as they are enlightening (Swyngedouw, 1997). Smith (1993) argues that scale is produced through and constitutive of social relationships, and Thrift (1995) goes so far to claim: "There is no such thing as scale" (p. 33). By forcing one to rethink how time and space are produced—that is, topologically rather than in terms of conventional Cartesian and Kantian views of space that have dominated geography—actor-network theory becomes "a machine for waging war on Euclideanism" (Law, as cited in Murdoch, 1998, p. 357).

One way in which space has been retheorized in terms of networks is commodity chains (Dicken et al., 2001). A commodity chain is a network of labor and production processes that gives rise to a commodity; it extends from the raw material through various stages in processing and delivery and ends in consumption. For example, the coffee commodity chain begins with the grower, typically an impoverished farmer, and extends through the processing plant, exporters, traders, roasting companies, retailers, and, finally, the consumer. Similarly, the meat commodity chain extends from ranchers to feedlots to packinghouses to cold storage to grocers and finally to consumers. The gold commodity chain begins with miners working under horrendous conditions in countries such as South Africa, through diamond cutters (usually in Europe), to retailers and consumers in the developed world, who celebrate gold's romantic allure without considering the conditions of its production (Hartwick, 1998). At each stage, the commodity is transformed in some way and value is added. The same company may control one or more stages in a commodity chain, depending on how vertically integrated or disintegrated the production process is. Because different nodes where these activities are carried out are spatially separated, commodity chains are geographical as well as economic and cultural phenomena.

Commodity chains are thus a means of depicting not only the ways economic activity reverberates through the production process, the linkages among different economic sectors, and flows of value over time and space but also the ways it overcomes the artificial separation between consumption and production. They allow one to see the commodity as more than just a thing; they reveal it to be an embodiment of processes at different spatial scales. Essentially, commodity chains are mechanisms that enable one to trace the impacts of consumption decisions back through the production and distribution process, broadening the scale of analysis from the local to the global. They track the commodity through complex, contingent lines of causality linking sellers and buyers across multiple spatial scales.

Over time, with the expansion of capitalism globally, commodity chains have become longer and longer (Gereffi & Korzeniewicz, 1994; Kaplinsky, 2000). This device affords an understanding, for example, of the ways in which globalization has unleashed a tidal wave of cheap imports that has propelled the high rates of consumer spending in societies such as that in the United States. By uniting consumption with production, it points to the sacrifices made by low-wage labor trapped in sweatshops in the developing world in order to provide American consumers with cheap goods. Such a perspective reveals consumption as being an economic, cultural, psychological, and environmental act that simultaneously reproduces both the world's most abstract space, the global economy, and the most intimate, the individual subject and body.

Concluding Thoughts

In the last several decades, most social sciences have grown by leaps and bounds in their degree of conceptual subtlety and theoretical sophistication. As modernist, teleological views of social change over time have come under intense criticism, alternative, poststructural views that take seriously the human capacity to change the world have risen in prominence. All social systems, structures, and relations are plastic and mutable, even if they are unintentional productions. Social constructivism has shed light on the deeply social, and hence changeable, nature of many phenomena previously taken-for-granted—poverty, gender, and "race," for example. So, too, have time and space been steadily repositioned from their status as containers "outside" society (a view that implicitly holds time above space in significance) to products that are generated "within" society (i.e., as outcomes of social action as much as shapers of it). Once one does away with convenient dualities like time versus space, global versus local, and individual versus society and views them as twin aspects of one underlying set of relations, it is quite possible to see how all of these categories are reproduced historically and geographically. I conclude that the fixed surfaces that were hegemonic during the long years of modernist thought in geography may be displaced by networks. This perspective rests upon complex, contingent human interactions, discards analytically misleading notions such as scale, and lends insight into the ways in which power-geometries underscore all aspects of social life in time and space.

References

Anderson, B. (1983). *Imagined communities: On the origins and spread of nationalism*. London: Verso.

Bennett, J. (1974). Counterfactuals and possible worlds. *Canadian Journal of Philosophy, 4*, 381–402.

Berger, J. (1974). *The look of things*. New York: Viking Press.

Bingham, N. (1996). Objections: From technological determinism towards geographies of relations. *Environment and Planning D: Society and Space, 14*, 635–657.

Borges, J. (1970). The garden of forking paths. In D. Yates & J. Irby (Eds.), *Labyrinths, selected stories and other writings* (pp. 27–41). London: Harmondsworth.
Braudel, F. (1992). *Civilization and capitalism, 15th–18th century: Vol. I. The structures of every-day life* (S. Reynolds, Trans.). Berkeley, CA: University of California Press.
Buttimer, A. (1976). Grasping the dynamism of lifeworld. *Annals of the Association of American Geographers, 66*, 277–292.
Carr (1961). *What is history?* New York: Vintage.
Castells, M. (1996). *The information age: Vol. 1. The rise of the network society*. Oxford, England: Blackwell.
Castells, M. (1997). *The information age: Vol. 2. The power of identity*. Oxford, England: Blackwell.
Cloke, P. & Johnston, R. (Eds.). (2005). *Spaces of geographical thought*. London: Sage.
Demandt, A. (1993). *History that never happened*. Jefferson, NC: McFarland.
Dicken, P., Kelly, P., Olds, K., & Yeung, H. (2001). Chains and networks, territories and scales: Towards an analytical framework for the global economy. *Global Networks, 1*, 89–112.
Duncan, J. & Ley, D. (1982). Structural Marxism and human geography. *Annals of the Association of American Geographers, 72*, 30–59.
Fukuyama, F. (1992). *The end of history and the last man*. New York: Free Press.
Gereffi, G. & Korzeniewicz, M. (Eds.) (1994). *Commodity chains and global capitalism*. Westport: Praeger.
Giddens, A. (1984). *The constitution of society: Outline of the theory of structuration*. Berkeley, CA: University of California Press.
Giddens, A. (1991). *Modernity and self-identity*. Stanford, CA: Stanford University Press.
Goodman, N. (1978). *Ways of worldmaking*. Indianapolis, IN: Hackett.
Gould, S. (1989). *Wonderful life: The Burgess shale and the nature of history*. New York: Norton.
Hartwick, E. (1998). Geographies of consumption: A commodity-chain approach. *Environment and Planning D: Society and Space, 16*, 423–437.
Harvey, D. (1990). Between space and time: Reflections on the geographical imagination. *Annals of the Association of American Geographers, 80*, 418–434.
Harvey, D. (2006). Space as a keyword. In N. Castree & D. Gregory (Eds.), *David Harvey: A critical reader* (pp. 270–294). Oxford, England: Blackwell.
Hawking, S. 1988. *A brief history of time* (10th ed.). New York: Bantam Books.
Johnston, R., Gregory, D., & Smith, D. (1994). *The dictionary of human geography*. Oxford, England: Blackwell.
Jones, J. & Hanham, R. (1995). Contingency, realism, and the expansion method. *Geographical Analysis, 27*, 185–207.
Kaplinsky, R. (2000). Globalization and unequalisation: What can be learned from value chain analysis? *Journal of Development Studies, 37*, 117–146.
Kvart, I. (1986). *A Theory of counterfactuals*. Indianapolis, IN: University of Indiana Press.
Latour, B. (1993). *We have never been modern*. London: Harvester Wheatsheaf.
Law, J. (1994). *Organising modernity*. Oxford, England: Blackwell.
Law, J. & Hassard, J. (Eds.) (1999). *Actor network theory and after*. Oxford, England: Blackwell.
Lewis, D. (1973). *Counterfactuals*. Oxford, England: Blackwell.
Lewis, D. (1986). *On the plurality of worlds*. Oxford, England: Blackwell.
Massey, D. (1993). Power geometry and a progressive sense of place. In J. Bird, B. Curtis, T. Putnam, G. Robertson, & L. Tickner (Eds.), *Mapping the futures: Local cultures, global change* (pp. 55–69). London: Routledge.
Massey, D. (1999). Imagining globalization: Power-geometries of time-space. In A. Brah, M. Hickman, and M. Mac an Ghail (Eds.), *Global futures: Migration, environment and globalization* (pp. 27–44). Basingstoke, England: Macmillan.
Massey, D. (2005). *For space*. London: Sage.
Murdoch, J. (1995). Actor-networks and the evolution of economic forms: Combining description and explanation in theories of regulation, flexible specialization and Networks. *Environment and Planning A, 27*, 731–757.

Murdoch, J. (1997). Towards a geography of heterogeneous associations. *Progress in Human Geography, 21*, 321–337.
Murdoch, J. (1998). The spaces of actor-network theory. *Geoforum, 29*, 357–374.
Nisbet, R. (1980). *History of the idea of progress*. New York: Basic Books.
Nozick, R. (1981). *Philosophical explanations*. Cambridge, MA: Harvard University Press.
Nute, D. (1975). Counterfactuals and the similarity of worlds. *Journal of Philosophy, 72*, 773–778.
Pred, A. (1984). Place as historically contingent process: Structuration and the time-geography of becoming places. *Annals of the Association of American Geographers, 74*, 279–297.
Pred, A. (1990). *Making histories and constructing human geographies: The transformation of practice, power relations, and consciousness*. Boulder, CO: Westview.
Prignogine, I. (1996). *The end of certainty: Time, chaos, and the new laws of nature*. New York: Free Press.
Rostow, W. (1971). *The stages of economic growth; a non-communist manifesto*. Cambridge, England: Cambridge University Press.
Sayer, A. (1992). *Method in social science: A realist approach* (2nd ed.). London: Routledge.
Serres, M. (1997). *The troubador of knowledge*. Ann Arbor, MI: University of Michigan Press.
Serres, M. & Latour, B. (1995). *Conversations on science, culture, and time*. Ann Arbor, MI: University of Michigan Press.
Smith, N. (1993). Homeless/global: Scaling places. In J. Bird, B. Curtis, T. Putnam, G. Robertson, and L. Tickner (Eds.), *Mapping the futures: Local cultures, global change* (pp. 87–119). London: Routledge.
Smith, S. (1984). Practicing humanistic geography. *Annals of the Association of American Geographers, 74*, 353–374.
Soja, E. (1989). *Postmodern geographies*. London: Verso.
Soja, E. (1993). Postmodern geographies and the critique of historicism. In J. Jones, W. Natter, & T. Schatzki (Eds.), *Postmodern contentions* (pp. 113–136). New York: Guilford.
Soja, E. (1996). *Thirdspace: Journeys to Los Angeles and other real-and-imagined places*. Oxford, England: Blackwell.
Swyngedouw, E. (1997). Neither global nor local: "Glocalization" and the Politics of Scale. In K. Cox (Ed.), *Spaces of globalization: Reasserting the power of the local* (pp. 137–166). New York: Guilford.
Thrift, N. J. (1983). On the determination of social action in space and time. *Environment and Planning D: Society and Space, 1*, 23–57.
Thrift, N. J. (1995). A hyperactive world. In R. Johnston, P. Taylor, & M. Watts (Eds.), *Geographies of global change* (pp. 18–35). Oxford, England: Blackwell.
Thrift, N. & Leyshon, A. (1994). A phantom state? The de-traditionalization of money, the international financial system and international financial centres. *Political Geography, 13*, 299–327.
Tobin, J. (1985). Neoclassical Theory in America: J. B. Clark and Fisher. *American Economic Review, 75*, 28–38.
Weber, M. (1958). *The Protestant ethic and the spirit of capitalism* (T. Parsons, Trans.). New York: Charles Scribner's Sons (Original work published 1905).

Chapter 15
Geophilosophy and Creative Milieus

Stephan Günzel

Milieu is a concept that stems from the natural sciences. It is regularly used to describe the state of being of a given setting, predominantly one involving fluids. Insofar as an autonomous subject is not imaginable within a milieu, this use gives a nonhumanistic, even antihumanistic, spin on the idea of milieu when it refers to the social sphere. The concept of milieu thus includes a world model similar to that encompassed by the concept of environment—but even more rigorous. Within an environment a person is still someone *in* it, with nature or the social world around it. By contrast, "in" a milieu (which literally means "middle" or "medium") implies that there is no outside and, hence, no inside: A milieu by definition is absolute, extensive, and embracing. It is therefore metasubjective and metaobjective at the same time. Nevertheless, it is not to be thought of as deterministic, for it is no more and no less than a medium, a possibility for events. Speaking of a "creative milieu" is rather tautological, for a milieu per se allows for things to be created and for changes to take place. When introduced into social analysis, however, the idea of creative milieu implies the existence of milieus that are not creative or that were not creative at a certain point in history *and* geography. In this regard a certain fallacy can be avoided. Even though a milieu is something that can be addressed in time and space, its impact (except in biology) cannot be traced back to the milieu's natural properties—only to other social and political properties operating at a given time and place. This characteristic is the main insight of geophilosophy as defined by the French philosophers Deleuze and Guattari (1991/1994), who ultimately follow the concept back to the German writer Friedrich Nietzsche. According to all three thinkers, geophilosophy can be summarized as the idea that philosophy or science in general needs a creative milieu to develop and that this milieu event, though it can be localized in time and space, is not explicable as being caused by the natural features of that specific environment. Identifying creative milieus within the history of science or a geography of ideas is thus not one topic among others but rather the central issue.

S. Günzel (✉)
Philosophische Fakultät/Medienwissenschaft, Ernst-Abbe-Platz 8, 07743 Jena, Germany
e-mail: StGuenzel@aol.com

P. Meusburger et al. (eds.), *Milieus of Creativity*, Knowledge and Space 2,

What Is Geophilosophy?

Two associations might be made with respect to the term geophilosophy. One is more or less esoteric and has to do with ecological thinking in the sense of "deep ecology" as fostered by Devall (1985). Even though the call to rethink the earth's value is attributable to geophilosophical motives, this notion is not vital. The second association with geophilosophy can be taken for a philosophical version of "geopolitics." The term has, in fact, been used in that sense to criticize implied political territoriality couched in a philosophical statement, as when Lyotard (1988/1990) analyzed Martin Heidegger's writings as a philosophy of the "blood-and-soil" type. Unlike these two interpretations (the ecological and the geopolitical), geophilosophy as meant in this chapter is concerned with the philosophy of geography, of geology, or both and with geography, geology, or both *as* philosophy. In the one sense, then, geophilosophy centers on philosophy as a philosophy of science; in the other sense, on an apparent nonphilosophical subject as philosophy. This relation can be thought of as dialectical: Insofar as a philosophy of a given science exists, this branch of science has an impact on philosophy.

Philosophy of Geosciences: Nietzsche and the Historical Paradigm

With respect to the science of geography the important historical period was the nineteenth century. In that epoch geography emancipated itself from the science of history, to which it was ancillary, and became established at universities in Europe as a subject in its own right. Geology, on the other hand, has a different origin, coming from seventeenth- and eighteenth-century geognosis. Nevertheless, geology was modernized at the same time that geography became a separate discipline. An important philosophical author relating to these two disciplines is Nietzsche, who is regarded primarily as an ecological or geopolitical thinker. The cliché attached to his name derives mainly from two things: the expression "will to power" and his figure named Zarathustra, who notoriously called upon his disciples to "remain faithful to the earth" (Nietzsche, 1883/1988, Vol. 4, p. 149 and 15). This popular perception of Nietzsche is limited by ideological appropriations of the early twentieth century. Today, his writings are becoming important in a totally different way (Günzel, 2003), for Nietzsche was not only a promoter of catchy phrases, he was also a well-informed reader of the scientific and nonscientific literature of his day. His library was a wide-ranging collection of nineteenth-century research, including anthropology, philology, and natural sciences such as astronomy, physics, chemistry, and, last but not least, geography. Nietzsche is thereby an outstanding witness to historical transformations of knowledge and a forerunner of the "spatial turn" in cultural and social sciences. He did not simply criticize the teleological paradigm of historical thinking; he criticized it with spatial figures of thought and both geographical and geological arguments.

Nietzsche Reading Ratzel

Beside other texts, Nietzsche's library contains a copy of Friedrich Ratzel's *Anthropogeographie* (1882), which Nietzsche bought and read soon after its publication. Most of the book is a description of geographical conditions in different areas of the world that Ratzel reckoned to be influential for cultural development. Within the book's 500 pages only a few passages offer theoretical reflections on what geography is. As one can tell from Nietzsche's marks, those pages were the ones he was interested in. No other pages were even touched (for they are not cut open). What one does encounter in this work are comments by Ratzel concerning the part geography plays in a philosophical perspective.

Ratzel concentrates mostly on Kant and Hegel. In Ratzel's opinion, Kant utterly ignored geography in his theoretical work (though he gave lectures on physical geography for most of his academic career). On this score Ratzel sees a certain contrast in Hegel. The introduction to Hegel's lectures on the philosophy of world history provides a sketch of geographical conditions that, in Hegel's eyes, function as a stage or scene, as a "theatre [*Schauplatz*] for history" (Hegel, 1837/1975, p. 97). According to Hegel, history as the diachronic aspect of absolute spirit courses not only through time but through space over the earth's surface as well. As he sees it, climatic conditions in particular necessarily cause history (i.e., the spirit's history) to leave out continents like Africa or regions like Siberia, where the climatic conditions are excessively harsh.

Ratzel (1882) finds that it seems outright "absurd" (p. 31) to assume that geography determines history in the way Hegel describes. For Ratzel, the problem comes in with Kant's (1784/1983) idea of history as a teleological development and the fulfillment of mankind's task, which Kant asserts to be transcendentally derivable from the concept of "freedom." To Kant, freedom was still only an idea or a hypothesis; but to Hegel, it is the very explanation of historical development: Hegel saw freedom as the driving force behind history, and history as a gradual evolvement of time *through space*. For this reason, geography, according to Hegel, can never be a subject of its own but is always taken into consideration with respect to history.

Nietzsche closely read those passages in which Ratzel criticizes German Idealism for its understanding of geography. Thus, Nietzsche's selective reading of Ratzel aims directly at the center of the philosophy of geography, for what is at stake in the mid-nineteenth century is the role of geography as science. Until Hegel, modern philosophy seems to have known mainly one possibility, geography as the maidservant of history, the paradigm of diacronicity. Teleological structures became *the* major obsession of scientific thought in the late eighteenth and early nineteenth centuries. By contrast, Nietzsche's own work intensified the implicit geographical judgments in historical thinking, as when he pointed out that historical development has to be folded back onto the situation of a particular culture (1874/1988, Vol. 1, pp. 243–334). But this should not lead to a geodeterministic reading of the origin of nations. Instead a geographical (or spatial) description would, according to Nietzsche, draw attention to the factual diversity of cultures.

Nietzsche and Geology

Today it is widely accepted that the modern age is the epoch "after history." In some cases this notion is a take on an apocalyptic or self-referential meaning in the manner of *posthistoire*-thinkers like Jean Baudrillard (1992/1994). But the core idea is not that crude. As theories of the spatial turn have shown (e.g., Soja, 1989), twentieth-century posthistorical theories in the human sciences *initially imply a* view of the local diversity that must be considered when one wishes to talk about global issues. In other words, the anthropology, history, and geography of the mid-nineteenth century were characterized by a top-down approach, which is based on transcendent presuppositions. Human sciences since the late twentieth century, however, have been characterized by a bottom-up approach, which is empirical.

Nietzsche is one of the first philosophers to question the primacy of history. To him, the bottom-up approach is identical with what he identifies as the "English method" (Nietzsche, 1887/1988, Vol. 5, p. 250) and what is related to both Hume's (1748/1963) skeptical philosophy and to Mill's (1843/2002) theory of induction. One of the major assumptions attacked by scientists from England was catastrophism. The catastrophist's model of the earth's development rested upon the idea of divine intervention throughout time. This position is still adopted, especially among creationists, to explain the existence of extinct species. Whereas Cuvier's catastrophism (Cuvier, 1829) was commonly accepted among German geognostics such as Abraham Gottlieb Werner (1750–1817) and Johann Wolfgang von Goethe (1749–1832), Charles Lyell (1797–1875) initiated a different understanding of geology: No flood, nor "revolutions" (Cuvier) are responsible for the changes of the earth's shape. Instead, Lyell (1990) argued that there are constant geological processes acting upon stone, no matter what period (Gould, 1987). After resigning from the University of Basel, Nietzsche caught up on natural sciences through his friend Paul Rée, who was versed in the latest British developments in those fields. Long before Nietzsche, Rée (1877/2003) was the one to make use of concepts from a natural science, stating in the preface to his book that the aim of his inquiry was to apply the geological method to moral science:

> Just as the geologist begins by seeking out and describing different formations and then inquires into the causes from which they have arisen, so too the author has begun by taking up moral phenomena from experience, and has then gone into the history of their beginning, as far as his abilities allowed. (p. 85)

This passage compares the work of the geologist with that of a moral philosopher, who looks at actual results and then does not trace them back to an absolute origin but rather explains them in terms of rules appropriate to that purpose.

Rée's essay was nothing less than the prefiguration of Nietzsche's later enterprise that Nietzsche called "genealogy." Indeed, *Genealogy of Morals* (Nietzsche, 1887/1988) was not a philosophy of moral behavior but rather a critical project intended to lay bare the natural laws underlying the development of any moral. In adopting the theory of Charles Darwin's colleague, the pre-historian John Lubbock (1870), Nietzsche assumed that morals emerged at the moment of settlement, about

10,000 years ago, when the value of an act was no longer judged primarily by its results but by its intent. To Nietzsche, the difference between the noble and the poor became *the* basis for moral judgment from that point onward. Nietzsche concluded that if morality came into being because of a natural development, then morality could eventually also be replaced by something else. In *Beyond Good and Evil* (1886/1988, Vol. 5, pp. 9–243), the central chapter of which is entitled "A Natural History of Morals," Nietzsche calls that prospective time the *aussermoralische Epoche*, the period beyond morality (pp. 50–51, §32). In even more general terms Nietzsche had looked for a philosophical concept that covers not only morality but any cultural aspect. At that point he discovers the concept of the "eternal recurrence (of the same)" (Nietzsche, 1884/1988, Vol. 4, pp. 270–277). This concept moves Nietzsche even closer to Lyell's geology than Rée is, for the eternal recurrence is a description of what actualism in geology calls upon the individual to do: to start from a present event and identify the natural laws behind phenomena valid in the present and the *past*. Nietzsche in his ethics is thus a uniformitarian in that he rejects the idea of an invariable ethic by endorsing a concept of morality that is subject to chance even to the point of having to come about through evolutionary processes rather than divine intervention.

Geosciences' Impact on Philosophy

In summary, the geophilosophical interpretation of Nietzsche's texts demonstrates a number of points. First, through a major thinker of the nineteenth century, one can identify the features of an early philosophy of two major earth sciences, geography and geology. Whereas Nietzsche's discussion of geography results in a critique of the dominance of historical thinking and teleological presuppositions, his engagement with geology results in an attempt to transfer a method from natural sciences to human sciences. There is no claim to return to the positivistic project of reducing human actions to a physicalistic explanation. Rather, it becomes obvious that the geosciences played an important role in the development of philosophical thought at a specific stage of such reflection. For the geosciences challenged the elementary concepts of history (geography as opposed to teleology) and morality (geology as opposed to invariability). Geography and geology are central, not marginal, to the human sciences. And this constellation still prevails. Experience since the late 1980s confirms that geography in particular has an impact on basic structures of knowledge. No matter what one's position within the spatial turn—be it culturo-materialistic (i.e., social aspects are structured spatially) or critical of "spatializing" the social realm (i.e., space is a product of human action)—geography and especially anthropogeography are the major fields channeling discussion today, just as in Nietzsche's time.

One could even argue that this capacity lies in the nature of geography, for geography is the structural, not the logical, opposite to philosophy and thus a constant challenge for philosophy. Whereas philosophy is a search for ways of systematic knowledge and for abstraction from the given, geography is a description of contingencies. To say it the other way round: Contingency is not a thread to geography, but it is to philosophy, for philosophy's traditional approach is mainly nonempirical and deductive (top-down). There are exceptions, a famous one being Hume (1748/1963). He considered his philosophy to be "mental geography" (p. 13), by which the philosopher takes no consideration of hidden forces behind the existing world but rather tries to describe what actually occurs. A modern inheritor of Hume's empiricism is the phenomenological movement of Edmund Husserl and Maurice Merleau-Ponty. Throughout phenomenology the task is to be nondeductive; there is a primacy of perception in relation to which all scientific arguments must be justified. Historically, it is striking how philosophy and geography are conjoined by a strong contrast that makes them a perfect couple. This bond finally could be the reason why phenomenology was especially suitable for social geography in the late twentieth century (e.g., Buttimer, 1993; Pickles, 1985; Relph, 1976; Seamon, 1979; Thrift, 1996).

Geography as Philosophy: The Milieu of Creativity

Geophilosophy is concerned not only with the philosophy of geography (or geology) but also with geography (or geology) *as* philosophy. As the preceding passages about Nietzsche show, the earth sciences have an effect on philosophy. The influence harks back to discussions during the Enlightenment, especially after the Lisbon earthquake of 1755. That event had a unique and strong impact on the theological concept of prophecy, as in the work of Voltaire. The dialectic relation between philosophy and geosciences has thus been a topic in French philosophy less than in German philosophy since the late twentieth century. Even though one can find prior uses of the term geophilosophy, it was not explicitly used as a philosophical term before Deleuze and Guattari (1991/1994) claimed that geophilosophy is the essential idea or the aim of philosophy itself.

The conception of those two authors is crucial for two reasons. First, they furnish the most explicit demonstration of what it means to look upon geography as philosophy (Bonta & Protevi, 2004). Second, they explain what geophilosophy is by linking their definition to a theory of creative milieus. Their key idea is that philosophy emerged from a milieu of creativity: Greece in the classical era. To Deleuze and Guattari (1991/1994), the ancient city-state of Athens offered three major components that allowed a development leading from the existence of individual philosophers (the Pre-Socratics) to the academic institution known as philosophy. The three components were (a) a taste for freedom of opinion, (b) friendship, and (c) immanence. According to Deleuze and Guattari, they not only determine what philosophy is (in the western sense) but also constitute major characteristics of democracy.

The first factor, the taste for freedom of opinion, is unconceivable in an empire. This factor conclusively shows that Plato was an atypical Greek. To Plato, opinion (*doxa*) is a threat to knowledge (*episteme*), whereas it was the very condition enabling his teacher, Socrates, to practice philosophy in the market. The *agora* was where philosophers were given a place for dispute. The second factor, friendship, relates not only to the social aspect but also to the fact that a philosopher in the literal sense is the friend of wisdom or truth. But in order to practice philosophy, one ultimately requires a certain milieu conducive to democratic governance. According to Deleuze and Guattari (1991/1994), the nature of this milieu is immanence, the third factor. As such, immanence does not originate in history but rather in geography: The strategic location of Athens afforded both distance from the empires of the east and the kind of access to the sea that fostered trade and a sufficient livelihood. It was this milieu of creativity in which philosophers from Africa and Asia were permitted to gather and form the institution of philosophy.

To Deleuze and Guattari (1991/1994), then, philosophy, science, and creativity emerge in a certain geographical milieu:

> Why philosophy in Greece at that moment? ... Geography is not confined to providing historical form with a substance and variable places. It is not merely physical and human but mental, like the landscape. Geography wrests history from the cult of necessity in order to stress the irreducibility of contingency. It wrests it from the cult of origins in order to affirm the power of a "milieu." (pp. 96–97)

To call a set of historical elements a milieu means that it was not the physical geography of Greece that determined that civilization's historical development (as Hegel would say) but rather that physical and economic conditions coincided with an intellectual attitude. The important point in Deleuze and Guattari's reasoning is that philosophy, in order to be creative, must be in keeping with those conditions and can be creative only when friendship and the freedom of opinion coexist with immanence. Today this immanence no longer resides in access to the sea. Instead, it is virtual and derives from the access to means of communication.

Conclusion

Philosophical reflections on creative milieus (Meusburger, 2000) can contribute in a number of ways to the present discussion of social geography. A concentration of innovative knowledge is always spatially located, just as it is temporal. The current consensus seems to be that most theories ascribing to the spatial turn refer to Foucault (1986) but do not trace the idea back to earlier reflections, especially Nietzsche's, which were crucial in Foucault's notion of spatial history (Elden, 2001). From a geophilosophical perspective, a purely historical description of the development of knowledge tends to revert to nineteenth-century historical thinking, which was as committed to geodeterministic argumentation as the subsequent geopolitics were. Creative milieus therefore have to be conceived of not as a result of space but rather as something that must coalesce in space or exist in a certain place in order to be operant. The main precondition to any creative milieu seems to be the same today as it was when the institutionalization of philosophy and science began in antiquity. The reason that philosophy could exist in Greece was that groups of philosophers were allowed to settle down, reflect, and teach within a protodemocratic system.

References

Baudrillard, J. (1994). *The illusion of the end* (C. Turner, Trans.). Cambridge, England: Polity (Original work published 1992).

Bonta, M. & Protevi, J. (2004). *Deleuze and geophilosophy: A guide and glossary*. Edinburgh, UK: Edinburgh University Press.

Buttimer, A. (1993). *Geography and the human spirit*. Baltimore, MD: John Hopkins University Press.

Cuvier, G. (1829). *A discourse on the revolutions of the surface of the globe, and the changes thereby produced in the animal kingdom*. London: Whittaker, Treacher & Arnot (Original work published 1825).

Deleuze, G. & Guattari, F. (1994). *What is philosophy?* (H. Tomlinson & G. Burchill, Trans.). London: Verso (Original work published 1991).

Devall, B. (1985). *Deep ecology*. Salt Lake City, UT: Smith.

Elden, S. (2001). *Mapping the present: Heidegger, Foucault and the project of a spatial history*. London: Continuum.

Foucault, M. (1986). Of other spaces (J. Miskowiec, Trans.). *Diacritics*, *16*, 22–27.

Gould, S. (1987). *Time's arrow—Time's cycle: Myth and metaphor in the discovery of geological time*. Cambridge, MA: Harvard University Press.

Günzel, S. (2003). Nietzsche's geophilosophy. *Journal of Nietzsche Studies*, *25*, 78–91.

Hegel, G. W. F. (1975). *Lectures on the philosophy of world history: Introduction* (H. B. Nisbet, Trans.). Cambridge, England: Cambridge University Press (Original work published 1837, this edition 1955).

Hume, D. (1963). *An enquiry concerning human understanding* (L. A. Selby-Bigge, Ed.). Oxford, England: Clarendon (Original work published 1748, this edition 1902).

Kant, I. (1983). Idea for a universal history with a cosmopolitan purpose (T. Humphrey, Trans.). In I. Kant (Ed.), *Perpetual peace and other essays on politics, history, and morals* (pp. 29–39). Indianapolis, IN: Hackett (Original work published 1784).

Lubbock, J. (1870). *The origin of civilisation and the primitive condition of man: Mental and social condition of savages*. London: Longmans/Green.

Lyell, C. (1990). *Principles of geology, being an attempt to explain the former changes of the Earth's surface by reference to causes now in operation* (3 Vols.). Chicago, IL: University of Chicago Press (Original work published 1830–1833).

Lyotard, J.-F. (1990). *Heidegger and 'the Jews'* (A. Michel & M. S. Roberts, Trans.). Minneapolis, MN: University of Minnesota Press (Original work published 1988).

Meusburger, P. (2000). The spatial concentration of knowledge: Some theoretical considerations. *Erdkunde*, *54*, 352–364.

Mill, J. S. (2002). *A system of logic: Ratiocinative and inductive, being a connected view of the principles of evidence, and the methods of scientific investigation*. Honolulu, HI: University Press of the Pacific (Original work published 1843, this edition 1891).

Nietzsche, F. (1988). *Kritische Studienausgabe* (G. Colli & M. Montinari, Ed.) (15 Vols). Berlin/New York: de Gruyter (Not yet fully translated as *The complete works* [E. Behler, Ed.]. Stanford, CA: Stanford University Press.

Pickles, J. (1985). *Phenomenology, science and geography: Spatiality and the human sciences*. Cambridge, England: Cambridge University Press.

Ratzel, F. (1882). *Anthropo-Geographie oder Grundzüge der Anwendung der Erdkunde auf die Geschichte* [Anthropogeography or principles of applying geography to history]. Stuttgart, Germany: Engelhorn.
Rée, P. (2003). The origin of the moral sensations (R. Small, Trans.). In R. Small (Ed.), *Basic writings* (pp. 83–167). Chicago, IL: University of Illinois Press (Original work published 1877).
Relph, E. (1976). *Place and placelessness*. London: Pion.
Seamon, D. (1979). *A geography of the lifeworld: Movement, rest and encounter*. London: Croom Helm.
Soja, E. (1989). *Postmodern geographies: The reassertion of space in critical social theory*. London: Verso.
Thrift, N. J. (1996). *Spatial formations*. London: Sage.

Abstracts of the Contributions

On the Psychology of Creativity

Joachim Funke

Abstract The chapter reviews psychological research in the field of creativity, beginning with an explanation of the available analytical methods. It shows the sources of psychological knowledge about this issue, explains what creative thinking looks like, and how it is manifested. It also deals with the determinants of creative thinking (person, environment, and product), the need for creative thinking, and empirical results about improvement of creativity.

Domain-Generality Versus Domain-Specificity of Creativity

Robert J. Sternberg

Abstract This chapter addresses the question of the domain-generality versus domain-specificity of creativity through a theoretical analysis of the construct. It is argued that creativity is neither domain-general nor domain-specific, but rather, has elements of each. We can increase its domain-generality by carrying with us the attitudes that support it across a variety of domains. Creativity is as much a decision about and an attitude toward life as it is a matter of ability. It is often obvious in young children, but it may be harder to find in older children and adults because their creative potential has been suppressed by a society that encourages intellectual conformity. The domain-specificity of creativity depends in part upon knowledge base, in part upon developed skills for accessing the knowledge base, but most importantly, on the decisions one makes regarding how to use the knowledge base. Knowledge is by no means sufficient for creativity. Indeed, the data suggest that knowledge can as well impede creative thinking as facilitate it. The important element of creativity is the socialization that prepares one to think "outside the box" and thus use one's knowledge in a creative fashion.

Scientific Creativity as a Combinatorial Process: The Chance Baseline

Dean Keith Simonton

Abstract The authors proposes the thesis that the key features of scientific creativity are explicable through combinatorial models, which cast light on most aspects of the phenomenon with the fewest possible assumptions and thus satisfy the law of parsimony (Ockham's razor). At the very least, combinatorial models provide a baseline for comparing alternative theories that use more assumptions to explain the same phenomena. The author begins with six core assumptions that specify how combinatorial creativity operates in the context of the individual scientist, the concepts and ideas that constitute the domain, and the colleagues and associates who define the field. These six assumptions lead to several implications for (a) scientific careers (individual variation and longitudinal change in output) and (b) scientific communities (specifically, the central attributes of multiple discovery and invention). The author elaborates the theory with a mathematical model that makes highly precise and empirically distinctive predictions. The chapter ends with a discussion of how the combinatorial models connect with other empirical findings regarding scientific creativity.

The Riddle of Creativity: Philosophy's View

Günter Abel

Abstract This chapter explores six aspects of creativity: (a) the distinction between mere novelty and genuine or radical creativity, (b) the difference between weak, strong, and moderate (intuitive) creativity, (c) the psychological features of creative individuals as opposed to philosophical "assumptions about creativity," (d) the use and understanding of symbolizing signs as a basic characteristic of the human mind, especially the creative mind, (e) the crucial relation between "creativity" and "rules," and (f) the relation between the creative mind and the human being's naturalistic and scientific world view.

Creativity: Multidimensional Associative or Chaotic Process? Methodological Comments on Creative Processes and Metaphors in Aesthetics and Innovation

Hans Lenk

Abstract Humans as the meta-interpreting, continuously symbolically transcending beings are the creative beings par excellence. Especially important in this process are creative metaphors consisting in seeing and establishing similarities and differentiations from different perspectives on diverse levels and overlapping strata. If stimulations towards new developments are based on transpositions to other perspectives and towards higher levels and strata, then we have a particularly creative, i.e. creativity-stimulating, "createaphor". "Createaphors" are also metaphors—however special ones which would overarch perspectives, bridge and transform as well as maintain tension within a stimulating play between similarities and dissimilarities.

Creativity is symbolic authentic activity, personal Eigen-activity: A philosophy of being creative would at the same time amount to a philosophy of an extended personal and authentic Eigen-activity of human, subjective, social or artificial interpreting systems. We could go as far as to ascribe to humans a createaphoric consciousness. We may understand the human being as the potential author and agent who is capable of creating createaphors. Creativity is a permanent and continuing creative process, a kind of an ongoing transformation of createaphors, the capacity and motivation to reach beyond old or "dying" metaphors by going on with the play of creativity.

Milieus of Creativity: the Role of Places, Environments and Spatial Contexts

Peter Meusburger

Abstract The environment is an important constituent of creativity that can either encourage or impair creative processes. Learning, problem-finding and problem-solving not only require cognitive skills but result from a complex and dynamic interaction between the actor and his or her surroundings. Viewing creativity from an interactional perspective accentuates the relevance of situational, contextual, and cultural determinants. Places, environments, spatial contexts, and spatial relations matter because a stimulating environment and talented individuals must come together and interact before a creative process can come into being. Because the generation, evaluation and adoption of creative ideas and products vary spatially, the spatial distribution of creative processes must not be ignored. Results and methodological problems of creativity research are discussed from a geographical perspective to clarify misunderstandings that might complicate transdisciplinary discourse between human geography and other social and behavioral sciences. The purpose is to learn how to conceptualize the relations between environment and agents without becoming trapped in geodeterminism.

Creativity, Intelligence, and Culture: Connections and Possibilities

James Kaufman

Abstract The topic of intelligence and culture is a widely studied and often highly controversial area. Despite obvious cognitive similarities, the relationship between *creativity* and culture has generated only a tiny fraction of the work on intelligence. This chapter discusses the relationship between creativity and intelligence and how creativity differs across ethnicities and cultures. The research indicates that creativity and intelligence, despite sharing so many similar abilities, show different patterns across cultures and ethnicities. The groups that receive the highest scores on intelligence and ability measures are not necessarily the same groups that receive the highest scores on creativity measures. The argument is not that any one culture is more creative or intelligent than another. Rather, if measures of intelligence show certain kinds of differences and measures of creativity show other differences, then using both intelligence and creativity measures would yield more information and present a more complete and more equitable picture of an individual than either type of measure alone would.

Exploring the Relationships Between Problem-Solving Style and Creative Psychological Climate

Scott G. Isaksen

Abstract This chapter outlines an ecological or systemic approach to improving the comprehension of the way people perceive their context and how it supports or inhibits their creativity. The author examines how individuals of differing problem-solving styles rate their working climates on nine dimensions of climate. Problem-solving styles were examined to assess individual differences on three dimensions of style. The Situational Outlook Questionnaire® was used to assess individual perceptions of the psychological climate. This measure also gave respondents opportunity to answer three open-ended questions in order to understand factors that support or hinder their creativity and suggested actions for improving the climate. Only 2 of 27 correlations were found to be significant (and the p values were low). Tests of equality of means indicated no significant differences between different problem-solving styles and results on the nine dimensions of climate. Despite the lack of meaningful or significant quantitative results, some meaningful differences were found with the qualitative and narrative data.

Creativity in Cross-Cultural Innovation Teams: Diversity and Its Implications for Leadership

Ricarda Bouncken

Abstract This study inquires into effects of national culture on team work and innovation in global teams. Results indicate that cultural values have unequal effects on teamwork and creativity in the innovation process. Interfering with communication and cohesiveness, inequalities of power distance usually hamper teamwork quality and creativity in global teams. Clear and direct communication emerges as another challenge facing global teams, for it proves to be important for the success of a project. The research also shows that personal self-reflection is one of the greatest advantages of working in a cross-cultural team, and, above all, that cross-cultural teams have a broader knowledge base than culturally homogeneous teams do. Lastly, leadership behavior is shown to be a means of mitigating the harm caused by certain components of cultural differences. A sensitive team leader, for example, can channel different behaviors during discussions, possibly increasing the creativity and innovativeness of global teams.

Space(s) of Innovation: Regional Knowledge Economies

Martina Fromhold-Eisebith

Abstract Economic geographers and regional scientists have long debated why phenomena of innovativeness, technology development, and the economic exploitation of knowledge are so unevenly distributed across space. Apparently, certain basic regional factors and the collaboration of colocating actors contribute greatly to the emergence and self-reinforcing dynamics of spatial concentrations of innovative firms. The chapter summarizes arguments of the academic discussion and discusses reasons for spatially distinct features of knowledge generation and commercialization. After clarifying logical relationships between the notions of spaces of innovation and milieus of creativity, the author presents assumptions about the nature of innovativeness and knowledge creation that bear on considerations of spatiality. She examines various reasons why innovative actors agglomerate and asks how local contexts sustain economic creativity. She approaches the conceptual dimension through ideal-type constellations, referring to clusters, innovative milieus, regional innovation systems, and learning regions. The methodological question of empirically capturing the role that regional contexts have for innovativeness is addressed, and major conclusions for further research are drawn.

The Unconscious City: How Expectancies About Creative Milieus Influence Performance

Jens Förster

Abstract People have expectations about cities (e.g., Paris = love; Amsterdam = creative), and those expectations are represented in memory. Thinking about a city can directly activate those expectancies and can influence the way a person thinks, feels, or behaves. Through specific mental procedures, the individual may start thinking in a creative manner when reminded of a city associated with creative thinking. Drawing on theories of social cognition, the chapter presents an experiment in which participants were first assessed on whether they thought that certain cities (e.g., Amsterdam, London, Cologne, and New York) are associated with creative thinking. Participants were then subliminally primed with the names of cities. Participants who had been reminded in this way of a city they associated with creativity performed more creatively on a subsequent creativity task than did those who had been reminded of a city they did not associate with creativity. Participants were not aware of such effects. The research demonstrates that expectations about cities may unconsciously affect thinking and performance. Limitations and boundary conditions for such effects are discussed.

Conceptual Spaces

Margaret A. Boden

Abstract Exploratory and transformational creativity are defined in terms of conceptual spaces in people's minds, with conceptual spaces being understood to mean culturally accepted styles of thinking in a particular area (e.g., in art, science, or cooking). Each style is seen as representing a space of possibilities—the set of structures (ideas) that can be generated through the stylistic rules concerned. In any interesting conceptual space, the possibilities are so numerous that no one person is likely to have visited (thought of) all points. Exploratory creativity involves moving through the space to find new ideas (i.e., previously unvisited locations) and discover the space's boundaries and limitations. If it becomes clear that no ideas unvisited points remain or (more likely) or no uninstantiated *kinds* of structure remain, the creative thinker may transform one or more dimensions of the space (one or more stylistic rules) so as to generate ideas that could not have been generated before. This transformational creativity typically elicits "impossibilist" surprise: the new idea seems (and was) impossible, and yet it happened. Computational psychology, and sometimes actual computer models, can help people understand how both types of creativity are possible.

Looking at the Present Through the Future: Science-Fiction Urbanism and Contingent and Relational Creativity

Rob Kitchin

Abstract This chapter explores ways in which the creative writing of several science-fiction authors emerges from diverse engagements with the world, how their fiction does work in the world, and how a recursive relationship has in some cases developed between novelists and people who read and act upon these stories. The empirical material illustrating this argument derives from a project that involved analysis of 34 novels and four collections of short stories with plots focusing on the development and use of cyberspatial, virtual reality, other information and communication technologies, and issues such as telemediation, computer intelligence, surveillance and governance, person–machine relations (cyborgs), and the changing nature of work and urbanism. Of particular interest was how the novelists engaged with notions of space and time and described the new geographies of the near future. The chapter inquires into how the authors of these stories imagine urban environments and the nature of future cities. Recognizing that fiction is a product of its place and time and that some of its ideas come from contemporary urban theory, the author argues that these novelists' visions of the near future serve as a powerful cognitive lens on urbanism in the present, extrapolating from spatial processes that are already at work. It is shown that urban theory has, in turn, drawn inspiration from these novels, creating a recursive relationship between novelists and urban theorists.

Teleology, Contingency, and Networks

Barney Warf

Abstract Inspired by structuration theory and philosophical realism, geographers and sociologists have increasingly come to emphasize the contingent nature of social reality (i.e., the manner in which it could be different).They have come to view time and space as social constructions rather than presocial givens. This chapter asserts that the theorization of social and spatial life necessitates rejection of the dichotomy between the real and the imaginary, the actual and the possible, the ontological and the epistemological. What is taken to be "real" is not simply what is observable or actual; it forms one outcome secreted from a broader universe of possibilities. It is argued that reality thus includes phenomena that never happened in fact, but *could* have happened plausibly as defined by theory. From this perspective, history and geography are the understanding not only of why things happen, but why they do not.

Geophilosophy and Creative Milieus

Stephan Günzel

Abstract The main focus in on two facets of geophilosophy. In one sense, geophilosophy depicts an aspect of philosophy of science, one that reflects the methods of geography in particular and earth sciences in general. In a second sense, the discipline deals with geography's emergence from an existence as subordinate subject matter in the science of history into the role of an independent discipline, a development through which it turned away from teleological argumentation. Accordingly, geophilosophy conveys the idea that modern philosophy itself is geographical in that it instills an understanding of concepts as syntactical elements arranged in a nondiachronic manner. The work of Friedrich Nietzsche is crucial to this perspective, for he explicitly criticized the teleological view of philosophy against the background of geography and geology. The final part of the article shows how the idea of a creative milieu can contribute to a reevaluation of philosophy's origin in geographical way.

The Klaus Tschira Foundation

Physicist Dr. h.c. Klaus Tschira established the Klaus Tschira Foundation (KTF) in 1995 as a not-for-profit organization conceived to support research in informatics, the natural sciences, and mathematics and to foster public understanding of these sciences. Klaus Tschira's commitment to this objective was honored in 1999 with the "Deutscher Stifterpreis," the prize awarded by the National Association of German Foundations. Klaus Tschira is a cofounder of SAP AG in Walldorf, one of the world's leading companies in the software industry.

The KTF provides support mainly for research in applied informatics, the natural sciences, and mathematics and to fund educational projects for students at public and private universities and schools. The resources are largely used for projects initiated by the foundation itself. It commissions research from institutions such as EML Research, founded by Klaus Tschira. The central goal of that organization for applied informatics is to develop new information-processing systems whose technology is perceived as user-friendly. In addition, the KTF invites applications for projects that are in line with the central concerns of the foundation.

The seat of the KTF is Villa Bosch in Heidelberg (Fig. 1), the former residence of Carl Bosch (1874–1940), the Nobel Prize Laureate for Chemistry. Carl Bosch, scientist, engineer, and businessman, joined BASF in 1899 as a chemist and became its CEO in 1919. In 1925 he was appointed CEO of the then newly created IG Farbenindustrie AG, and in 1935 he became chairman of the supervisory board of this chemical conglomerate. In 1937 Bosch was elected president of the Kaiser Wilhelm Gesellschaft (later renamed as the Max Planck Gesellschaft), the premier scientific society in Germany. Bosch's work combined chemical and technological knowledge at its best. Between 1908 and 1913, together with Paul Alwin Mittasch, he solved numerous problems in the industrial synthesis of ammonia, drawing on a process discovered earlier by Fritz Haber (Karlsruhe), who won the Nobel Prize for Chemistry in 1918. The Haber-Bosch process, as it is known, quickly became the most important method of producing ammonia—and remains so to this day. Bosch's research also influenced high-pressure synthesis of other substances. He was awarded the Nobel Prize for Chemistry in 1931, together with Friedrich Bergius.

In 1922 BASF erected a spacious country mansion and ancillary buildings in Heidelberg-Schlierbach for its CEO, Carl Bosch. The villa is situated in a small park on the hillside above the Neckar river and within walking distance from the famous Heidelberg Castle. As a fine example of the style and culture of the 1920s, Villa Bosch is considered one of the most beautiful buildings in Heidelberg and has been declared a protected cultural site. After World War II, it served as a domicile for high-ranking military staff of the United States Army. Thereafter, a local enterprise used the villa as its headquarters for several years. In 1967 Süddeutsche Rundfunk, a broadcasting company, established its Heidelberg studio there. Klaus Tschira bought Villa Bosch as a future home for his planned foundations toward the end of 1994 and had the building restored and modernized. Combining the historic ambience of the 1920s with the latest infrastructure and technology, Villa Bosch reopened in new splendor in mid-1997, ready for fresh challenges. The former garage, located 300 meters west of the villa, now houses the Carl Bosch Museum Heidelberg, founded and managed by Gerda Tschira and dedicated to the memory of the Nobel laureate, his life, and his achievements.

This book is the result of a symposium entitled "Milieus of Creativity," which took place at Villa Bosch, September 17–20, 2006 (Fig. 2).

For further information contact:

Klaus Tschira Foundation gGmbH
Villa Bosch
Schloss-Wolfsbrunnenweg 33
D-69118 Heidelberg, Germany
Tel.: (+49) 6221-533-101
Fax: (+49) 6221-533-199
beate.spiegel@ktf.villa-bosch.de

Public relations:
Renate Ries
Tel.: (+49) 6221-533-214
Fax: (+49) 6221-533-198
renate.ries@ktf.villa-bosch.de

http://www.ktf.villa-bosch.de

Fig. 1 Villa Bosch (© Peter Meusburger, Heidelberg)

Fig. 2 Participants of the symposium "Milieus of Creativity" at Villa Bosch in Heidelberg, September 17–20, 2006 (© Thomas Bonn, Heidelberg)

Index

Breinigsville, PA USA
24 February 2010

233101BV00007B/71/P

9 781402 098765